SCIENTIFIC
AND
TECHNOLOGICAL
THINKING

SCIENTIFIC AND TECHNOLOGICAL THINKING

Edited by

Michael E. Gorman
University of Virginia

Ryan D. Tweney
Bowling Green State University

David C. Gooding
University of Bath

Alexandra P. Kincannon
University of Virginia

Psychology Press
Taylor & Francis Group

New York London

First published by Lawrence Erlbaum Associates, Inc.
Lawrence Erlbaum Associates, Inc., Publishers
10 Industrial Avenue
Mahwah, New Jersey 07430

First issued in paperback 2012

This edition published 2012 by Psychology Press

Psychology Press
Taylor & Francis Group
711 Third Avenue
New York, NY 10017

Psychology Press
Taylor & Francis Group
27 Church Road
Hove, East Sussex BN3 2FA

Psychology Press is an imprint of Taylor and Francis, an informa group company

Cover design by Kathryn Houghtaling Lacey

Library of Congress Cataloging-in-Publication Data

Scientific and technological thinking / edited by Michael E. Gorman
... [et al.].
    p. cm.
Includes bibliographical references and index.
ISBN 0-8058-4529-1 (cloth : alk. paper)
ISBN 978-0-415-65415-9 (Paperback)
1. Creative ability in technology. 2. Creative thinking.   I. Gorman, Michael E., 1952–
T49.5.N475   2004
153.3'5—dc22   2003063115
                                                                  CIP

Contents

Contributors   vii

Preface   ix

1   Editors' Introduction   1
Michael E. Gorman, Ryan D. Tweney, David C. Gooding,
and Alexandra P. Kincannon

2   Interpreting Scientific and Engineering Practices: 17
Integrating the Cognitive, Social, and Cultural Dimensions
Nancy J. Nersessian

3   Causal Thinking in Science: How Scientists and Students   57
Interpret the Unexpected
Kevin N. Dunbar and Jonathan A. Fugelsang

4   A Framework for Cognitive Studies of Science and Technology   81
David Klahr

5   Puzzles and Peculiarities: How Scientists Attend   97
to and Process Anomalies During Data Analysis
Susan Bell Trickett, Christian D. Schunn, and J. Gregory Trafton

6   On Being and Becoming a Molecular Biologist: 119
Notes From the Diary of an Insane Cell Mechanic
Jeff Shrager

7   Replicating the Practices of Discovery: Michael Faraday   137
    and the Interaction of Gold and Light
    Ryan D. Tweney, Ryan P. Mears, and Christiane Spitzmüller

8   How to Be a Successful Scientist   159
    Paul Thagard

9   Seeing the Forest for the Trees: Visualization, Cognition,   173
    and Scientific Inference
    David C. Gooding

10 Problem Representation in Virginia Woolf's Invention   219
    of a Novelistic Form
    Maria F. Ippolito

11 What's So Hard About Rocket Science?   259
    Secrets the Rocket Boys Knew
    Gary Bradshaw

12 A Systems-Ordered World   277
    Thomas P. Hughes

13 Levels of Expertise and Trading Zones: Combining Cognitive   287
    and Social Approaches to Technology Studies
    Michael E. Gorman

14 Technology at the Global Scale: Integrative Cognitivism   303
    and Earth Systems Engineering and Management
    Brad Allenby

15 The Future of Cognitive Studies of Science and Technology   345
    Michael E. Gorman, Ryan D. Tweney, David C. Gooding,
    and Alexandra P. Kincannon

    Author Index   353

    Subject Index   361

Contributors

Brad Allenby, Environment, Health & Safety, AT&T, Bedminster, New Jersey; Darden Graduate School of Business and School of Engineering, University of Virginia; Theological Seminary, Princeton University

Gary Bradshaw, Department of Psychology, Mississippi State University

Kevin N. Dunbar, Department of Psychological and Brain Sciences, Department of Education, Dartmouth College

Jonathan A. Fugelsang, Department of Psychological and Brain Science, Dartmouth College

David C. Gooding, Science Studies Centre, Department of Psychology, University of Bath

Michael E. Gorman, Department of Science, Technology and Society, School of Engineering and Applied Science, University of Virginia

Thomas P. Hughes, Professor Emeritus, University of Pennsylvania

Maria F. Ippolito, Department of Psychology, University of Alaska Anchorage

Alexandra P. Kincannon, Department of Psychology, University of Virginia

David Klahr, Department of Psychology, Carnegie Mellon University

Ryan P. Mears, Department of Psychology, Bowling Green State University

Nancy J. Nersessian, Program in Cognitive Science, College of Computing & School of Public Policy, Georgia Institute of Technology

Christian D. Schunn, Learning Research & Development Center, University of Pittsburgh

Jeff Shrager, Department of Plant Biology, Carnegie Institute of Washington; Institute for the Study of Learning and Expertise, Stanford University

Christiane Spitzmüller, Department of Psychology, University of Houston

Paul Thagard, Department of Philosophy, University of Waterloo

J. Gregory Trafton, Naval Research Laboratory

Susan B. Trickett, Department of Psychology, George Mason University

Ryan D. Tweney, Department of Psychology, Bowling Green State University

Preface

This volume is the product of a workshop on cognitive studies of science and technology that was held at the University of Virginia in March 2001. The goal of the workshop was to assemble a diverse group from a variety of fields, including graduate students and junior and senior faculty, to discuss the latest research and to generate ideas for the future of "Cognitive Studies of Science and Technology." The workshop was made possible through the generous support of the National Science Foundation, the Strategic Institute of the Boston Consulting Group, and the National Collegiate Inventors and Innovators Alliance.

The chapters in this volume (except chap. 14) are authored by workshop participants. They describe recent developments and discuss ongoing issues in the study of the cognitive processes of discovery and invention. Building on our workshop discussions, we have developed a conceptual framework that we hope will help to clarify the current state of the field and to spawn new ideas for future investigation. For readers interested in the original papers and authors, and the lively discussion that occupied much of the workshop, all of this material was recorded live in digital format. It can be shared, with permission of the original participants, via the workshop Web site at http://repo-nt.tcc.virginia.edu/cogwkshop/ . For a brief description of our deliberations, see the following chapter: Gorman, M. E., Kincannon, A., and Mehalik, M. M. (2001). Spherical horses and shared toothbrushes: Lessons learned from a workshop on scientific and technological thinking. In K. P. Jantke and A. Shinohara (Eds.), *Discovery science* (pp. 74–86). Berlin, Germany: Springer-Verlag.

1

Editors' Introduction

Michael E. Gorman
University of Virginia

Ryan D. Tweney
Bowling Green State University

David C. Gooding
University of Bath

Alexandra P. Kincannon
University of Virginia

At the turn of the 21st century, the most valuable commodity in society is knowledge, particularly new knowledge that may give a culture, a company, or a laboratory an adaptive advantage (Christensen, 1997; Evans & Wurster, 2000; Nonaka & Takeuchi, 1995). Turning knowledge into a commodity poses two dangers. One is to increase the risk that one culture, company, or group can obtain an unfair advantage over others. The other is to impose a one-dimensional, goal-driven view of something that, as the chapters in this volume will show, is subtle, complex, and diverse as to its motivations and applications. To be sure, knowledge about the cognitive processes that lead to discovery and invention can enhance the probability of making valuable new discoveries and inventions. However, if made widely available, this knowledge could ensure that no particular interest group "corners the market" on techno-scientific creativity. It would also facilitate the development of business strategies and social policies based on a genuine understanding of the creative process. Furthermore, through an understanding of principles underlying the cognitive processes related to discovery, educators can use these principles to teach students effective problem-solving strategies as part of their education as future scientists.

A special focus of this volume is an exploration of what fine-grained case studies can tell one about cognitive processes. The case study method is normally associated with sociology and anthropology of science and technology; these disciplines have been skeptical about cognitive explanations. If there is a well-established eliminativism among neuroscientists (Churchland, 1989), there is also a social eliminativism that seeks to replace cognitive accounts with sociological accounts (Woolgar, 1987). In these socio-anthropological case studies, interactions, inscriptions, and actions are made salient. The private mental processes that underlie public behavior have to be inferred, and most anthropologists and sociologists of science are trained to regard these cognitive processes as epiphenomenal. By contrast, intellectual or internalist studies by historians of science go into reasoning processes in detail (Drakes, 1978; Mayr, 1991; Westfall, 1980). Historians of science have produced a large number of studies that describe processes of experimentation, modeling, and theory construction. These studies can be informative about reasoning and inference in relation to declarative knowledge, experiential knowledge and experimental data (Galison, 1997; Gooding, 1990; Principe, 1998), visualization (Rudwick, 1976; Tweney & Gooding, 1991; Wise, 1979), and the dynamics of consensus formation through negotiation and other forms of personal interactions (Rudwick, 1976). However, historians generally have no interest in identifying and theorizing about general as opposed to personal and culture-specific features of creative processes. Thus, very few historical studies have combined historical detail with an interest in general features of the creative process (for exceptions, see Bijker, 1995; Carlson & Gorman, 1990; Giere, 1992; Gooding, 1990; Gruber, 1974; Law, 1987; Miller, 1986; Tweney & Gooding, 1991; Wallace & Gruber, 1989).

Despite the usefulness of fine-grained case studies, cognitive psychologists have traditionally lamented their lack of rigor and control. How can one identify general features, let alone develop general principles of scientific reasoning, from studies of a specific discovery, however detailed they may be? One answer is to develop the sorts of computational models preferred by cognitive scientists (Shrager & Langley, 1990). One classic example of this kind of modeling is, of course, Kulkarni and Simon's (1988) simulation of a historical account by Larry Holmes (1989) dealing with Hans Krebs's discovery of the ornithine cycle (see also Langley, Simon, Bradshaw, & Zytkow, 1987). These models can be abstracted from historical cases as well as current, ethnographic ones. However, it is important to remember that such models are typically designed to suit the representational capabilities of a particular computer language. Models derived from other domains—such as information theory, logic, mathematics, and computability theory—can become procrustean beds, forcing the territory to fit the researcher's preconceived map (Gorman, 1992; Tweney, 1990).

The tension among historical, sociological, and cognitive approaches to the study of science is given thorough treatment in the chapter 2, by Nersessian. She distinguishes between good old-fashioned artificial intelligence, represented by computer programs whose programmers claim they discover scientific laws, and social studies of science and technology, represented by detailed or "thick" descriptions of scientific practice. Proponents of the former approach regard cognition as the manipulation of symbols abstracted from reality; proponents of the latter see science as constructed by social practices, not reducible to individual symbol systems. Nersessian describes a way for cognitive studies of science and technology to move beyond these positions, taking what she calls an environmental perspective that puts cognition in the world as well as in the brain. Ethnography can be informative about cognitive matters as well as sociological ones, as Nersessian's study of biomedical engineering laboratories shows, and historical research helps make the cognitive practices intelligible.

In order to ground cognitive theory in experimental data, psychologists have conducted reasoning experiments, mainly with college students but also with scientists. These experiments allow for controlled comparisons, in which all participants experience the same situation except for a variable of interest that is manipulated. A psychologist might compare the performance of scientists and students on several versions of a task. Consider, for example, a simple task developed by Peter Wason (1960). He showed participants in his study the number triplet "2, 4, 6" and asked them to propose additional triplets in an effort to guess a rule he had in mind. In its original form, the experimenter's rule was always "any three increasing numbers," which proved hard for most participants to find, given that the starting example suggests a much more specific rule (e.g., "three even numbers"). Each triplet can be viewed as an experiment, which participants used to generate and test hypotheses. For Wason, participants seemed to manifest a *confirmation bias* that made it hard to see the need to disconfirm their own hypotheses. Later research suggested a much different set of explanations (see Gorman, 1992, for a review). Tasks such as this one were designed to permit the isolation of variables, such as the effect of possible errors in the experimental results (Gorman, 1989); however, they are not representative of the complexities of actual scientific practice. The idealization of reasoning processes in scientific and technological innovation (Gorman, 1992; Tweney, 1989), and of scientific experiments in computer models (Gooding & Addis, 1999), is a critical limitation of this kind of research.

What makes models powerful and predictive is their selectivity. A good model simplifies the modeled world so as to make it amenable to one's preferred methods of analysis or problem solving. However, selectivity can make one's models limited in scope and, at worst, unrealistic. Insofar as this is a problem, it is often stated as a dichotomy between abstraction and realism, as for example by the mathematician James Gleick: "The choice is always the same. You can make your model more complex and more faithful to reality, or you can make it simpler and easier to handle" (Gleick, 1987, p. 278). David Gooding illustrated this problem at a workshop with a joke that could become a central metaphor for science and technology studies:

> A millionaire with a passion for horse racing offered a large prize—enough to buy a few Silicon Graphics machines—to anyone who could predict the outcome of any horse race. Three scientists took up the challenge, a physiologist, a geneticist and a theoretical physicist. One year later the three scientists announced their results. Here's what each reported:
>
> *The Physiologist:* "I have analysed oxygen uptake, power to weight ratios, dietary intake and metabolic rates, but there are just too many variables. I am unable to predict which horse will win."
>
> *The Geneticist:* "I have examined blood lines, breeding programs and all the form books, but there are just too many uncertainties. I cannot predict who will win any race."
>
> *The Physicist:* "I have developed a theoretical model of the dynamics of horse racing, and have used it to write a computer program that will predict the outcome of any horse race to 7 decimal places. I claim the prize. But—there is one proviso. The model is only valid for a perfectly spherical horse moving through a vacuum"[1]

Experimental simulations of scientific reasoning using tasks like Wason's (1960) 2–4–6 task are abstractions just like the spherical horse: They achieve a high degree of rigor and control over participants' behavior but leave out many of the factors that play a major role in scientific practice. Gooding (in press) argues that in the history of any scientific field there is a searching back and forth between models and real world complexities, to achieve an *appropriate level of abstraction*—not overly simple, capturing enough to be representative or valid, yet not so complex as to defeat the problem-solving strategies of a domain. Gooding's chapter for this volume (chap. 9) shows how scientists use visualization in creating models that enable them to negotiate the tension between simplicity and solvability on the one hand and complexity and real world application on the other. He argues that, like any other scientific discipline, cognitive studies of science and technology must find appropriate abstractions with which to describe, investigate, model, and theorize about the phenomena it seeks to explain.

---

[1]This anecdote is attributed to Adam Katalsky in *New Scientist*, December 19–26, 1992.

To compare a wide range of experiments and case studies conducted in different problem domains, one needs a general framework that will establish a basis for comparison. As Chris Schunn noted in the workshop on cognitive studies of science and technology that inspired this volume (see Preface), models, taxonomies, and frameworks are like tooth-brushes—no one wants to use anyone else's. In science and technology studies, this has been the equivalent of the "not invented here" syndrome. This usually reflects the methodological norms of a discipline, such as the sociological aversion to cognitive processes, which is reminiscent of behavioral psychology's rejection of mental processes as unobservables. One strategy for achieving de facto supremacy is to assume, even if one cannot demonstrate it, that one's own "toothbrush" is superior to any other (Gorman, 1992).

DISCOVERY

Sociological and historical studies of the resolution of scientific controversies have shown that the supremacy of a particular theory, technology, or methodological approach involves negotiation. Because no method is epistemologically neutral, this negotiation often focuses on the validity of the method(s) of establishing facts and of making inferences from them (Galison, 1997). Therefore, rather than promoting the investigative potential of a single method, we advocate approaches that address both Gooding's problem of abstraction and Schunn's problem of shareable frameworks. Dunbar and Fugelsang develop one such approach in their contribution (chap. 3) to this volume. This approach combines experiments, modeling and case studies in a complementary manner. They develop the distinction (first made by Bruner, Goodnow, & Austin, 1956) between *in vitro* studies of scientific thinking (which involve abstract tasks like Wason's [1960] 2–4–6 task) and *in vivo* studies (which involve observing and analyzing scientific practice). Dunbar (1999) used an *in vitro* task to study how participants reasoned about a genetic control mechanism and conducted *in vivo* studies of molecular biology laboratories. In chapter 3, Dunbar and Fugelsang label four more approaches in the same style:

1. *Ex vivo* research, in which a scientist is taken out of her or his laboratory and investigated using *in vitro* research, by presenting problems similar to those he or she would use in his or her research.
2. *In magnetico* research, using techniques such as magnetic resonance imaging to study brain patterns during problem solving, including potentially both *in vitro* and *in vivo* research.
3. *In silico* research, involving computational simulation and modeling of the cognitive processes underlying scientific thinking, including

the good old-fashioned artificial intelligence work cited by Nersessian and alternatives.

4. *Sub specie historiae* research, focusing on detailed historical accounts of scientific and technological problem solving. These *in historico* studies can serve as data for *in silico* simulations.

Later chapters offer a variety of examples of *sub specie historiae* and *in vivo* studies, with references to the other types of research noted earlier.

In chapter 4, Klahr takes a framework that he was involved in developing and stretches it in a way that makes it useful for organizing and comparing results across chapters in this volume. His idea is that discovery involve searches in multiple problem spaces (Klahr & Dunbar, 1988; Simon, Langley, & Bradshaw, 1981). For example, a scientist may have a set of possible experiments she might conduct, a set of possible hypotheses that might explain the experimental results, and a set of possible sources of experimental error that might account for discrepancies between hypotheses and results. Klahr adds two other dimensions: (a) whether a space is general or domain specific and (b) whether the search is conducted by individuals, dyads, or teams. This framework leads to interesting questions, such as: Under what circumstances does a mismatch between current hypothesis and current experimental result lead to a search of a space of possible errors in the experiment? and When does it trigger a search for a new hypothesis? Empirical studies can provide specific answers to these questions.

We can now suggest one possible framework to support a comparative analysis of different kinds of study. We can combine Klahr's idea of multiple search spaces with Dunbar's distinction among six types of methodology to produce a multidimensional matrix that allows us to organize and compare the research studies reported in this volume. For example, the studies involving the introduction of error into the 2–4–6 task involve an *in vitro* methodology, three types of general problem spaces, and are conducted on individuals (Gorman, 1989). Tweney and his colleagues have used scientists as experimental participants in a selection task (Tweney & Yachanin, 1985) and in "artificial universe" studies (Mynatt, Doherty, & Tweney, 1978). Similarly, specific computational simulations could be put in the *in silico* row, with the problem spaces they modeled as columns. Historical case studies could be placed in the *sub specie historiae* row, with their problem spaces across the top; these would either be domain specific or, in the case of individuals who move across domains in the course of their careers, person specific. Consider, for example, Herbert Simon, whose career included original contributions to economics, computer science, psychology, and cognitive studies of science. There will, of course, be cases where methods are mixed within the same study, as Dunbar and Fugelsang do when they combine *in magnetico* with *in vitro* techniques.

At this point, gaps in the matrix draw our attention to ways of extending research. These gaps, together with studies we consider relevant but that do not fit the framework, suggest how the framework needs to be developed. They also help identify limitations in the framework itself, as we discuss in chapter 15. Similarly, new studies will add new problem spaces, especially because discovery and invention involve the creation of new problem spaces. Furthermore, as the "thick" historical and ethnographic studies show, much scientific and technological thinking is not conducted "over a defined problem space but across a domain consisting of, say, apparatus, external texts (books, articles, tables, etc.), internal memory (reflecting the scientist's domain knowledge), and a symbol system that may need to be altered or created, rather than merely rearranged" (Tweney, 2001, p. 154). We emphasize, therefore, that we use problem spaces as an organizational heuristic rather than as an attempt to prescribe an ontology of resources and procedures for discovery.

Trickett, Schunn, and Trafton report in chapter 5 a domain-specific *in vivo* study. They show that, to resolve anomalies, two astronomers and a physicist search both hypothesis and data spaces. The data consist of images of galaxies in the astronomy case and of various representations of the relation between a model and data in the physics case. In both cases, the researchers had to decide which kinds of visualizations worked best; therefore, they searched through problem-specific spaces of visualizations. In both cases, the scientists generated new visualizations and studied existing ones more closely. This study suggests the need to incorporate visualizations into our framework. As Gooding argues in chapter 9, visualizations are used across a wide range of contexts; in particular, they are used to generate phenomenological descriptions, proto-explanations, dynamical models, and in the context of communicating about results. It follows that ways of dealing with different modes of representation should be included in each of the spaces in the Simon–Klahr–Dunbar scheme, in addition to considering that researchers might search a space of possible visualizations on certain kinds of problems.

Trickett et al.'s study (chap. 5) also uses a dyad as a unit of analysis. Two astronomers worked together on identifying and accounting for anomalies in their visualizations. Collaboration was not the focus of the study, however; the same analytic framework was used for the physicist working alone and the astronomers working together. Trickett et al. point out the need for future research to determine whether two or more scientists working together are more likely to notice anomalies than individuals working in relative isolation.

Some of the most successful scientists and inventors kept notebooks, which they used to enhance their problem-solving strategies. Shrager's chapter, 6, provides a contemporary example of how record keeping can support learning. Shrager describes how he gathered data on the process by which he became a molecular biologist. His is therefore a reflexive *in vivo* study. Shrager kept both a laboratory notebook and a protocol book; he remarked at the workshop (see Preface), "If you lose your lab notebook, you're hosed." So, another kind of potential search space is the notes a scientist or inventor keeps on experiments, hypotheses, new designs, and so on—notes that are incredibly valuable when it comes to conducting further research and in establishing priority for an invention or discovery. Gorman and his colleagues have described the way in which Bell made a major improvement on his telephone by searching his own notebook for promising results and discovering one that led to an improved telephone design (Gorman, Mehalik, Carlson, & Oblon, 1993). Tweney and Gooding (1991) showed how Faraday used records of his work to monitor his research stratagems in order to evaluate and refine his research program. However, Shrager also kept a different kind of diary that included his observations on his own learning process. At the end of his chapter, he speculates that this kind of diary might be useful both for science and technology studies scholars and in the education of scientists and engineers.[2]

Notebooks and diaries are more than an external memory aid that scientists and inventors can search; they also create a space of reflection and transformation. Bell, for example, sprinkled his notebook with reflections on his style of invention, gradually realizing that his strength was not in the hands-on component of inventing but in developing the theoretical principles underlying the transmission of speech. For Bell, this "theory" was not a formal hypothesis but a powerful mental model he described and deployed (Gorman, 1997). Similarly, Shrager's diary includes reflections on how the problem-solving style he has to develop for molecular biology differs from the style he uses in repairing cars.

Faraday's extensive laboratory investigations are recorded in his laboratory diary, and partial records also survive in the form of material artifacts. These include many instruments and objects with which Faraday "fixed" phenomena. Chapter 7, by Tweney, Mears, and Spitzmüller, is an *in historico* study of Faraday's investigation of the fine structure of matter. Faraday's studies of gold colloids and thin gold films have left a rich source of information about his methods and a way of recovering the phenomena that Faraday observed. Faraday designed experiments to test hypotheses, kept extensive notes, and struggled for appropriate ways to produce phenomena that would help him visualize the interaction of light and matter at the atomic level. In other words, he created artifacts in an effort to answer questions about phenomena. These artifacts constitute a space of negotiation

---

[2]Gorman and Shrager are currently exploring these possibilities in a pilot research project involving graduate students entering nanotechnology and systems engineering.

that is closely related to the visual space explored by scientists in Trickett et al.'s and Gooding's chapters, because they also attempted to improve visualizations, just as Faraday struggled to find the right physical representations of his growing understandings. Tweney et al.'s account reveals the way in which Faraday's private speculations were eventually translated into public documents and artifacts intended to demonstrate phenomena, rather than simply explore them. Note also that Tweney et al.'s replications of Faraday's procedures constitute something like an attempt to merge the *in historico* method with other methods—a kind of real-world version of *in silico* investigations. It is important to emphasize that these replications add a necessarily very personal dimension to the analysis of Faraday's research. This personal dimension is also examined in Gooding's chapter, and it corresponds to the early, exploratory stages of the discovery process, as does Shrager's diary analysis.

Thagard's chapter, 8, provides a useful end-point to a series of chapters devoted to science. He took a list of successful habits for scientists generated by workshop members (see Preface) and compared them with the recommendations of three eminent scientists. In the workshop itself, Thagard recommended that the participants consider a space of questions a scientist might pursue. Finding an interesting problem that one can solve is not a trivial matter, not least because interest and importance do not necessarily go hand in hand with solvability. Answers to many of the questions that Darwin and Faraday articulated near the start of their careers emerged over several decades (Gruber, 1974; Tweney & Gooding, 1991); others eluded them entirely. Harder still is to find a problem that has the potential to make a breakthrough. As Thagard notes, Herb Simon recommended going off the beaten path to find such problems. He had a particular knack for cultivating collaborators as he charged into the unknown. Similarly, James Watson advocated taking risks, but having very bright colleagues and mentors on whom to fall back.

Thagard's chapter highlights the importance of collaboration, an activity partly captured by having a dyadic category in one's evolving framework. However, within this dyadic category there ought to be an analysis of different kinds of collaborations. For example, some stay within a domain, and others stretch across disciplines. In some collaborations the work of each participant is easily distinguished from the other, and in others the whole is greater than the sum of the parts. Many collaborations involve entire teams of researchers, stretching beyond the dyadic category.

In chapter 10, Ippolito uses insights from the literature on scientific thinking to help readers understand Virginia Woolf's development as a writer, including her role in creating a new kind of stream-of-consciousness novel. Woolf kept a diary of reflections on her own process and generalizations about what it took to be a writer. She also fished wherever she went for interesting details about other human beings, and noted them—creating a space of "observations" where she conducted perceptual rehearsal, gaining "practiced familiarity with certain classes of problems" and from which she constructed representations she hoped to share with her readers. The process of creating these representations was rigorous and transformative. As Woolf (1926, p. 135) wrote, "Life is subjected to a thousand disciplines and exercises. It is curbed; it is killed. It is mixed with this, stiffened with that, brought into contrast with something else; so that when we get our scene ... a year later the surface signs by which we remembered it have disappeared." Ippolito compares Woolf's process to the way in which scientists such as Faraday, Newton, Maxwell, and Einstein developed sophisticated representations that led to discoveries.

INVENTION

Is the construction of the extended narrative of a novel more like an invention than a discovery? Chapters 11–14 are an attempt to understand the kind of thinking that goes into the development of new technologies.

In chapter 11, Bradshaw focuses on the Rocket Boys described by Homer Hickham in his book of the same name. In his earlier work, Bradshaw invoked a problem-space framework to explain why the Wright Brothers succeeded; while their competitors search only through a design space, the Wrights considered both design and function spaces. Bradshaw's analysis of the Rocket Boys includes a larger array of more specific problem spaces, such as types of fuel, fins, and alternatives for the nozzle geometry. Bradshaw maintains that this decomposition of the design space into more specific problem spaces is one reason the boys succeeded,[3] but they could not test all possible variations arising from the combinations of these multiple factors. Instead, the boys developed a shared mental model of rocket design. This model evolved as they studied sources and interacted with experts, such as their teachers. The boys also took good notes on their experiments and results and found a good division of labor within the team.

Hughes's chapter, 12, illustrates one of the key problems facing inventors: integrating components that have been shown to work into a larger system that also performs as specified. This is known as the problem of *decomposition*. In his earlier work, Hughes has shown how Edison was an inventor of systems (Hughes, 1977); in his more recent work, Hughes has shown how systems engineers create not just new artifacts but also the systems in which these artifacts will play a role (Hughes, 1998). Hughes's work

---

[3]In their use of this strategy, the boys unwittingly followed in the footsteps of Nobel laureate John Bardeen, who was taught decomposition by Eugene Wigner: "The first step was to decompose the problem, either into smaller problems with less scope or into simpler problems that contained the essence of the larger problem" (Hoddeson & Daitch, 2002, p. 54).

suggests the importance of including in one's framework the extent to which scientists and inventors are focusing on redesigning systems.

Chapter 12 reminds readers that new technological systems evolve through the work of many different actors. These occupy a space in which methods, techniques, meanings, and even the very basis for communication are negotiated: a trading zone (Fujimura, 1992; Galison, 1996; Star & Griesemer, 1989). Gorman lays out in chapter 13 a framework for studying and understanding the kinds of trading zones that emerge around technological systems. The effectiveness of the trading zone depends, in part, on the nature of the network linking participants. A network represents both the connectivity or patterns of communication between actors and the power or authority structure. What Gorman calls a *State 1* is a structure dominated by one party or by the viewpoint of a single person or group. A *State 2* is a zone where participants are relatively equal and communicate via a creole that is compatible with several interpretations and approaches to a problem. A *State 3* occurs when participants develop a shared mental model. Therefore, Gorman suggests that a developed cognitive approach to science and technology should include trading zones, especially when considering research that depends on dyads, teams, and entire systems.

Because one of our goals in this volume is to establish that cognitive studies of science and technology have relevance outside of academia, a representative from industry who also has strong academic credentials has contributed a chapter to this volume. Brad Allenby is the creator of Earth Systems Engineering Management (ESEM), which involves a new way of thinking about environmental issues that incorporates insights from the literature on science and technology studies. Human technological systems have been transforming nature for thousands of years, but the pace has accelerated. Allenby argues in chapter 4 that there is no point in trying to undo these effects; instead, scientists' responsibility is to manage these complex systems intelligently. To succeed, ESEM will require the formation of State 3 networks regarding environmental systems, because the complexity of these systems and their vulnerability to unexpected perturbations require a continuous dialogue among stakeholders. ESEM is one kind of activity that requires searches in multiple spaces at different levels and particular attention to how these spaces combine at the systems level.

SUMMARY

Table 1.1 shows the chapters in this volume organized in a matrix that focuses on research methodology on one axis and problem space on another, in context with a few other seminal studies, shown in italics. The studies in this volume are mostly *in vivo* and *sub specie historiae,* as one of our goals was to demonstrate the value of these approaches in combination with others.

TABLE 1.1

Studies in This Volume Classified by Research Methodology
and Problem Space

| Problem Spaces | Research Methodology | | | | | |
| | Sub Specie | | | | | |
	In Vitro	In Vivo	Ex Vivo	Historiae	In Silico	In Magnetico
Experiment	Big Trak[a]	Chapter 3				
Hypothesis/ model	Chapter 3					
Anomalies		Chapter 5				
Visualizations		Chapter 5		Chapters 7, 9		
External memory aids		Chapter 6		Chapter 10		
Questions		Chapter 8			Chapter 8	
Design				Chapter 11		
Function				Chapter 11		

[a]Shrager and Klahr (1986).

Most of the studies described in this book were also domain specific, although at different levels and with different types of specificity. The Big Trak study (Shrager & Klahr, 1986) is noted for comparison as an example of an *in vivo* general problem-solving study not linked to a specific domain.

Table 1.2 shows another dimension on which studies could be classified: whether a group or a system is the primary focus of the analysis. Of course, there could be finer gradations in group size, perhaps following the social evolutionary framework that distinguishes among group, tribal, and larger organizational structures (Caporael, 1997). All these levels of human interaction depend on a close coupling between human and nonhuman actants, as Nersessian, and Tweney et al., and Allenby point out in chapters 2, 7, and 14. The systems level reflects a scale-up in complexity as well as in the number of actors and actants. One could use Gorman's three network states to distinguish among types of trading zones.

These tables are meant to be provocative rather than comprehensive, and we invite readers to incorporate additional studies and develop new

TABLE 1.2

Chapters That Treat Dyads, Teams, or Systems as the Primary Unit of Analysis

Dyad	Team	System
Trickett et al.'s astronomers	Nersessian's biomedical engineering laboratories	Hughes
Thagard	Dunbar's molecular biology laboratories	Gorman
	Bradshaw's Rocket Boys	Allenby

categories. The point is to find heuristics for comparing cognitive studies of science and technology. In the case of this volume, our heuristic framework highlights the focus on detailed case studies of domain-specific problem solving. In chapter 15, we reconsider this framework and point the way toward future research.

REFERENCES

Bijker, W. E. (1995). *Of bicycles, bakelites and bulbs: Toward a theory of sociotechnical change*. Cambridge, MA: MIT Press.

Bruner, J., Goodnow, J., & Austin, G. (1956). *A study of thinking*. New York: Wiley.

Caporael, L. (1997). Evolution of truly social cognition: The core configurations model. *Personality and Social Psychology Review, 1*, 276–298.

Carlson, W. B., & Gorman, M. E. (1990). Understanding invention as a cognitive process: The case of Thomas Edison and early motion pictures, 1888–1891. *Social Studies of Science, 20*, 387–430.

Christensen, C. M. (1997). *The innovator's dilemma: When new technologies cause great firms to fail*. Boston: Harvard Business School Press.

Churchland, P. M. (1989). *Mind and science: A neurocomputational perspective*. Cambridge, MA: MIT Press.

Drakes, S. (1978). *Galileo at work: His scientific biography*. Chicago: University of Chicago Press.

Dunbar, K. (1999). How scientists build models: *In vivo* science as a window on the scientific mind. In L. Magnani, N. Nersessian, & P. Thagard (Eds.), *Model-based reasoning in scientific discovery* (pp. 85–100). New York: Kluwer Academic.

Evans, P., & Wurster, T. S. (2000). *Blown to bits: How the new economics of information transforms strategy*. Boston: Harvard Business School Press.

Feynman, R., & Leighton, R. (1985). *Surely you're joking, Mr. Feynman*. New York: Norton.

Fujimura, J. (1992). Crafting science: Standardized packages, boundary objects and "translation." In A. Pickering (Ed.), *Science as practice and culture* (pp. 168–213). Chicago: University of Chicago Press.

Galison, P. (1996). *Computer simulations and the trading zone*. Stanford, CA: Stanford University Press.

Galison, P. (1997). *Image & logic: A material culture of microphysics*. Chicago: University of Chicago Press.

Giere, R. N. (Ed.). (1992). *Cognitive models of science* (Vol. 15). Minneapolis: University of Minnesota Press.

Gleick, J. (1987). *Chaos: Making a new science*. Harmondsworth, England: Penguin.

Gooding, D. (1990). Mapping experiment as a learning process: How the first electromagnetic motor was invented. *Science, Technology and Human Values, 15*, 165–201.

Gooding, D. C. (in press). Varying the cognitive span: Experimentation, visualization and digitization. In H. Radder (Ed.), *Scientific experimentation and its philosophical significance*. Pittsburgh, PA: University of Pittsburgh Press.

Gooding, D. C., & Addis, T. (1999). A simulation of model-based reasoning about disparate phenomena. In P. Thagard (Ed.), *Model-based reasoning in scientific discovery* (pp. 103–123). New York and London: Kluwer Academic/Plenum.

Gorman, M. E. (1989). Error, falsification and scientific inference: An experimental investigation. *Quarterly Journal of Experimental Psychology, 41A*, 385–412.

Gorman, M. E. (1992). *Simulating science: Heuristics, mental models and technoscientific thinking*. Bloomington: Indiana University Press.

Gorman, M. E. (1997). Mind in the world: Cognition and practice in the invention of the telephone. *Social Studies of Science, 27*, 583–624.

Gorman, M. E., Mehalik, M. M., Carlson, W. B., & Oblon, M. (1993). Alexander Graham Bell, Elisha Gray and the speaking telegraph: A cognitive comparison. *History of Technology, 15*, 1–56.

Gruber, H. E. (1974). *Darwin on man: A psychological study of scientific creativity*. London: Wildwood House.

Hoddeson, L., & Daitch, V. (2002). *True genius: The life and science of John Bardeen*. Washington, DC: Joseph Henry.

Holmes, F. L. (1989). *Hans Krebs: The formation of a scientific life, 1900–1933* (Vol. I). Oxford, England: Oxford University Press.

Holmes, F. L. (1989). Antoine Lavoisier and Hans Krebs: Two styles of scientific creativity. In D. B. Wallace & H. E. Gruber (Eds.), *Creative people at work: Twelve cognitive case studies* (pp. 44–68). Oxford, England: Oxford University Press.

Hughes, T. P. (1977). Edison's method. In W. B. Pickett (Ed.), *Technology at the turning point*. San Francisco: San Francisco Press.

Hughes, T. P. (1998). *Rescuing Prometheus*. New York: Pantheon.

Klahr, D., & Dunbar, K. (1988). Dual space search during scientific reasoning. *Cognitive Science, 12*, 1–48.

Kulkarni, D., & Simon, H. A. (1988). The processes of scientific discovery: The strategies of experimentation. *Cognitive Science, 12*, 139–175.

Langley, P., Simon, H. A., Bradshaw, G. L., & Zytkow, J. M. (1987). *Scientific discovery: Computational explorations of the creative processes*. Cambridge, MA: MIT Press.

Law, J. (1987). Technology and heterogeneous engineering. In W. E. Bjiker, T. P. Hughes, & T. J. Pinch (Eds.), *The social construction of technological systems* (pp. 111–134). Cambridge, MA: MIT Press.

Mayr, E. (1991). *One long argument*. London: Penguin.

Miller, A. (1986). *Imagery in scientific thinking*. Cambridge, MA: MIT Press.

Mynatt, C. R., Doherty, M. E., & Tweney, R. D. (1978). Consequences of confirmation and disconfirmation in a simulated research environment. *Quarterly Journal of Experimental Psychology, 30*, 395–406.

Nonaka, I., & Takeuchi, H. (1995). *The knowledge-creating company: How Japanese companies create the dynamics of innovation.* New York: Oxford University Press.

Principe, L. M. (1998). *The aspiring adept: Robert Boyle and his alchemical quest: Including Boyle's "lost" Dialogue on the transmutation of elements.* Princeton, NJ: Princeton University Press.

Rudwick, M. J. S. (1976). The emergence of a visual language for geology, 1760–1840. *History of Science, 14,* 149–195.

Shrager, J., & Klahr, D. (1986). Instructionless learning about a complex device: The paradigm and observations. *International Journal of Man–Machine Studies, 29,* 153–189.

Shrager, J., & Langley, P. (1990). *Computational models of scientific discovery and theory formation.* San Mateo, CA: Morgan Kaufmann.

Simon, H. L., Langley, P. W., & Bradshaw, G. (1981). Scientific discovery as problem solving. *Synthese, 47,* 1–27.

Star, S. L., & Griesemer, J. R. (1989). Institutional ecology, "translations" and boundary objects: Amateurs and professionals in Berkeley's museum of vertebrate zoology, 1907–39. *Social Studies of Science, 19,* 387–420.

Tweney, R. D. (1989). Faraday's discovery of induction: A cognitive approach. In F. A. S. L. James (Ed.), *Faraday rediscovered* (pp. 189–209). New York: American Institute of Physics.

Tweney, R. D. (1990). Five questions for computationalists. In P. Langley (Ed.), *Computational models of scientific discovery and theory formation* (pp. 471–484). San Mateo, CA: Morgan Kaufmann.

Tweney, R. D. (2001). Scientific thinking: A cognitive–historical approach. In T. Okada (Ed.), *Designing for science: Implications for everyday, classroom, and professional settings* (pp. 141–173). Mahwah, NJ: Lawrence Erlbaum Associates, Inc.

Tweney, R. D., & Gooding, D. C. (1991). *Michael Faraday's chemical notes, hints, suggestions and objects of pursuit of 1822.* London: Peregrinus/IEE.

Tweney, R. D., & Yachanin, S. A. (1985). Can scientists assess conditional inferences? *Social Studies of Science, 15,* 155–173.

Wallace, D. B., & Gruber, H. E. (1989). *Creative people at work: Twelve cognitive case studies.* New York: Oxford University Press.

Wason, P. C. (1960). On the failure to eliminate hypotheses in a conceptual task. *Quarterly Journal of Experimental Psychology, 12,* 129–140.

Westfall, R. S. (1980). *Never at rest: A biography of Isaac Newton.* Cambridge, MA: Cambridge University Press.

Wise, M. N. (1979). The mutual embrace of electricity and magnetism. *Science, 203,* 1310–1318.

Woolf, V. (1926). Life and the novelist. In *Collected essays* (Vol. 2, pp. 131–136). New York: Harcourt, Brace & World.

Woolgar, S. (1987). Reconstructing man and machine: A note on sociological critiques of cognitivism. In W. E. Bjiker, T. Hughes, & T. Pinch (Eds.), *The social construction of technological systems* (pp. 311–328). Cambridge, MA: MIT Press.

2

Interpreting Scientific and Engineering Practices: Integrating the Cognitive, Social, and Cultural Dimensions

Nancy J. Nersessian
Georgia Institute of Technology

Cognitive studies of science and technology ("cognitive studies") participate in two interdisciplinary fields: (a) cognitive science and (b) science and technology studies (STS). My analysis starts from issues about how cognitive studies are situated with respect to the social and cultural research programs in STS. As we will see, these issues have implications for how cognitive studies are situated within cognitive science as well. Within STS there is a perceived divide between cognitive accounts and social and cultural ("sociocultural"[1]) accounts of knowledge construction, evaluation, and transmission. Sociocultural accounts are dominant and have tended to claim that cognitive factors are inconsequential to interpreting these practices. Scientists are seen as having interests and motivations and as being members of cultures, but cognition remains, in effect, "black boxed." Cog-

---

[1] I categorize social and cultural accounts together here as *sociocultural* as a matter of convenience. "Social" and "cultural" are, of course, not coextensive notions, and analyses of these dimensions of scientific practice are quite diverse in the literature.

nitive studies accounts, for their part, have paid deference to the importance of the social and cultural dimensions of practice but have not, by and large, made these dimensions an integral part of their analyses. The situation has fostered a perception of incompatibility between cognitive and sociocultural accounts. One clear indication of this perception is the now-expired infamous "ten-year moratorium" on cognitive explanations issued first in 1986 by Bruno Latour and Stephen Woolgar (1986, p. 280; Latour, 1987, p. 247), by which time, they claimed, all pertinent aspects of science would be explained in terms of sociocultural factors. Perceptions to the contrary, any such divide is artificial. Producing scientific knowledge requires the kind of sophisticated cognition that only rich social, cultural, and material environments can enable. Thus, the major challenge for interpreting scientific and engineering knowledge-producing practices is to develop accounts that capture the fusion of the social–cognitive—cultural dimensions in these.

I argue in this chapter that the perception stems not from a fundamental incompatibility between cognitive and sociocultural accounts of science and technology but rather arises from the fact that integration has been hampered by implicit and explicit notions of "cognition" used on both sides of the perceived divide. Implicit echoes of Cartesian dualism underlie the anticognitive stance in sociocultural studies, leading to sociocultural reductionism. On this side, Cartesianism is rejected as untenable but, rather than developing an alternative theory to encompass cognitive explanatory factors, these are rejected outright. Within cognitive studies, these echoes are more explicit in their association with the traditional cognitive science view of cognition connected with GOFAI ("Good Old Fashioned AI" [coined by Haugeland, 1985]). The founding "functionalist" assumption of AI, that has in turn dominated cognitive science, is that thinking or intelligence is an abstractable structure that can be implemented in various media, including computers and humans. Cognitive reductionism identifies cognition with symbol processing that, in humans, takes place within an individual mind. Research in cognitive studies of science supports the position that important aspects of the representational and reasoning practices of scientists and engineers cannot be explained without invoking cognitive structures and processes. However, this large body of research, especially "in vivo" (coined by Dunbar, 1995) observational studies and "cognitive–historical" (coined by Nersessian, 1992; see also Nersessian, 1995b) studies, has led equally to recognizing that the social, cultural, and material environments in which science is practiced are critical to understanding scientific cognition (see, e.g., Dunbar 1995; Giere, 1988, 2002; Gooding, 1990; Gorman, 1997; Gorman & Carlson, 1990; Kurz & Tweney, 1998; Nersessian, 1984, 1995, 2002b; Thagard, 2000; Tweney, 1985, 2002). Accommodating these insights requires inserting a third approach to interpreting science and engineering practices—one that can serve as a *via media* in that it is nonreductive. The main purpose of this chapter, and an important part of the agenda for this volume, is to theorize cognition in relation to context or environment.

One route to attaining integration is to reconceptualize "cognition" by moving the boundaries of representation and processing beyond the individual so as to view scientific and engineering thinking as a complex system encompassing cognitive, social, cultural, and material aspects of practice. This direction is being pursued for accounts of mundane cognition in contemporary cognitive science, where proponents of such accounts refer to them as *embodied* and *embedded*. These accounts challenge central assumptions of GOFAI, and so the research is creating controversy within the field of cognitive science. To date, it has played little role in either cognitive or sociocultural studies of science. Accounts within this emergent research paradigm, which I call *environmental perspectives*, seek to provide explanations of cognition that give substantial roles to bodily and sociocultural factors. Advocates of environmental perspectives argue that the traditional symbol-processing view has mistaken the properties of a complex *cognitive system*, comprising both the individual and the environment, for the properties of an individual mind. They aim to develop an analytical framework in which cognitive processes are not separated from the contexts and activities in which cognition occurs. In this chapter I argue that a promising path to integration of cognitive and sociocultural dimensions of scientific and engineering practices lies in developing studies that both use the research of environmental perspectives on the social–cognitive–cultural nexus and contribute to its development.

THE CARTESIAN ROOTS OF COGNITIVE AND SOCIAL REDUCTIONISM IN STS

What, besides a penchant for rhetorical flourish, could explain such a pronouncement as the 10-year moratorium? One can agree that scientists are human in that they have interests, motivations, and sociocultural loci in conducting research. However, they also have sophisticated cognitive capabilities that historical records and contemporary practices provide strong evidence that they use in doing science. The roots of the position expressed in the 10-year moratorium pronouncement are complex in 20th-century intellectual history in that they arise as a reaction against a mix of issues, including the history of ideas approach to the history of science, the internal–external distinction in history and in sociology of science, the perceived hegemony of philosophical accounts of scientific knowledge, and the logicist "rules and representations" account of thinking of GOFAI analyses of science in early cognitive science. My concern here is with the Cartesian thread that runs through all of these.

The vision of early cognitive studies of science grew out of Herbert Simon's (Simon, Langley, & Bradshaw, 1981) important idea that scientific discovery involves problem-solving processes that are not different in kind from the problem-solving processes used in mundane circumstances. Coupled with the functionalist assumption of GOFAI, this insight led to attempts to abstract problem solving heuristics, and implement them in AI "scientific discovery" programs capable of making important scientific discoveries, such as was claimed for Kepler's laws (Langley, Simon, Bradshaw, & Zytkow, 1987) and the Krebs cycle (Kulkarni & Simon 1988). Those who dismiss cognitive explanations countered that when one studies, for example, the practices of high energy particle physicists, knowledge is produced not by what goes on in the mind of a solitary problem solver but by a "network" (Latour, 1987) or "mangle" (Pickering, 1995) of humans, machines, social arrangements, and cultures. Most researchers in contemporary cognitive studies would agree. Discovery programs are post hoc reconstructions. Once a solution is known, there are other ways to derive it. Once the data are known, a discovery program using good heuristics, such as BACON, can derive Kepler's laws. Later programs, such as KEKADA, used significant historical research (Holmes, 1980) to build systems that use many of the heuristics employed by Krebs, and, in this case, novel possible routes to the answer were also "discovered." However, what is missing from these computational accounts are the constructive processes of knowledge development, which are much more complex than simply using the appropriate heuristics. Why someone decides to collect such data, how data are selected as salient, what kinds of experimental devices and instruments are used and constructed for collection and analysis and how these are manipulated, how serendipity can play a role, and so forth, are all critical to constructing the knowledge that makes for a so-called "scientific discovery." However, discovery programs make up only a small fraction of the research in cognitive studies. The nonreductive nature of the social, cultural, and material environment is clear and agreed on in numerous cognitive studies accounts, such as those referenced earlier.

In my own research on Maxwell and the construction of the field concept, for example, I have repeatedly argued that even if one focuses on Maxwell's reasoning processes it matters a great deal to understanding how he derived the mathematical equations that Maxwell was trained in the Scottish geometrical (physical and visual) approach to using mathematics; was trained in Cambridge, England, as a mathematical physicist; was located in a milieu that valued Faraday's theoretical speculations as well as his experimental results, and included teachers and colleagues such as Thomson and his penchant for analogical models; and that he was located in Victorian Britain where, among other factors, there was widespread cultural fascination with machines and mechanisms (Crosbie Smith & Wise, 1989; Davies, 2000; Nersessian, 1984, 1992, 2002b; Siegel, 1991). These sociocultural factors, taken together with cognitive factors, help to explain the nature of the theoretical, experimental, and mathematical knowledge and the methodological practices with which Maxwell formulated the problem and approached its solution. They are reflected in Maxwell's reasoning through mechanical models in deriving the equations, and one cannot understand his construction of these equations without taking these factors into account. Continental physicists working on electromagnetism at the time, such as Ampère, used quite different practices and drew from fundamentally different theoretical assumptions and mathematical and physical representational structures (see, e.g., Hoffman, 1996). Differences in sociocultural factors figure into why members of these communities were not able to derive the field equations. However, one also cannot explain the practices of either community without taking human cognition into account.

Why, then, are cognitive accounts that underscore the importance of sociocultural dimensions not seen as compatible with, or complementary to, sociocultural accounts? One likely issue is that many, though not all, of the cognitive analyses have individual scientists and inventors at their focus. These individuals, though, are conceived as engaging in a sociocultural activity. A Maxwell wrestling alone in his study with a problem is still engaged in a sociocultural process that includes the factors discussed earlier. To find the root of the conflict one needs to consider the issue of what notions of cognition inform the cognitive and the sociocultural sides of the debate.

Cognitive Reductionism

I will begin with the cognitive side, because these accounts make explicit use of cognitive science research. Cognitive studies accounts have been constructed largely without directly challenging the assumptions underlying the traditional cognitive science view of cognition, and this view contains vestiges of a Cartesian mind–body dualism. To connect this analysis with the discussion of environmental perspectives presented in the ENVIRONMENTAL PERSPECTIVES ON COGNITION section, it is useful to focus on the assumptions of the traditional view that are highlighted by these critics. On the traditional view, the cognitive system comprises the *representations* internal to an individual mind and the internal computational *processes* that operate on these. On the functionalist assumption of that view, thinking is "disembodied" in that it is independent of the medium in which it is implemented. Also, although the environment is represented in the content of thinking through being represented in memory, cognitive processing is independent of the social, cultural, and material environment, and thus cognition is not "embedded." Recently, these founding assumptions of cognitive science were reiterated and elaborated on by Alonso Vera and Herbert Simon (1993) in response to criticisms arising from within cognitive science.

In their article, Vera and Simon (1993) argued that the characterization of the traditional view by its critics, as outlined earlier, is a caricature, or at least rests on a misunderstanding of the original claims. They contended that the traditional view does not deny the importance of embodiment and sociocultural context to cognition—indeed, Simon's (1981, pp. 63–66) early "parable of the ant" recognizes that the complexity in the ant's behavior arises from acting in the environment. Rather, the claim is that what is important about the environment for thinking processes is abstracted through perception and represented in memory by the symbols generated by the cognitive system. The unit of analysis in studying cognition is a "physical symbol system" (see also Simon & Newell, 1972). A physical symbol system has a memory capable of storing and retaining symbols and symbol structures and a set of information processes that form structures as a function of sensory stimuli. In humans, and any natural or artificial physical symbol system with sensory receptors and motor action, sensory stimuli produce symbol structures that cause motor actions and modify symbol structures in memory. Thus, a physical symbol system can interact with the environment by (a) receiving sensory stimuli from it and converting these into symbol structures in memory and (b) acting upon it in ways determined by the symbol structures it produces, such as motor symbols. Perceptual and motor processes connect symbol systems with the environment and provide the semantics for the symbols. Clearly, then, Vera and Simon claimed, cognition is embodied and embedded but also takes place within the individual physical symbol system.

Granting the subtleties of Vera and Simon's (1993) rearticulation of the traditional view, one can see that it still complies with the Cartesian characterization. First, cognition is independent of the medium in which is it implemented. The physical nature of the patterns that constitute symbols is irrelevant. The processing algorithms are media independent. It makes no difference whether the medium is silicon or organic or anything else. So, 'mind' and 'medium' are independent categories. Second, the social and cultural environments in which cognition occurs are treated as abstract content on which cognitive processes operate. These dimensions are examined only as sociocultural knowledge residing inside the mind of a human individual or internal to other physical symbol systems.

Sociocultural Reductionism

Turning now to sociocultural studies, the conception of cognition that pervades this side of the perceived divide is largely implicit. It rests on folk notions that are uninformed by research in cognitive science, or even just in psychology. The best way to understand why these accounts reject the explanatory significance of factors pertaining to human cognition is to see the rejection as stemming from a tacit understanding of cognition that also retains vestiges of Cartesian dualism. The mind–body, individual–social, and internal–external dichotomies associated with Cartesianism are all in play on the sociocultural side as well, only this time they provide justification for rejecting cognitive explanatory factors—that is, rejecting these distinctions provides the grounds for rejecting cognitive explanations. As Latour (1999) argued, a cognitive explanation is tantamount to maintaining the epistemological position that the source of knowledge is ideas internal to the mind, where "mind" is a ghostly presence in a physical vessel. Cognitive explanations are cast out in a reactionary response to seeing dualism and GOFAI as providing the only possible ways of understanding 'mind' and 'cognition.' Reductionism is thus taken in the other direction. Sociocultural studies replace cognitive reductionism with sociocultural reductionism. Banishing cognitive explanatory factors amounts to "throwing out the baby with the bath water."

First, cognition is thrown out because it is identified with internal mental processes. Second, there is a disconnect between cognition and behavior. Actions are seen as resulting from the social, cultural, and material environments in which they occur, and from motivations and interests, which are customarily considered noncognitive factors. Cognition is "black boxed" and not part of the explanatory mix in analyzing knowledge construction. Third, the individual is held to be the wrong unit of analysis. In the "actor network," agency is not located specifically in humans. All actors—human and artifactual—are on equal footing. Cognition is rejected as an explanatory category because, traditionally, it belongs to individuals conceived as loci of solitary mental processing, independent of cultures and communities. These are all indications that an implicit belief that Cartesianism is "the only game in town" underlies sociocultural reductionism.

Rapprochement

Vestiges of Cartesianism on both sides of the divide in STS have been serving to create it. On the one hand, the traditional GOFAI account has not received explicit challenge from researchers in cognitive studies of science and engineering. On the other hand, a Cartesian conception of cognition serves as a basis for rejecting the relevance of cognitive explanatory factors by sociocultural studies. What is needed, instead, is a way of theorizing the cognitive, social, and cultural aspects of practice in relation to one another. Progress toward an integrative account is being hampered by assumptions from which research on both sides of the divide, in fact, points away. On the one side, the best way of reading the cumulative results of observational and cognitive–historical research in cognitive studies is as providing a challenge to the notion that the social, cultural, and material worlds of practice can be reduced to a few parameters in a traditional account of cognition. On the other side, the moratorium has ended. Indeed, even Latour (1999) has made good on his original promise (Latour 1987, p. 247) to "turn to the mind" if anything remained to be explained after the 10-year period. He has turned to the mind in order to discuss the relativism and realism debate in the "science wars," but what he says is pertinent here (Latour, 1999). Latour traced the roots of this debate to the Cartesian "mind-in-a-vat" that places the world external to mind and has that mind trying to understand the world by looking out from the vessel in which it resides (1999, pp. 4–10). He argued that research in sociocultural studies has established that knowledge production lies not within the mind but in the rich social, cultural, and material worlds of practices. Thus, the way forward is for mind to "*reconnect* through as many relations and vessels as possible with the rich vascularization that makes science flow" (1999, p. 113). Others in sociocultural studies are also moving toward accounts that can be read as taking note of cognition, such as Peter Galison's (1997) concern with the "image" and "logic" traditions in the material culture of particle physicists, Karin Knorr Cetina's (1999) recent analysis of scientific practices as part of "epistemic cultures," and Hans-Jörg Rheinberger's (1997) analysis of experimentation in molecular biology as producing "epistemic things." The time is ripe for rapprochement. Combined, research on the cognitive and sociocultural sides shows the divide to be artificial. There is a need for a new account of the social–cognitive–cultural nexus adequate to interpret scientific and engineering practices.

Within contemporary cognitive science there is movement toward an understanding of cognition, where "cognition refers not only to universal patterns of information transformation that transpire inside individuals but also to transformations, the forms and functions of which are shared among individuals, social institutions, and historically accumulated artifacts (tools and concepts)" (Resnick, Levine, & Teasley, 1991, p. 413). These accounts were not developed in response to the issues within STS discussed earlier, but I believe they offer significant groundwork for thinking about the integration problem. In the next section I present a brief analysis that weaves together significant threads of this research.

ENVIRONMENTAL PERSPECTIVES ON COGNITION

Some time ago, several cognitive scientists began expressing dismay with the "cognitive paradigm" as it had developed thus far and began calling for what they saw as a fundamental revisioning of the notion of cognition. Donald Norman (1981) posed the challenge:

> The human is a social animal, interacting with others, with the environment and with itself. The core disciplines of cognitive science have tended to ignore these aspects of behavior. The results have been considerable progress on some fronts, but sterility overall, for the organism we are analyzing is conceived as pure intellect, communicating with one another in logical dialog, perceiving, remembering, thinking when appropriate, reasoning its way through well-formed problems that are encountered in the day. Alas the description does not fit actual behavior. (p. 266)

Traditional cognitive science research attempts to isolate aspects of cognition to study it on the model of physics—the "spherical horses" approach.[2] Although traditional studies are still the mainstay of cognitive science, over the last 20 years significant investigations of cognition in authentic contexts of human activity such as learning and work have become numerous. These examinations range from studies of the effects of sociocultural milieu on categorization, conceptualization, and reasoning, to primate studies relating the emergence of culture and the evolution of human cognition, to neuroscience studies examining the potential of the human brain to be altered by the sociocultural environment of development. These various research thrusts can be characterized as attempts to account for the role of the environment (social, cultural, and material) in shaping and participating in cognition. Many of these analyses make *action* the focal point for understanding human cognition. Human actors are construed as thinking in complex environments; thus these analyses have emphasized that cognition is "embodied" (see, e.g., Barsalou, 1999; Glenberg, 1997; Glenberg & Langston, 1992; Johnson, 1987; Lakoff, 1987; Lakoff & Johnson, 1998) and "embedded," which, variously, are construed as "distributed" (see, e.g., Hutchins, 1995; Norman, 1988; Zhang, 1997; Zhang & Norman, 1995), "encultured" (see, e.g., Donald, 1991; Nisbett, Peng, Choi, & Norenzayan, 2001; Shore, 1997; Tomasello, 1999), or "situated" (see, e.g., Clancey, 1997; Greeno, 1989a, 1998; Lave, 1988; Suchman, 1987).

---

[2]As noted by the editors of this volume, two significant metaphors pervaded the workshop on Cognitive Studies of Science and Technology. "Spherical horses" comes from a joke told by David Gooding: A multimillionaire offered a prize to whomever could predict the outcome of a horse race: a stockbreeder, a geneticist, or a physicist. The stockbreeder said there were too many variables, the geneticist said the prediction could not be made about any horse in particular, and the physicist claimed the prize: physics could make the prediction accurately to many decimal places, provided the horse were conceived as perfectly spherical and moving through a vacuum. "Shared toothbrushes" came from an observation made by Christian Schunn that, as with toothbrushes, no academic wants to use someone else's theoretical framework.

In contrast to the traditional cognitive science construal of the environment as mental content on which cognitive processes operate, these perspectives maintain that cognitive processes cannot be treated separately from the contexts and activities in which cognition occurs. For example, in arguing for a distributed notion of cognition, Edwin Hutchins (1995) contended that rather than construing culture as content and cognition as processing, what is required is for "cognition" and "culture" to be seen as interrelated notions construed in terms of *process*. Such construal leads to a shift in theoretical outlook from regarding cognitive and sociocultural *factors* as independent variables to regarding cognitive and sociocultural *processes* as integral to one another. The environmental perspectives maintain that the traditional view has mistaken the properties of a complex *cognitive system*, comprising individuals and environment, for the properties of an individual mind. The main points of contention are *not* whether the environment can be accommodated but rather *whether accounting for environmental factors requires altering fundamental notions of the structures and processes that make up cognition and of the methods through which to investigate cognition*. The argument is about the very nature of cognition and how to investigate it.

Broadly characterized, the challenges posed by the environmental perspectives to the traditional cognitive science view center on three interrelated questions: (a) What are the bounds of the cognitive system? (b) what is the nature of the processing used in cognition? and (c) what kinds of representations—internal and external—are used in cognitive processing? The literature of environmental perspectives is by now quite extensive, so it will not be possible to lay out any position in detail. Also, the research that falls under this label is wide ranging, and there is as yet not much dialogue among areas. What I present here is a way to read a cross-section of the literature so as to highlight features of research I see as most pertinent to the project of reconceptualizing the social–cognitive–cultural nexus in STS. I begin by discussing the "situative perspective" (Greeno, 1998) and then link aspects of other perspectives to this discussion.

Situated and Distributed Cognition

Much of the impetus for developing theories of *situated cognition* has come from studies conducted by cognitive anthropologists and sociologists concerned with learning and with work practices. Jean Lave, for instance, has attempted to explain ethnographical studies that establish striking disparities between mathematical problem-solving competency in the real world and in school learning environments. In real world environments, such as supermarkets (Lave, 1988) and Brazilian street markets (Carraher, Carraher, & Schliemann, 1983), adults and children exhibit high levels of competence in solving mathematics problems that are structurally of the same kind as those they fail at solving in standard school and test formulations. Lave (1988) argued that the way to explain the disparities is to construe the relation between cognition and action as an interactive process in which the resources available in a specific environment play an essential role. Cognition is a relation between individuals and situations and does not just reside in the head. Explanations of human cognition in the situative perspective use the notion of *attunement to constraints and affordances*, adapted from Gibson's (1979) theory of perception. On the situative adaptation, an *affordance* is a resource in the environment that supports an activity, and a *constraint* is a regularity in a domain that is dependent on specific conditions.

The structure of an environment provides the constraints and affordances needed in problem solving, including other people, and these cannot be captured in abstract problem representations alone. In traditional cognitive science, problem solving is held to involve formulating in the abstract the plans and goals that will be applied in solving a problem. However, ethnographical studies of work environments by Lucy Suchman (1987) led her to argue that, contrary to the traditional cognitive science view, plans and goals develop in the context of actions and are thus *emergent* in the problem situation. Problem solving requires improvisation and appropriation of affordances and constraints in the environment, rather than mentally represented goals and plans specified in advance of action.

Within the situative perspective, analysis of a cognitive system, which James Greeno (1998) called an *intact activity system*, can focus at different levels: (a) on the individual, now conceptualized as an embodied, social, tool-using agent; (b) on a group of agents; (c) or on the material and conceptual artifacts of the context of an activity, or on any combination of these. In all cases, the goal is to understand cognition as an interaction among the participants in, and the context of, an activity. Cognition thus is understood to comprise the interactions between agents and environment, not simply the possible representations and processes in the head of an individual. In this way, situated cognition is *distributed*.

As with the situative perspective, the *distributed cognition* perspective contends that the environment provides a rich structure that supports problem solving. An environment does not, however, just supply scaffolding for mental processes, as the traditional view maintains. Rather, aspects of the environment are integral to the cognitive system and thus enter essentially into the analysis of cognition. To accommodate this insight, an account of cognitive processing needs to incorporate the *salient* resources in an environment in a nonreductive fashion. Salient resources are, broadly characterized, factors in the environment that can affect the outcome of an activity, such as problem solving. These cannot be determined *a priori* but need to be judged with respect to the instance. For ship navigators, for example, the function of a specific instrument would be salient to piloting the ship, but the material from which the instrument is made usually would not. For physicists, sketching on a blackboard or whiteboard or piece of paper is likely irrelevant to solving a problem, but sketching on a computer screen might be salient, because the computer adds resources that can affect the outcome. On the other hand, sketching on a board usually takes place when others are present and possibly assisting in the problem solving, and sketching on paper is often done for oneself, and so other details of a case could change what is considered salient.

Determining the *cognitive artifacts* within a specific system is a major part of the analytical task for advocates of the distributed perspective. Hutchins (e.g., 1995) has studied the cognitive functions of artifacts used in modern navigation, such as the alidade, gyrocompass, and fathometer. Various kinds of external representations are candidate cognitive artifacts, and much research has focused on visual representations, especially diagrams. Jiajie Zhang and Donald Norman (Zhang, 1997; Zhang & Norman, 1995), for example, have studied problem solving with isomorphic problems to ascertain potential cognitive functions of different kinds of visual representations. They found that external representations differentially facilitate and constrain reasoning processes. Specifically, they argue that diagrams can play more than just a supportive role in what is essentially an internal process; these external representations also can play a direct role in cognitive processing, without requiring the mediation of an internal representation of the information provided in them. The external representation can change the nature of the processing task, as when the tic-tac-toe grid is imposed on the mathematical problem of "15." One way this research contributes to breaking down the external–internal distinction is by expanding the notion of memory to encompass external representations and cues; that is, specific kinds of affordances and constraints in the environment are construed, literally, as memory in cognitive processing. Thus, Zhang and Norman (1995) argue that analysis of cognition in situations of problem solving with diagrams needs to be at the level of the cognitive system that comprises both the mental and diagrammatic representations.

Research in the situative and distributed perspectives largely consists of observational case studies in which ethnographic methods are used. Although these studies focus on details of particular cases and often provide "thick descriptions" of these (Geertz, 1973), their objective differs from sociocultural studies in STS that aim mainly to ferret out the specific details of a case. The aim of the cognitive science research discussed here is to understand the nature of the regularities of cognition in human activity. Hutchins framed that objective succinctly:

There are powerful regularities to be described at the level of analysis that transcends the details of the specific domain. It is not possible to discover these regularities without understanding the details of the domain, but the regularities are not about the domain specific details, they are about the nature of cognition in human activity. (Hutchins, as quoted in Woods, 1997, p. 117)

Currently there are many research undertakings that share the situated cognition and distributed cognition objective of furthering an account that construes cognition and environment in relation to one another. Research in all environmental perspectives areas is very much research in progress, so it tends to focus internally to an area, without much interaction across them. In the remainder of this section I provide a brief tour through significant research programs that, when considered as comprising a body of interconnected research, offer a substantially new way of understanding human cognition and of thinking about the social–cognitive–cultural nexus in science and engineering practices.

Embodied Mental Representation

Individual human agents are parts of cognitive systems, and an accounting of the nature of their mental representations and processes is an outstanding research problem for environmental perspectives. Some research in distributed cognition makes use of mainstream notions of mental representation, such as mental models and concepts. The most radical proponents of the situative perspective, however, go so far as to contend that mental representations play no role in cognitive processes. Driving a car around a familiar campus provides an example of an activity that might not require use of a mental map of the campus; the affordances and constraints in the environment could suffice for navigating to one's office. However, it is difficult to see how complex problem-solving practices, such as those in science and engineering, could simply make use of environmental affordances and constraints. A more moderate position, such as the one articulated by Greeno (1989b), maintains that although not all cognitive practices need to use mental representations, and not all information in a system needs to be represented mentally, some kinds of practices might use them. Scientific and engineering problem-solving practices are prime candidates for practices that use mental representations. However, it is unlikely that environmental perspectives can simply adopt traditional cognitive science understandings of these representations.

In thinking about the human component of a cognitive system, a line of research that examines the implications of the *embodied* nature of human cognition potentially can be appropriated. Embodied cognition focuses on the implications of the interaction of the human perceptual system with the environment for mental representation and processing. Proponents contend that there is empirical evidence that perceptual content is retained in mental representations and that perceptual and motor processes play a significant role in cognitive processing (see, e.g., Barsalou, 1999; Craig, Nersessian, & Catrambone, 2002; Glenberg, 1997; Johnson, 1987; Kosslyn, 1994; Lakoff, 1987). Psychologist Lawrence Barsalou (1999) formulated a theory of "perceptual symbol systems" that calls into question the traditional understanding of mental representation as *amodal,* or composed of symbols that are arbitrary transductions from perception. He argued, rather, that there is an extensive experimental literature that can be read as supporting the contention that mental representations retain perceptual features, or are *modal.* On Barsalou's account, cognitive processing uses "perceptual symbols," which are neural correlates of sensory experiences. These representations possess *simulation* capabilities; that is, perceptual and motor processes associated with the original experiences are re-enacted when perceptual symbols are used in thinking. One implication of this account is that situational information should be retained in concept representations, and there is abundant evidence from psychological experiments supporting this (Yeh & Barsalou, 1996). Thus, affordances and constraints of situational information can be at play even in using conceptual understanding in activities, such as in problem solving.

One highly influential account of the embodied nature of mental representation has been provided by George Lakoff and Mark Johnson, who argue that mental representations arise through metaphorical extension from bodily experiences. All representations, no matter how abstract, they contend, can be shown to derive from fundamental kinesthetic *image schemas* that structure experience prior to the formation of conceptual representations. An example of an image schema that pervades human thinking is the *"container"* schema with *"in"* and *"out"* as primary reference points to the human body (Lakoff, 1987, p. 252). The notion of being trapped in a marriage and getting out of it reflects this image schema. Another is the more complex *"force"* schema, with *interaction, directionality, path, origin,* and *degree* as dimensions of fundamental bodily interactions in the world (Johnson, 1987, pp. 41–42). One uses this schema when, for example, talking of having writer's block. Conceptual structures are cast as developing out of such schemas and thus as being meaningful in terms of these. Lakoff and Johnson argue that metaphorical extension is a universal cognitive mechanism that can accommodate observed individual and cultural variability in conceptual structure.

Cognition and Culture

In *Culture in Mind*, anthropologist Bradd Shore (1997) addressed the problem of the role of universal cognitive mechanisms in the development of mental representations, the content of which are culturally variable and context relative. His approach to the problem draws on ethnographic studies of various cultural groups to examine the interplay between the "cultural affordances" offered by local sociocultural structures and the universal cognitive processes involved in meaning making in the creation of "cultural models" exhibited in local practices. Cultural models have two dimensions: (a) the publicly available, or "instituted" form, such as in rituals and games, and (b) the mental construct or "mental model" that individuals create and use to understand the world. The instituted forms are not simply "faxed" to the mind but "undergo a variety of transformations as they are brought to mind" (Shore, 1997, p. 52). Shore's account of the transformative processes of constructing mental models uses the notion of meaning construction as involving processes of metaphorical extension, developed by Lakoff and Johnson. Shore concluded that although there are possibly an infinite variety of cultural models, the relations between culture and cognition are governed by such universal cognitive mechanisms.

Comparative studies between humans and other primates in primatology research and in the area of cognitive development have led Michael Tomasello (1999; Tomasello & Call, 1997), among others, to contend that cognition is inherently cultural. He argues that culture is central to the development of uniquely human cognitive abilities, both phylogenetically and ontogenetically. The question of the origins of these unique abilities is a key problem for understanding cognitive development. From the perspective of biological evolution, the time span is just too short to account for the vast cognitive differences that separate humans from the primates closest to them genetically, the chimpanzees. On the basis of experimental and observational studies of ontogenesis in human children and in other primates, Tomasello posits that the development of the uniquely human cognitive abilities began with a small phylogenetic change in the course of biological evolution: the ability to see conspecifics as like oneself and thus to understand the intentionality of one's actions. This change has had major consequences in that it enabled processes of imitation and innovation that allow for the accumulation of culture through transmission—what Tomasello (1999) calls *cultural evolution*.

According to the account Tomasello (1999) developed, cultural evolution is the engine of cognitive evolution; that is, he claims that the expansion of cognitive capacities in the human primate has occurred as an adaptation to culture. It is significant, then, that this account theorizes culture not as something added to accounts of cognition—culture is what makes human cognition what it is. Human cognition and culture have been co-evolving. The cultural tools of each generation (including language development) are left behind for the next generation to build upon. Tomasello (1999) called this the "rachet effect." Regardless of the fate of his claim about the root of this ability lying in a *uniquely* human ability to understand conspecifics as intentional beings (recent work shows that other primates, and dogs, might also possess the ability; Agnetta, Hare, & Tomasello, 2000; Tomasello, Call, & Hare, 1998), humans are unique in the way they pass on and build on culture. In ontogenesis, children absorb the culture and make use of its affordances and constraints in developing perspectively based cognitive representations. Tomasello (1999) argued that language development plays a crucial role in creating cognitive capacities in the processes of ontogenesis. This view parallels the early speculations of Lev Vygotsky (1978), whose work has influenced the development of the situative perspective discussed earlier. Vygotsky argued that cognitive development is sociocultural in that it involves the internalization of external linguistic processes.

Another influential comparative account that examines the relations between culture and the development of human cognitive capacities is offered by the evolutionary psychologist Merlin Donald (1991), who uses a wide range of evidence from anthropology, archaeology, primatology, and neuroscience to argue his case. One aspect of this account reinforces the notion that not all cognitive processing need be of internal representations. External representations are indispensable in complex human thinking, and their development has been central to the processes of cultural transmission. Donald's analysis of the evolutionary emergence of distinctively human representational systems starts from the significance of *mimesis*—or re-creation, such as using the body to represent an idea of the motion of an airplane—in the developments of such external representations as painting and drawing (40,000 years ago), writing (6,000 years ago) and phonetic alphabets (4,000 years ago). He argues for a distributed notion of memory as a symbiosis of internal and external representation on the basis of changes in the visuo–spatial architecture of human cognition that came about with the development of external representation. On this account, affordances and constraints in the environment are, *ab initio*, part of cognitive processing.

Research into the relations between culture and cognition, together with neuroscience research into cognitive development, can be construed as moving beyond the old nature–nurture debate and developing an *interactionist* approach. It attempts to provide an account of how evolutionary endowment and sociocultural context act together to shape human cognitive development. In support of this conception, neuroscience studies of the impact of sociocultural deprivation, enrichment, and trauma on brain structure and processes lead to a conception of the brain as possessing significant cortical plasticity and as a structure whose development takes place in response to the sociocultural environment as well as to genetic inheritance and biological evolution (see, e.g., Elman et al., 1998; van der Kolk, McFarlane, & Weisaeth, 1996).

Finally, in so connecting cognition and culture, this body of research indicates that human cognition should display both universal and culturally specific characteristics. Tomasello (1999, pp. 161–163) discussed some of the universal learning abilities, such as those connected with language learning; these include the ability to understand communicative intentions, to use role reversal to reproduce linguistic symbols and constructions, and to use linguistic symbols for contrasting and sharing perspectives in discourse interactions. Recent investigations by Richard Nisbett and his colleagues (Nisbett et al., 2001) provide evidence of culturally specific features of cognition. Their research examined learning, reasoning, problem solving, representation, and decision making for such features. This research was also inspired by the substantial body of historical scholarship that maintains that there were systematic cultural differences between ancient Greek and Chinese societies, especially concerning what Nisbett et al. (2001) call the "sense of personal *agency*" (p. 292). Nisbett et al. hypothesized that these kinds of differences between so-called Eastern and Western cultures, broadly characterized as holistic versus analytic thinking (p. 293), should be detectable in a wide range of cognitive processes, such as categorization, memory, covariation detection, and problem solving.

The comparative contemporary cultures in Nisbett et al.'s (2001) study are those whose development has been influenced either by ancient China (China, Japan, and Korea) or by ancient Greece (western Europe and North America). In a series of experiments with participants from east Asian and Western cultures, and participants whose families had changed cultural location, Nisbett et al. examined explanations, problem solving, and argument evaluation. Some significant systematic differences were found along the five dimensions they identified in the ancient cultures (in the order Eastern vs. Western): (a) focusing on continuity versus on discreteness, (b) focusing on field versus on object, (c) using relations and similarities versus using categories and rules, (d) using dialectics in reasoning versus using logical inference from assumptions and first principles, and (e) using experienced-based knowledge in explanations versus using abstract analysis. Although Nisbett et al.'s grouping of very diverse cultures into such gross categories as "Eastern" and "Western" is problematic, the general results are intriguing and promise to lead to further research into the issue of culturally specific features of cognition.

Environmental Perspectives and the Integration Problem

Situating the problem of interpreting scientific and engineering practices with respect to the framework provided by environmental perspectives on cognition affords the possibility of analyzing the practices from the outset as bearing the imprint of human cognitive development, the imprint of the sociocultural histories of the localities in which science is practiced, and the imprint of the wider societies in which science and technology develop. The implications of the growing body of environmental-perspectives research for the project of constructing integrative accounts of knowledge-producing practices in science and engineering are extensive. Working them out in detail is beyond the scope of any one chapter. One approach to exploring the implications would be to recast some of the analyses in the literatures of both cognitive studies and sociocultural studies of science and engineering in light of it. Here, for example, I am thinking of such research as by Cetina, Galison, Giere, Gooding, Gorman, Latour, Rheinberger, Tweney, and myself, cited earlier.

Another approach would be to undertake new research projects that aim from the outset at integration. In the next section I offer my current research project on interpreting knowledge-producing practices in biomedical engineering research laboratories as an exemplar of an integrative approach. This project combines ethnographic studies with cognitive–historical analyses to examine reasoning and representational practices. My colleagues and I are examining these research practices at all of the levels of analysis noted by Greeno (1998) for situated cognitive systems: at the level of researchers as individual, embodied, social, tool-using agents; at the level of groups of researchers; at the level of the material and conceptual artifacts of the context of laboratory activities; and at various combinations of these.

RESEARCH LABORATORIES AS EVOLVING DISTRIBUTED COGNITIVE SYSTEMS

Science and engineering research laboratories are prime locations for studying the social–cognitive–cultural nexus in knowledge-producing practices. Extensive STS research has established that laboratory practices are located in rich social, cultural, and material environments. However, these practices make use of sophisticated cognition in addressing research problems. In this section I discuss some features of my current research project that has among its aims the interpretation of reasoning and representational practices used in problem solving in biomedical engineering (BME) laboratories. The research both appropriates and contributes to research within the environmental perspectives discussed in the previous section. My colleagues and I do not adopt or apply any particular theory but rather use a cross-section of that thinking about the nature of cognition as a means of framing our investigation into these research practices. We are influenced also by research on both sides of the supposed divide in STS. As a contribution to STS, specifically, we aim to develop analyses of the creation of BME knowledge in which the cognitive and the sociocultural dimensions are integrated analytically from the outset. Our focus is on the cognitive practices, but we analyze cognition in BME laboratories as situated in localized reasoning and representational practices. This is collaborative research that would not be possible without an interdisciplinary team.[3] The case study has been underway for less than 2 years, so the analysis presented here is preliminary. Nevertheless, it provides a useful exemplar of how integration might be achieved.

We have begun working in multiple sites, but here I discuss a specific tissue engineering laboratory, Laboratory A, that has as its ultimate objective the eventual development of artificial blood vessels. The daily research is directed toward solving problems that are smaller pieces of that grand objective. Our aim is to develop an understanding of (a) the nature of reasoning and problem solving in the laboratory; (b) the kinds of representations, tools, forms of discourse, and activities used in creating and using knowledge; (c) how these support the ongoing research practices; and (d) the nature of the challenges faced by new researchers as they are apprenticed to the work of the laboratory.

We conceive of and examine the problem-solving activities in Laboratory A as *situated* and *distributed*. These activities are situated in that they lie in localized interactions among humans and among humans and technological artifacts. They are distributed in that they take place across systems of humans and artifacts. BME is an *interdiscipline* in that melding of knowledge and practices from more than one discipline occurs continually, and significantly new ways of thinking and working are emerging. Most important for our purposes is that innovation in technology and laboratory practices happens frequently, and learning, development, and change in researchers are constant features of the laboratory environment. Thus, we characterize the laboratory as comprising "evolving distributed cognitive systems." The characterization of the cognitive systems as *evolving* adds a novel dimension to the existing literature on distributed cognition, which by and large has not examined these kinds of creative activities.

Investigating and interpreting the cognitive systems in the laboratory has required innovation, too, on the part of our group of researchers studying the laboratory. To date, ethnography has been the primary method for investigating situated cognitive practices in distributed systems. As a method

[3]This research is conducted with Wendy Newstetter (co-PI), Elke Kurz-Milcke, Jim Davies, Etienne Pelaprat, and Kareen Malone. Within this group of cognitive scientists we have expertise in ethnography, philosophy of science, history of science, psychology, and computer science. We thank our research subjects for allowing us into their work environment and granting us numerous interviews, and we gratefully acknowledge the support of the National Science Foundation Research on Learning and Education Grant REC0106773.

it does not, however, suffice to capture the critical *historical* dimension of the research laboratory: the evolution of technology, researchers, and problem situations over time that are central in interpreting the practices. To capture the evolving dimension of the laboratory, we have developed a mixed-method approach that includes both ethnography and cognitive–historical analysis.

A Mixed-Method Approach to Investigating Evolving Distributed Cognitive Systems

None of the conceptions of distributed cognition in the current literature account for systems that have an evolving nature. In Hutchins's (1995) studies of distributed cognition in work environments—for instance, the cockpit of an airplane or on board a ship—the problem-solving situations change in time. The problems faced, for example, by the pilot change as she is in the process of landing the plane or bringing a ship into the harbor. However, the nature of the technology and the knowledge that the pilot and crew bring to bear in those processes are, by and large, stable. Even though the technological artifacts have a history within the field of navigation, such as the ones Hutchins documented for the instruments aboard a ship, these do not change in the day-to-day problem-solving processes on board. Thus, these kinds of cognitive systems are dynamic but largely *synchronic*. In contrast, we are studying cognition in innovative, creative settings, where artifacts and understandings are undergoing change over time. The cognitive systems of the BME research laboratory are, thus, dynamic and *diachronic*. Although there are loci of stability, during problem-solving processes the components of the systems undergo development and change over time. The technology and the researchers have evolving, *relational* trajectories that must be factored into understanding the cognitive system at any point in time. To capture the evolving dimension of the case study we have been conducting both cognitive–historical analyses of the problems, technology, models, and humans involved in the research and ethnographic analyses of the day-to-day practices in the laboratory.

Ethnographic analysis seeks to uncover the situated activities, tools, and interpretive frameworks used in an environment that support the work and the ongoing meaning-making of a community. Ethnography of science and engineering practices aims to describe and interpret the relations between observed practices and the social, cultural, and material contexts in which they occur. Our ethnographic study of the BME laboratory develops traces of transient and stable arrangements of the components of the cognitive systems, such as evidenced in laboratory routines, the organization of the workspace, the artifacts in use, and the social organization of the laboratory at a particular point in time, as they unfold in the daily research activities and

ground those activities. Ethnographic studies of situated sociocultural practices of science and engineering are abundant in STS (see, e.g., Bucciarelli,
1994; Latour & Woolgar, 1986; Lynch, 1985). However, studies that focus on
situated *cognitive* practices are few in number in either STS or in cognitive science. Furthermore, existing observational (Dunbar, 1995) and ethnographic
studies (see, e.g., Goodwin, 1995; Hall, Stevens, & Torralba, in press; Ochs &
Jacoby, 1997) of scientific cognition lack attention to the historical dimension that we find important to our case study.

Cognitive–historical analysis enables one to follow trajectories of the human and technological components of a cognitive system on multiple levels, including their physical shaping and reshaping in response to problems,
their changing contributions to the models that are developed in the laboratory and the wider community, and the nature of the concepts that are at
play in the research activity at any particular time.[4] As with other cognitive–historical analyses, we use the customary range of historical records to
recover how the representational, methodological, and reasoning practices
have been developed and used by the BME researchers. The practices can
be examined over time spans of varying length, ranging from shorter spans
defined by the activity itself to spans of decades or more. The aim of cognitive–historical analysis is to interpret and explain the generativity of these
practices in light of salient cognitive science investigations and results
(Nersessian, 1992, 1995b). Saliency is determined by the nature of the
practices under scrutiny. In this context, the objective of cognitive–historical analysis is not to construct an historical narrative; rather, the objective is
to enrich understanding of cognition in context by examining how
knowledge-producing practices originate, develop, and are used in science
and engineering domains.

In STS there is an extensive literature in the cognitive studies area that
uses cognitive–historical analysis. My own studies and those of many others
have tended to focus on historical individuals, including Faraday, Maxwell,
and Bell, and on developing explanatory accounts of concept formation,
concept use, and conceptual change (Andersen, 1996; Chen, 1995;
Gooding, 1990; Gorman, 1997; Gorman & Carlson, 1990; Nersessian,
1985, 1992, 2002b; Tweney, 1985). Many of these studies have argued for
the significance of the material context of concept formation, with special
focus on a wide range of external representations in interpreting concept
formation practices, such as Gooding's (1990) study of how Faraday's concept of "electromagnetic rotations" emerged through complex interactions
with sketches on paper and prototype apparatus, my own on the generative

---

[4]For a comparison of cognitive–historical analysis to other methodologies—laboratory
experiments, observational studies, and computational modeling—used in research on scientific discovery, see Klahr and Simon (1999).

role of the lines-of-force diagram on the development of his field concept (Nersessian, 1984, 1985), and Tweney's recent work on various physical manipulations of microscope slides in Faraday's developing understanding of gold (Tweney, 2002; chap. 7, this volume). They have also shown the importance of sociocultural context, as, for example, in Gooding's (1989) account of the origins of Faraday's lines-of-force concept in the material and communicative strategies of other practitioners and in my (Nersessian, 1984, 1992, 2002b) discussions of the context of Maxwell's modeling practices in mathematizing the electromagnetic field concept as noted in the section titled THE CARTESIAN ROOTS OF COGNITIVE AND SOCIAL REDUCTIONISM IN STS. When studying contemporary science and engineering, what ethnography adds is the possibility of examining both the social, cultural, and material contexts of as they currently exist and the practices as they are enacted.

In our study of BME practices thus far, the analyses are focused on the technological artifacts that push BME research activity and are shaped and reshaped by that activity. Ethnographic observations and interviews have indicated the saliency of specific artifacts in the social–cognitive–cultural systems of the laboratory, as I discuss in the section titled "THE BME LABORATORY AS AN EVOLVING DISTRIBUTED COGNITIVE SYSTEM." These artifacts become and remain part of the laboratory's history. The cognitive–historical analyses focus on reconstructing aspects of the laboratory's history. How the members of the laboratory appropriate the history and use and alter the artifacts in their daily research in turn become the focus of our ethnographic analyses. We aim to construct an account of the *lived relation* that develops between the researchers and specific artifacts, rather than an account of the developing knowledge about these artifacts *per se*. By focusing on the lived relations we mean to emphasize the activity of the artifacts in a relational account of distributed cognitive systems. These lived relations have cognitive, social, and cultural dimensions. Combining cognitive–historical analysis with ethnography allows examination of these relationships *in situ*, as they have developed—and continue to develop—in time. It is important that developing a relationship with an artifact entails appropriating its history, which chronicles the development of the problem situation, including what is known about the artifact in question. The researchers, for instance, include postdoctoral fellows, PhD students, and undergraduates, all of whom have learning trajectories. These trajectories, in turn, intersect with the developmental trajectories of the diverse technological artifacts and of the various social systems within the laboratory.

Users of an artifact often redesign it in response to problems encountered, either of a technical nature or to bring it more in accord with the *in vivo* model. To begin research, a new participant must first master the relevant aspects of the existing history of an artifact necessary to the research and then figure out ways to alter it to carry out her project as the new research problems demand, thereby adding to its history. For example, one highly significant technological artifact in Laboratory A is the *flow loop*, an engineered device that emulates the shear stresses experienced by cells within blood vessels (see Fig. 2.1). A PhD student we interviewed discussed how the previous researcher had modified the block to solve some technical problems associated with bacterial contamination—a constant problem in this line of research. The flow loop, as inherited by the new student, had previously been used on smooth muscle cells. The new student was planning to use the flow loop to experiment with vascular constructs of endothelial cells that are thicker than the muscle cells and not flat. Because the vascular constructs are not flat, spacers need to be used between the block and the glass slides in order to improve the flow pattern around the boundary to bring the *in vitro* model more in accord with the *in vivo* model. To begin that research she, together with another new student, had to re-engineer the flow loop by changing the width of the flow slit to hold the spacers.

Making sense of the day-to-day cognitive practices in a BME laboratory and constructing cognitive histories of artifacts are prima facie separate tasks. However, that the research processes in the distributed cognitive systems of Laboratory A evolve at such a fast pace necessitates going back and forth between the two endeavors. The ethnographic observations of the development, understanding, and use of particular artifacts by various laboratory members, as well as ethnographic interviews, have enabled us to conjoin the cognitive–historical study of laboratory members, laboratory objects, and the laboratory itself with an eye on the perception of these entities by the laboratory members themselves.

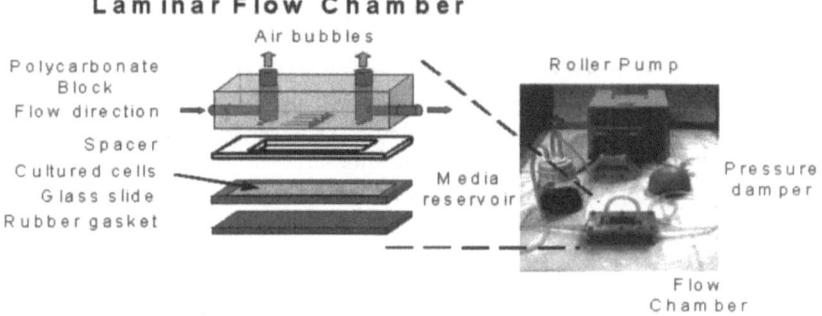

FIG. 2.1.   Diagram and photograph of a flow loop.

The BME Laboratory as an Evolving
Distributed Cognitive System

Laboratory A applies engineering principles and methods to study living cells and tissues with the goal of eventual development of artificial blood vessels to implant in the human cardiovascular system. The laboratory members come from predominantly engineering backgrounds. They tend to seek biological knowledge on an as-needed basis. Biological knowledge is embedded in the artifacts the researchers construct for simulation purposes and other model-based reasoning they use in the course of research. Early experimentation in this area was conducted by the principal investigator (PI) and others with animals *in vivo* and *ex vivo* (substitutes implanted but not kept within the body). However, *in vivo* research has many limitations, such as that one cannot test the strength of various kinds of scaffolding for blood vessels. The research has now moved *in vitro*, through the design of facsimiles of relevant aspects of the *in vivo* environment. These technological facsimiles are locally constructed sites of *in vitro* experimentation.

A major research goal is to optimize *in vitro* models so as to move closer and closer to *in vivo* situations. When used within the human body, the bioengineered substitutes must replicate the functions of the tissue being replaced. This means that the materials used to "grow" these substitutes must coalesce in a way that mimics the properties of native tissues. It also means that the cells embedded in the scaffolding material must replicate the capabilities of native cells so that the higher level tissue functions can be achieved. Moreover, the cells must be readily available. This requires developing methods and technologies for ensuring cell growth, proliferation, and production.

In vitro research in Laboratory A starts with culturing blood vessel cells, smooth muscle cells, and endothelial cells. Cells are embedded in various scaffolding materials and stimulated in environments that mimic certain aspects of the current understanding of flow processes in an effort to improve them—for example, making them proliferate or making them stronger. A significant part of creating artificial blood vessels is to have them withstand the mechanical forces associated with blood flow through vessels *in vivo*. Much of the technology created by the laboratory serves this purpose. Cells are stimulated in the *in vitro* simulation environments, and various instruments are used to extract and process information, most often pertaining to stress and strain, such as measures of elasticity (linear modulus), shear stress, ultimate tensile stress, toughness (the amount of energy it takes to break a construct), and cell volume and health under mechanical stimulation.

There are many dimensions along which to develop the analysis of the laboratory as an evolving distributed cognitive system and of the systems within it. In the following sections, I focus on our recasting of some traditional cognitive science interpretive notions by which we are attempting to break down the internal–external distinction—a major impediment to integrating cognitive and sociocultural dimensions of scientific and engineering practices. In these analyses it is important to keep in mind that (a) our use of the notion of "distributed cognitive system" to understand the problem-solving practices within the BME laboratory is for analytical purposes and is not intended to reify the systems and (b) what a system encompasses both in space and in time—that is, its "boundaries"—is, in our analysis, relative to a problem-solving process.

The Laboratory as "Problem Space."   The laboratory, as we construe it, is not simply a physical space existing in the present but rather a *problem space*, constrained by the research program of the laboratory director, that is reconfiguring itself almost continually as the research program moves along and takes new directions in response to what occurs both in the laboratory and in the wider community of which the research is a part. At any point in time the laboratory-as-problem-space contains resources for problem solving that comprise people, technology, techniques, knowledge resources (e.g., articles, books, artifacts, the Internet), problems, and relationships. Construed in this way, the notion of "problem space" takes on an expanded meaning from that customarily used in the traditional cognitive science characterization of problem solving as a search through an *internally* represented problem space. Here the problem space comprises both. Researchers and artifacts move back and forth between the wider community and the physical space of the laboratory. Thus the problem space has permeable boundaries.

For instance, among the most notable and recent artifacts (initiated in 1996) in Laboratory A are the tubular-shaped, bioengineered cell-seeded vascular grafts, locally called *constructs* (see Fig. 2.2). These are physical models of native blood vessels engineered to eventually function as viable implants for the human vascular system. The endothelial cells the laboratory uses in seeding constructs are obtained by researchers traveling to a distant medical school and bringing them into the problem space of the laboratory. On occasion, the constructs or substrates of constructs travel with laboratory members to places outside of the laboratory. Recently, for example, one of the graduate students has been taking substrates of constructs to a laboratory at a nearby medical school that has the elaborate instrumentation to perform certain kinds of genetic analysis (microarrays). This line of research is dependent on resources that are currently available only outside Laboratory A in the literal, spatial sense. The information produced in this locale is brought into the problem space of the laboratory by the researcher and figures in the further problem-solving activities of the laboratory.

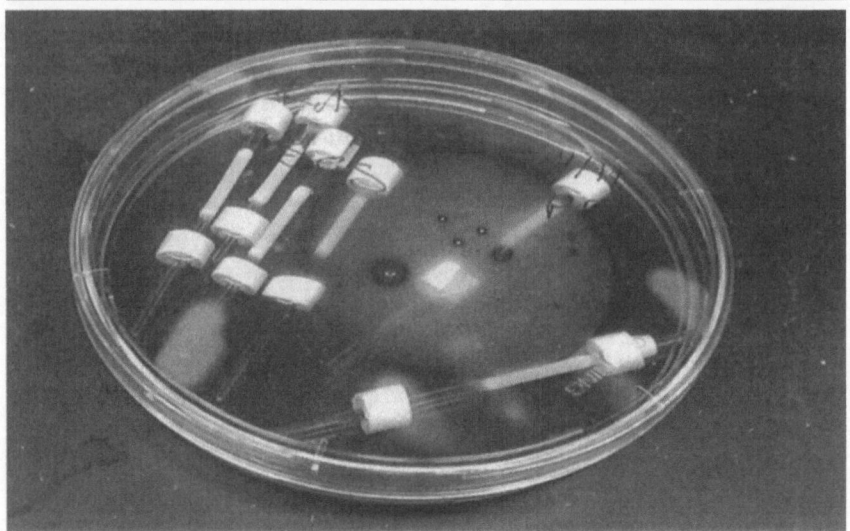

FIG. 2.2.   Photograph of a Dish of Vascular Constructs.

Following Hutchins (1995), my colleagues and I analyze the cognitive processes implicated in a problem-solving episode as residing in a *cognitive system* comprising both one or more researchers and the *cognitive artifacts* involved in the episode (see also Norman, 1991). In line with his analysis, a *cognitive system* is understood to be sociotechnical in nature, and *cognitive artifacts* are material media possessing the cognitive properties of generating, manipulating, or propagating representations.[5] So, right from the outset, the systems within the laboratory are analyzed as social–cognitive–cultural in nature. Determining the cognitive artifacts within any cognitive system involves issues of agency and intention that are pressing questions for cognitive science research, both in the development of the theoretical foundations of distributed cognition and in relation to a specific case study. On our analysis, not all parts of the cognitive system are equal. Only the researchers have agency and intentions, which enable the cognitive activities of specific artifacts.

Our approach to better understanding such issues is to focus on the technology used in experimentation. During a research meeting with the laboratory members, including the PI, we asked them to sort the material artifacts in the laboratory according to categories of their own devising and rank the importance of the various pieces to their research. Their classification in terms of "devices," "instruments," and "equipment" is represented in Table

---

[5]For related notions in the STS literature, see also Rheinberger (1997) on "epistemic things" and Tweney (2002) on "epistemic artifacts."

2.1. Much to the surprise of the PI, the newer PhD students initially wanted to rank some of the equipment, such as the pipette, as the most important for their research, whereas for the PI and the more senior researchers deemed the devices the laboratory engineers for simulation purposes as most important to the research. Additional ethnographic observations have led us to formulate working definitions of the categories used by Laboratory A's researchers. *Devices* are engineered facsimiles that serve as *in vitro* models and sites of simulation;[6] *instruments* generate measured output in visual, quantitative, or graphical form; and *equipment* assists with manual or mental labor.

Distributed Model-Based Reasoning. As noted earlier, an *in vivo–in vitro* division is a significant component of the cognitive framework guiding practice in Laboratory A. Because the test bed environment for developing artificial blood vessels cannot be the human body in which they will ultimately be implanted, the BME researchers have to design facsimiles of the *in vivo* environment where the experiments can be conducted. These devices pro-

TABLE 2.1
Sorting of Laboratory Artifacts by the Laboratory Members

	Ontology of Artifacts	
Devices	*Instruments*	*Equipment*
Flow loop	Confocal	Pipette
Bioreactor	Flow cytometer	Flask
Equi-biaxial strain	Mechanical tester	Water bath
Construct	Coulter counter	Refrigerator
	"Beauty and beast"	Sterile hood
	LM 5 (program)	Camera
	computer	

---

[6] We are using the term "device" because this is how the researchers in the laboratory categorized the *in vitro* simulation technology. This notion differs from the notion of "inscription devices" that Latour and Woolgar (1986, p. 51) introduced and that has been discussed widely in the STS literature. The latter are devices for literally creating figures or diagrams of phenomena. The former are sites of *in vitro* simulation, and further processing with instruments is necessary to transform the information provided by these devices into visual representations or quantitative measures.

vide locally constructed sites of experimentation where *in vitro* models are used to screen and control specific aspects of the *in vivo* phenomena they want to examine. Devices, such as the construct, the flow loop, and the bioreactor (discussed later), are constructed and modified in the course of research with respect to problems encountered and changes in understanding. Studying the devices underscores how the kinds of systems we are investigating diverge from those investigated by Hutchins (1995). The devices are not stable technological artifacts but have a history within the research of the laboratory. For example, the flow loop was first created in the research of the PI of this laboratory to simulate "known fluid mechanically imposed wall sheer stress"—in other words, to perform as a model of hemodynamics.[7] We have traced aspects of its development since 1985. The constructs were first devised in Laboratory A in 1996 as an important step in the overall objective of creating vascular substitutes for implantation. They afford experimentation not only on cells but also on structures more closely related to the *in vivo* model. The bioreactor, although having a longer and more varied history outside the laboratory, first made its appearance in this laboratory in conjunction with the tubular constructs and was not used anywhere before for that purpose. The current smooth muscle constructs are not strong enough to withstand the mechanical forces in the human (or animal) cardiovascular system. The *bioreactor* is used to stimulate the cells mechanically with the objective of changing their mechanical properties. The *equi-biaxial strain*, which simulates blood vessel expansion and contraction, is the newest device created specifically for this laboratory and is just starting to be used.

The cognitive artifacts in the distributed systems in the laboratory cut across the categories, although most are devices or instruments. Analysis of the ethnographic data has focused our attention on the devices, all of which we classify as cognitive artifacts. Devices instantiate models of the cardiovascular system and serve as *in vitro* sites of experimentation with cells and constructs under conditions simulating those found in the vascular systems of organisms. It is in relation to the researcher's intent of performing a simulation with the device to create new situations that parallel potential real world situations, and the activity of the device in so doing, that qualifies a device as a cognitive artifact within the system. For example, as a device, the flow loop *represents* blood flow in the artery. In the process of simulation, it *manipulates* constructs that are *representations* of blood vessel walls. After

---

[7]Although some of the material we quote from comes from published sources, given the regulations governing confidentiality for human subjects research, if the authors are among subjects we are not able to provide citations to that material here. It seems that the possibility of conducting historical research in conjunction with human subjects research was not anticipated! Laboratory A researchers are given an alias, "A plus a number," e.g., A10.

being *manipulated*, the constructs are then removed and examined with the aid of instruments, such as the confocal microscope, which *generates* images for many color channels, at multiple locations, magnifications, and gains. These *manipulations* enable the researchers to determine specific things, such as the number of endothelial cells and whether the filaments align with the direction of flow, or to simply explore the output, just "looking for stuff." Thus, the *representations generated* by the flow loop *manipulations* of the constructs are *propagated* within the cognitive system.

Devices perform as models instantiating current understanding of properties and behaviors of biological systems. For example, the flow loop is constructed so that the behavior of the fluid is such as to create the kinds of mechanical stresses experienced in the vascular system. However, devices are also systems themselves, possessing engineering constraints that often require simplification and idealization in instantiating the biological system they are modeling. For example, the flow loop is "a first-order approximation of a blood vessel environment ... as the blood flows over the lumen, the endothelial cells experience a shear stress ... we try to emulate that environment. But we also try to eliminate as many extraneous variables as possible" (laboratory researcher A10). So, as with all models, devices are idealizations.

The bioreactor provides an example of how the devices used in the laboratory need to be understood both as models of the cardiovascular system and as a systems in themselves. The bioreactor is used for many purposes in the field but, as used in Laboratory A, it was re-engineered for "mimicking the wall motions of the natural artery" (see Fig. 2.3). It is used to expose the constructs to mechanical loads in order to improve their overall mechanical properties. The researchers call this process *mechanical conditioning*—or, as one researcher put it, "exercising the cells." This preferably is done at an early stage of the formation of the construct, shortly after seeding the cells onto a prepared tubular silicon sleeve. *In vivo*, arterial wall motion is conditioned on pulsatile blood flow. With the bioreactor, though, which consists of a rectangular reservoir containing a fluid medium (blood-mimicking fluid) in which the tubular constructs are immersed and connected to inlet and outlet ports off the walls of the reservoir, "fluid doesn't actually move," as one laboratory member put it, "which is somewhat different from the actual, uh, you know, real life situation that flows." The sleeves are inflated with pressurized culture medium, under pneumatic control (produced by an air pump). The medium functions as an incompressible fluid, similar to blood. By pressurizing the medium within the sleeves, the diameter of the silicon sleeve is changed, producing strain on the cells, similar to that experienced *in vivo*. The bioreactor is thus a functional model of pulsatile blood flow, and needs to be understood by the researcher as such.

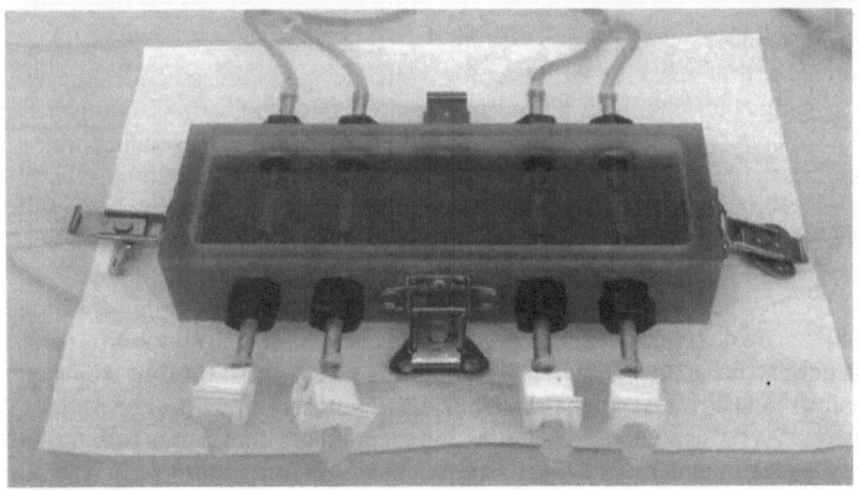

FIG. 2.3.   Photograph of a Bioreactor.

Distributed Mental Modeling.   Significant to our reconceiving the internal–external distinction is that the problem space comprises mental models
and physical artifacts together with a repertoire of activities in which
simulative model-based reasoning assumes a central place. Many instances
of model-based reasoning in science and engineering use external representations that are constructed during the reasoning process, such as diagrams,
sketches, and physical models. In line with the discussion of such representations in the ENVIRONMENTAL PERSPECTIVES ON COGNITION
section, these can be seen to provide constraints and affordances essential
to problem solving that augment those available in whatever internal representations are used by the reasoner during the process. In this way, "cognitive capabilities" are understood to encompass more than "natural"
capabilities. The devices used in Laboratory A are physical models used in
the problem solving. Within the cognitive systems in the laboratory, then,
devices instantiate part of the current community model of the phenomena
and allow simulation and manipulation. The intent of the simulation is to
create new situations *in vitro* that parallel potential *in vivo* situations.

One researcher we interviewed called the processes of constructing and
manipulating these *in vitro* sites "putting a thought into the bench top and
seeing whether it works or not." These instantiated "thoughts" allow researchers to perform controlled simulations of an *in vivo* context—for example, of the local forces at work in the artery. The "bench top," as one
researcher explained, is not the flat table surface but comprises all the lo-
cales where experimentation takes place. In previous research, I
(Nersessian, 1999, 2002a) have characterized the reasoning involved in

simulative model-based reasoning as a form of dynamic mental modeling, possibly using iconic representations. There the focus was on thought experiments, and that analysis used the notion of a mental model in the traditional manner as referring to an internal object of thought. In the current research, I am expanding the notion of simulating a mental model to comprise both what are customarily held to be the internal thought processes of the human agent and the processing of the external device. Simulative model-based reasoning involves a process of coconstructing the "internal" researcher models of the phenomena and of the device and the "external" model that is the device, each incomplete. Understood in this way, simulating the mental model would consist of processing information both in memory and in the environment; that is, the mental modeling process is distributed in the cognitive system.[8]

Cognitive Partnerships. Our account of the distributed cognitive systems in the laboratory characterizes cognition in terms of the lived relationships among the components of these systems, people, and artifacts. In Laboratory A these relationships develop in significant ways for the individual laboratory members and for the community as a whole. Newcomers to the laboratory, who are seeking to find their place in the evolving system, initially encounter the cognitive artifacts as materially circumscribed objects. For example, one new undergraduate who was about to use the *mechanical tester*, an instrument for testing the strength of the constructs (see Fig. 2.4), responded to our query about the technology she was going to use in her research project:

A2: I know that we are pulling little slices of the construct—they are round, we are just pulling them. It's the machine that is right before the computer in the lab. The one that has the big "DO NOT TOUCH" on it.

I: Is it the axial strain (mechanical tester)?

A2: I know it has a hook on it and pulls.

The novice researcher can describe the mechanical tester at this point in time as nothing more than parts. Another example is provided by the sorting task recounted in the section titled "The Laboratory as 'Problem Space,'" where novice researchers saw the equipment as more important to

---

[8]Of course, I use the term "mental" provocatively here, as a rhetorical move to connect the discussion with aspects of the traditional notion of mental modeling and extend the notion for use in the distributed-cognition framework. For related attempts to reconceive mental modeling, see Greeno (1989b) on the relation between mental and physical models in learning physics and Gorman (1997) on the relation between mental models and mechanical representations in technological innovation.

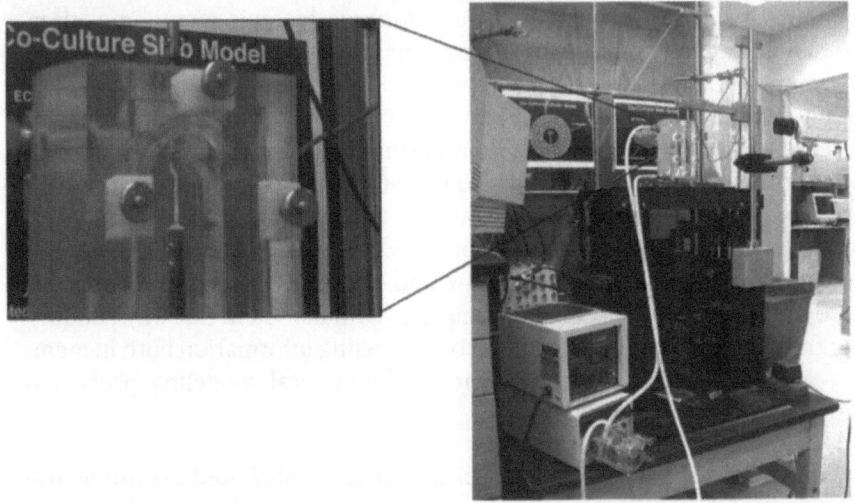

FIG. 2.4.   Photograph of the mechanical tester with close-up of hook mechanism.

their research than the simulation devices. We propose that growing cognitive membership in the laboratory involves a gradual process of coming to understand these objects as devices—as objects with evolving trajectories, constructed and used to respond to problems, to help answer questions, and to generate new ones. Thus, we find that one cannot divorce research from learning in the context of the laboratory and that learning involves building relationships with people and with artifacts.

We characterize the relationships between the various technological artifacts in the cognitive system and the researchers as *cognitive partnerships*. These partnerships provide the means for generating, manipulating, and propagating representations within the distributed cognitive systems of this research laboratory. Over time, understandings are constructed, revised, enhanced, and transformed through partnerships between the researchers and the artifacts in the community. As relationships change, so too do knowledge and participation.

The cognitive partnerships transform both researcher and artifact. A researcher who some months earlier was a newcomer and who saw the artifacts as just many kinds of machines and objects piled on shelves and on the bench top now can see a device as an *in vitro* site for "putting a thought [his/her thought] into the bench top and seeing whether it works or not." During the problem-solving processes involved in instantiating a thought and seeing if it works, devices are re-engineered, as exemplified above with the flow loop. Re-engineering is possible because the researcher with a developed partnership appropriates and participates in the history of a device. A senior PhD researcher, at that point in time, considered the "resident expert" on the bioreactor, was able easily to reconstruct some of his lived relationship with it and some of its history within this laboratory:

> I: Do you sometimes go back and make modifications? Does that mean you have some generations of this?

> … Uh yes I do. The first generation and the second generation or an offshoot I guess of the first generation. Well the first one I made was to do mechanical loading and perfusion. And then we realized that perfusion was a much more intricate problem than we had—or interesting thing to look at—than we had guessed. And so we decided okay we will make a bioreactor that just does perfusion on a smaller scale, doesn't take up much space, can be used more easily, can have a larger number of replicates, and so I came up with this idea.

He continued by pulling down previous versions of bioreactor (made by earlier researchers as well) and explaining the modifications and problems for which design changes were made. His account suggests a developed partnership.

Furthermore, in developed partnerships potential device transformations can be envisioned, as with one undergraduate research scholar we interviewed about the bioreactor:

> A16: I wish we could accomplish—would be to actually suture the actual construct in there somehow. To find a way not to use the silicon sleeve.… That would really be neat. Um, simply because the silicon sleeves add the next level of doubt. They're—they are a variable thing that we use, they're not always 100% consistent. Um the construct itself is not actually seeing the pressure that the sleeve does. And because of that you know, it doesn't actually see a—a pressure. It feels the distension but it doesn't really feel the pressure. It doesn't have to withstand the pressure. That's the whole idea of the sleeve. And so, um, I think that it would provide a little bit more realism to it. And uh, because that also, a surgeon would actually want to suture the construct into a patient. And um, because of that you're also mimicking the patient as well—if you actually have the construct in the path. I think another thing is to actually have the flow because um, so this flow wouldn't be important with just the sleeve in there. But if you had the construct in contact with the—with the liquid that's on the inside, you could actually start to flow media through there.

In this case an undergraduate student has been transformed over the course of several semesters to a BME researcher, contributing to immediate research goals; who transforms artifacts in his immediate research; who understands the outstanding problems and objectives; and who can envision how a device might change from a functional model to a model more closely paralleling the *in vivo* situation to push the research along. At this point in evolution, thinking is taking place through the cognitive partnering of the researcher and device. In their established form, relationships with artifacts entail cognitive partnerships that live in *interlocking* models performing internally as well as externally.

Implications of the Exemplar for Integration

Our approach to interpreting the knowledge-producing practices in the laboratory contributes to the project of developing means of interpreting cognitive, social, and cultural dimensions of practice in relation to one another. By starting from the perspective that cognition is embedded in complex environments, the laboratory's innovative problem-solving practices are interpreted as social–cognitive–cultural from the outset. The mixed methodology enables both thick descriptions of specific systems and hypotheses about "the nature of cognition in human activity" that go beyond the specifics of the laboratory under study. Consider the outline of our analysis of the flow loop. It is a major cognitive tool developed and used in the model-based reasoning in this laboratory. It is a significant cultural artifact, originating in the research program of the PI and then passed down through generations[9] of researchers, enabling each to build on the research of others, while sometimes being re-engineered as an artifact in the service of model-based reasoning. It is a locus for social interaction, such as that involved in learning and didactical interaction between mentor and apprentice. At one point it served as the vehicle for initiation into the community of practice, although at present cell culturing serves this purpose, because the problem situation has evolved, and now the flow loop is no longer the only experimental device. On the one hand, the histories of the lived relations among the flow loop and researchers can be developed into thick social–cognitive–cultural descriptions of the evolving systems of the laboratory. On the other hand, understanding the role of the flow loop as a device—a cognitive artifact for performing simulative model-based reasoning in the problem-solving activities within the distributed cognitive systems of the laboratory—leads to hypotheses about the nature of reasoning and representation. The mixed methodology facilitates the capture and analysis of the dynamics of the interplay among the cognitive, social, and cultural dimensions of scientific and engineering practices.

---

[9]Approximately 5 years marks a generational shift in this research, although different generations of researchers are in the laboratory at any one time.

CONCLUSIONS

The reductionism of the physical symbol system notion does not do justice to the practices of science and engineering, such as the complex relationship with the material environment, the highly distributed nature of reasoning in laboratory environments and elsewhere, and the extensive use of external representations in reasoning and communicating. These aspects of practice need to be factored into an account of cognition as more than simply content over which internal cognitive processes operate and as doing more than just providing scaffolding for cognition. The environmental perspectives on cognition provide a framework within which to do this. At the same time, studying reflexively the cognitive practices of scientists and engineers contributes to the task of developing that account of cognition.

STS accounts that see cognition as inconsequential in creating knowledge also do not do justice to these practices. Moreover, even if we start from the perspective that cognition is distributed within a system, there is always at least one human in the knowledge-making system, and often an individual plays a pivotal role: Maxwell's equations were formulated by Maxwell (in original form, of course). So the contribution of the individual human component in the system needs also to be understood: Internal representations and processes are still important. However, they need to be understood as inherently integrated with the "external" environment. Again, environmental perspectives, viewed in the interrelated way of the ENVIRONMENTAL PERSPECTIVES ON COGNITION section, assist in developing a framework in which to do this. The analysis presented in the RESEARCH LABORATORIES AS EVOLVING DISTRIBUTED COGNITIVE SYSTEMS section is my current way of approaching integration.

Integrating the cognitive and the sociocultural will have major implications for STS. Likewise, implications from studying cognition with respect to scientific and engineering practices stand to have a major impact on cognitive science. Take the following as one example. The physical symbol system notion assumes that cognitive processes are universal and the same through recorded history, so there is thought to be no need to contextualize or historicize theories of problem solving, learning, and decision making. In this cognitive science has modeled itself on physics—the phenomena to be explained are the "spherical horses." From the perspective of sociocultural analyses, scientific knowledge-producing practices have changed with changes in cultural assumptions, including values, and with developments in such things as instrumentation and mathematical representational systems. These changes are traditionally accommodated as changes in *what* scientists think about—that is, the content of representations changes culturally and historically—and not as changes in *how* scientists think, that is, in the nature of cognitive representations and processing. But if we reconceptualize "cognition," moving the boundaries beyond the individual to complex systems encompassing salient aspects of the environments of practice—that is, conceptualize cognition as distributed and situated in the environment and as lying in the interactions among parts of the system—what are the implications of these historical sociocultural changes for understanding scientific cognition—or, for that matter, ordinary cognition?

At this stage in the project of integration we are left with many unresolved issues. What is sure is that interpreting scientific and engineering practices requires s shift from looking at cognitive factors and sociocultural factors as independent variables in explanations of these practices to regarding cognitive processes as inherently sociocultural, and vice versa. Doing this requires rethinking foundational and methodological issues in cognitive science and in STS together—with the goal of creating "shared toothbrushes"—and we are only at the beginning of this process.

ACKNOWLEDGMENTS

I thank Elke Kurz-Milcke, Thomas Nickles, and the editors of this volume for their comments on earlier versions of this chapter. I appreciate also the comments made on an earlier version of this chapter by the participants in the "Cognitive Studies of Science" workshop organized by the Danish Graduate Research School in Philosophy, History of Ideas, and History of Science, especially Hanne Andersen, Ronald Giere, and Thomas Söderqvist. Finally, I gratefully acknowledge the support of the National Science Foundation in carrying out this research: Science and Technology Studies Scholar's Award SBE9810913 and Research on Learning and Education Grant REC0106773.

I dedicate this chapter to the memory of Frederic Lawrence Holmes, friend, mentor, and inspiration.

REFERENCES

Agnetta, B., Hare, B., & Tomasello, M. (2000). Cues to food location that domestic dogs (*Canis familiaris*) of different ages do and do not use. *Animal Cognition, 3*, 107–112.

Andersen, H. (1996). Categorization, anomalies, and the discovery of nuclear fission. *Studies in the History and Philosophy of Modern Physics, 27*, 463–492.

Barsalou, L. W. (1999). Perceptual symbol systems. *Behavioral and Brain Sciences, 22*, 577–609.

Bucciarelli, L. L. (1994). *Designing engineers.* Cambridge, MA: MIT Press.

Carraher, T. D., Carraher, D. W., & Schliemann, A. D. (1983). Mathematics in the streets and schools. *British Journal of Developmental Psychology, 3*, 21–29.

Cetina, K. K. (1999). *Epistemic cultures: How the sciences make knowledge.* Cambridge, MA: Harvard University Press.

Chen, X. (1995). Taxonomic changes and the particle-wave debate in early nineteenth-century Britain. *Studies in the History and Philosophy of Science, 26*, 251–271.

Clancey, W. J. (1997). *Situated cognition: On human knowledge and computer representations*. Cambridge, England: Cambridge University Press.

Craig, D. L., Nersessian, N. J., & Catrambone, R. (2002). Perceptual simulation in analogical problem solving. In L. Magnani & N. Nersessian (Eds.), *Model-based reasoning: Science, technology, values* (pp. 169–187). New York: Kluwer Academic/Plenum.

Davies, G. E. (2000). *The democratic intellect*. Edinburgh, Scotland: Edinburgh Press.

Donald, M. (1991). *Origins of the modern mind: Three stages in the evolution of culture and cognition*. Cambridge, MA: Harvard University Press.

Dunbar, K. (1995). How scientists really reason: Scientific reasoning in real-world laboratories. In R. J. Sternberg & J. E. Davidson (Eds.), *The nature of insight* (pp. 363–395). Cambridge, MA: MIT Press.

Elman, J. L., Bates, E. A., Johnson, M., Karmiloff-Smith, A., Parisi, D., &. Plunkett, K. (1998). *Rethinking innateness: A connectionist perspective on development*. Cambridge, MA: MIT Press.

Galison, P. (1997). *Image and logic: A material culture of microphysics*. Chicago: University of Chicago Press.

Geertz, C. (1973). *The interpretation of cultures*. New York: Basic Books.

Gibson, J. J. (1979). *The ecological approach to visual perception*. Boston: Houghton Mifflin.

Giere, R. N. (1988). *Explaining science: A cognitive approach*. Chicago: University of Chicago Press.

Giere, R. N. (2002). Scientific cognition as distributed cognition. In P. Carruthers, S. Stich, & M. Siegal (Eds.), *The cognitive basis of science* (pp. 285–299). Cambridge, England: Cambridge University Press.

Glenberg, A. M. (1997). What memory is for. *Behavioral and Brain Sciences, 20*, 1–55.

Glenberg, A. M., & Langston, W. E. (1992). Comprehension of illustrated text: Pictures help to build mental models. *Journal of Memory and Language, 31*, 129–151.

Gooding, D. (1989). "Magnetic curves" and the magnetic field: Experimentation and representation in the history of a theory. In D. Gooding, T. Pinch, & S. Schaffer (Eds.), *The uses of experiment* (pp. 183–244). Cambridge, England: Cambridge University Press.

Gooding, D. (1990). *Experiment and the making of meaning: Human agency in scientific observation and experiment*. Dordrecht, The Netherlands: Kluwer.

Goodwin, C. (1995). Seeing in depth. *Social Studies of Science, 25*, 237–274.

Gorman, M. (1997). Mind in the world: Cognition and practice in the invention of the telephone. *Social Studies of Science, 27*, 583–624.

Gorman, M. E., & Carlson, W. B. (1990). Interpreting invention as a cognitive progress: The case of Alexander Graham Bell, Thomas Edison, and the telephone. *Science, Technology, and Human Values, 15*, 131–164.

Greeno, J. G. (1989). A perspective on thinking. *American Psychologist, 44*, 134–141.

Greene, J. G. (1989b). Situations, mental models, and generative knowledge. In D. Klahr & K. Kotovsky (Eds.), *Complex information processing* (pp. 285–318). Hillsdale, NJ: Lawrence Erlbaum Associates.

Greene, J. G. (1998). The situativity of knowing, learning, and research. *American Psychologist, 53*, 5–24.

Hall, R., Stevens, R., & Torralba, T. (2002). Disrupting representational infrastructure in conversation across disciplines. *Mind, Culture, and Activity, 9*, 179–210.

Haugeland, J. (1985). *Artificial intelligence: The very idea*. Cambridge, MA: MIT Press.

Hoffman, J. R. (1996). *Andre-Marie Ampere*. Cambridge, England: Cambridge University Press.

Holmes, F. L. (1980). Hans Krebs and the discovery of the ornithine cycle. *Federation Proceedings, 39*, 216–225.

Hutchins, E. (1995). *Cognition in the wild*. Cambridge, MA: MIT Press.

Johnson, M. (1987). *The body in the mind: The bodily basis of meaning, imagination, and reason*. Chicago: University of Chicago Press.

Klahr, D., & Simon, H. A. (1999). Studies of scientific discovery: Complimentary approaches and divergent findings. *Psychological Bulletin, 125*, 524–543.

Kosslyn, S. M. (1994). *Image and brain*. Cambridge, MA: MIT Press.

Kulkarni, D., & Simon, H. A. (1988). The processes of scientific discovery: The strategy of experimentation. *Cognitive Science, 12*, 139–175.

Kurz, E. M., & Tweney, R. D. (1998). The practice of mathematics and science: From calculus to the clothesline problem. In M. Oaksford & N. Chater (Eds.), *Rational models of cognition* (pp. 415–438). Oxford, England: Oxford University Press.

Lakoff, G. (1987). *Women, fire, and dangerous things: What categories reveal about the mind*. Chicago: University of Chicago Press.

Lakoff, G., & Johnson, M. (1998). *Philosophy in the flesh*. New York: Basic Books.

Langley, P., Simon, H. A., Bradshaw, G. L., & Zytkow, J. M. (1987). *Scientific discovery: Computational explorations of the creative processes*. Cambridge, MA: MIT Press.

Latour, B. (1987). *Science in action*. Cambridge, MA: Harvard University Press.

Latour, B. (1999). *Pandora's hope: Essays on the reality of science studies*. Cambridge, MA: Harvard University Press.

Latour, B., & Woolgar, S. (1986). *Laboratory life: The construction of scientific facts*. Princeton, NJ: Princeton University Press.

Lave, J. (1988). *Cognition in practice: Mind, mathematics, and culture in everyday life*. New York: Cambridge University Press.

Lynch, M. (1985). *Art and artifact in laboratory science: A study of shop work and shop talk in a research laboratory*. London: Routledge and Kegan Paul.

Nersessian, N. J. (1984). *Faraday to Einstein: Constructing meaning in scientific theories*. Dordrecht, The Netherlands: Martinus Nijhoff/Kluwer Academic.

Nersessian, N. (1985). Faraday's field concept. In D. C. Gooding & F. A. J. L. James (Eds.), *Faraday rediscovered: Essays on the life & work of Michael Faraday* (pp. 337–406). London: Macmillan.

Nersessian, N. (1992). How do scientists think? Capturing the dynamics of conceptual change in science. In R. Giere (Ed.), *Minnesota studies in the philosophy of science* (pp. 3–45). Minneapolis: University of Minnesota Press.

Nersessian, N. (1995). Opening the black box: Cognitive science and the history of science. *Osiris, 10*, 194–211.

Nersessian, N. (1999). Model-based reasoning in conceptual change. In L. Magnani, N. J. Nersessian, & P. Thagard (Eds.), *Model-based reasoning in scientific discovery* (pp. 5–22). New York: Kluwer Academic/Plenum.

Nersessian, N. (2002a). The cognitive basis of model-based reasoning in science. In P. Carruthers, S. Stich, & M. Siegal (Eds.), *The cognitive basis of science* (pp. 133–153). Cambridge, England: Cambridge University Press.

Nersessian, N. (2002b). Maxwell and the "method of physical analogy": Model-based reasoning, generic abstraction, and conceptual change. In D. Malament (Ed.), *Reading natural philosophy: Essays in the history and philosophy of science and mathematics* (pp. 129–163). Lasalle, IL: Open Court.

Nisbett, R., Peng, K., Choi, I., & Norenzayan, A. (2001). Culture and systems of thought: Holistic v. analytic cognition. *Psychological Review, 108*, 291–310.

Norman, D. A. (1981). *Perspectives on cognitive science*. Hillsdale, NJ: Lawrence Erlbaum Associates.

Norman, D. A. (1988). *The psychology of everyday things*. New York: Basic Books.

Norman, D. A. (1991). Cognitive artifacts. In J. M. Carroll (Ed.), *Designing interaction* (pp. 17–38). Cambridge, England: Cambridge University Press.

Ochs, E., & Jacoby, S. (1997). Down to the wire: The cultural clock of physicists and the discourse of consensus. *Language in Society, 26*, 479–505.

Pickering, A. (1995). *The mangle of practice: Time, agency, and science*. Chicago: University of Chicago Press.

Resnick, L. B., Levine, J., & Teasley, S. (Eds.). (1991). *Perspectives on socially shared cognition*. Washington, DC: American Psychological Association.

Rheinberger, H.-J. (1997). *Toward a history of epistemic things: Synthesizing proteins in the test tube*. Stanford, CA: Stanford University Press.

Shore, B. (1997). *Culture in mind: Cognition, culture and the problem of meaning*. New York: Oxford University Press.

Siegel, D. (1991). *Innovation in Maxwell's electromagnetic theory*. Cambridge, England: Cambridge University Press.

Simon, H. A. (1981). *The sciences of the artificial* (2nd ed.). Cambridge, MA: MIT Press.

Simon, H. A., Langley, P. W., & Bradshaw, G. L. (1981). Scientific discovery as problem solving. *Synthese, 47*, 1–27.

Simon, H. A., & Newell, A. (1972). *Human problem solving*. Englewood Cliffs, NJ: Prentice Hall.

Smith, C., & Wise, M. N. (1989). *Energy and empire: A biographical study of Lord Kelvin*. Cambridge, England: Cambridge University Press.

Suchman, L. A. (1987). *Plans and situated actions: The problem of human–machine communication*. Cambridge, England: Cambridge University Press.

Thagard, P. (2000). *How scientists explain disease*. Princeton, NJ: Princeton University Press.

Tomasello, M. (1999). *The cultural origins of human cognition*. Cambridge, MA: Harvard University Press.

Tomasello, M., & Call, J. (1997). *Primate cognition*. Oxford, England: Oxford University Press.

Tomasello, M., Call, J., & Hare, B. (1998). Five primates follow the visual gaze of conspecifics. *Animal Behavior, 55*, 1063–1069.

Tweney, R. D. (1985). Faraday's discovery of induction: A cognitive approach. In D. Gooding & F. A. J. L. James (Eds.), *Faraday rediscovered* (pp. 189–210). New York: Stockton.

Tweney, R. D. (2002). Epistemic artifacts: Michael Faraday's search for the optical effects of gold. In L. Magnani & N. J. Nersessian (Eds.), *Model-based reasoning: Science, technology, values* (pp. 287–304). New York: Kluwer Academic/Plenum.

van der Kolk, B., McFarlane, A. C., & Weisaeth, L. (Eds.). (1996). *Traumatic stress: The effects of overwhelming experience on mind, body, and society*. New York: Guildford.

Vera, A., & Simon, H. (1993). Situated cognition: A symbolic interpretation. *Cognitive Science, 17*, 4–48.

Vygotsky, L. S. (1978). *Mind in society: The development of higher psychological processes*. Cambridge, MA: Harvard University Press.

Woods, D. D. (1997). Towards a theoretical base for representation design in the computer medium: Ecological perception and aiding human cognition. In J. Flack, P. Hancock, J. Cairn, & K. Vincente (Eds.), *The ecology of human–machine systems* (pp. 157–188). Hillsdale, NJ: Lawrence Erlbaum Associates.

Yeh, W., & Barsalou, L. W. (1996). The role of situations in concept learning. *Proceedings of the 18th annual conference of the Cognitive Science Society* (pp. 460–474). Mahwah, NJ: Lawrence Erlbaum Associates.

Zhang, J. (1997). The nature of external representations in problem solving. *Cognitive Science, 21,* 179–217.

Zhang, J., & Norman, D. A. (1995). A representational analysis of numeration systems. *Cognition, 57,* 271–295.

3

Causal Thinking in Science: How Scientists and Students Interpret the Unexpected

Kevin N. Dunbar
Jonathan A. Fugelsang
Dartmouth College

Scientists have attempted to delineate the key components of scientific thinking and scientific methods for at least 400 years (e.g., Bacon, 1620/ 1854; Galilei, 1638/1991; Klahr, 2000; Tweney, Doherty, & Mynatt, 1981). Understanding the nature of the scientific mind has been an important and central issue not only for an understanding of science but also what it is to be human. Given the enduring and wide-ranging interest in the scientific mind, there has been a multiplicity of ways used to investigate the genesis of scientific concepts, theories, and hypotheses. Most important to bear in mind is that many of the methods that have been used to understand science have been tied to changes in their respective fields of study. In philosophy, for example, the switch from an analytical to a more historically based approach resulted in major shifts in an understanding of science (Callebaut, 1992). Likewise, in psychological studies of scientific thinking there has been continuous discourse between advocates of naturalistic studies of human behavior and advocates of highly controlled experiments (Dunbar, 2000; Dunbar & Blanchette, 2001; Tweney et al., 1981). In fact, naturalistic versus highly constrained or controlled investigations has been a central issue that cuts across many disciplines (Lave, 1998; Suchman, 1987). Numerous books, articles, and conferences have been held that espouse one way of conducting research as being better or more "authentic" than others. Over the past decade, we have taken a more unifying approach by combining naturalistic with experimental methods. We have conducted what we have called *in vivo* naturalistic studies of scientists thinking and reasoning "live" in their laboratories (Dunbar, 1993b, 1995, 1997, 2002). We have also conducted what we have called *in vitro* research, where we have used controlled experiments to investigate specific aspects of scientific thinking (e.g., Baker & Dunbar, 2000). The topics that we chose to investigate in our laboratory were the ones that we identified in our *in vivo* studies. Using this approach, we were able to discover new aspects of analogical thinking that clarify both our understanding of scientific thinking in general and analogy in particular (Blanchette & Dunbar, 2000, 2001, 2002; Dunbar & Blanchette, 2001).

In this chapter, we discuss our three-pronged approach to understanding scientific thinking. First, using *in vivo* methods, we investigate reasoning about expected and unexpected findings and find that causal thinking is crucial. Second, using *in vitro* methods, we investigate the ways that students reason causally about expected and unexpected findings. Third, using *in magnetico* methods (functional magnetic resonance imaging [fMRI]) we are currently conducting brain imaging studies of scientific thinking (Fugelsang & Dunbar, 2003). We argue that at the heart of these analyses are causal reasoning strategies that people use to understand the unexpected, anomalous, or serendipitous.

A TAXONOMY OF RESEARCH ON SCIENCE

There are numerous ways of categorizing research on scientific thinking, such as the taxonomies proposed by Klahr and Simon (1999) and by Feist and Gorman (1998). All the taxonomies capture important ways of understanding scientific thinking. We have used a biological taxonomy to understand science that bears some similarity to other taxonomies but is based on the premise that understanding scientific thinking is the same as understanding any biological system. Consequently, many of the methods and tools that are used to understand any biological system can be used to understand the scientific mind. Our taxonomy can be seen in Table 3.1. Borrowing from biology, our initial categorization of methods used to investigate the scientific mind focused on *in vivo* and *in vitro* methods (Dunbar, 1993b, 1995, 1997, 1999b, 2000, 2002). We characterized *in vivo* research as investigating scientists as they think and reason "live." We have focused on scientists reasoning at laboratory meetings, but, as a number of chapters in this volume show, there are other possible ways of investigating

TABLE 3.1

Methods for Discovering the Workings of the Scientific Mind

Method	Description	Advantage	Disadvantage	Examples
In *vivo*	Investigate scientists as they think and reason live about their research.	Captures what scientists really do in a naturalistic context.	Does not give access to the nonverbalized cognitive activities. Time consuming.	Dunbar (1993a, 1995, 1997, 2000, 2002).
Ex *vivo*	a. Protocol studies of scientists working on a scientific problem. b. Interview scientists about how they do their research. c. Diary studies of scientists.	Can study a scientist directly. Protocols give details of working memory. Interviews show real issues for scientists. Likewise, diaries reveal important factors for scientists.	Scientists can present biased interpretations of themselves. Scientists may leave out many of the details of what they have done.	Thagard (1999), Nersessian (1992), chaps. 6 and 7 this volume.
In *vitro*	Conduct controlled experiments on an aspect of scientific thinking.	Can conduct detailed investigation that precisely investigates aspects of scientific thinking.	Task is highly constrained. May not tap what scientists really do.	Dunbar (1993), Klahr & Dunbar (1988), Klayman & Ha (1987, 1988), Tweney et al. (1981), Wason (1968).
In *magnetico*	Scan brains of people as they reason scientifically using functional magnetic resonance imaging.	Can reveal what brain sites are involved in scientific thinking and how they interact.	Task can be highly constrained. Very expensive.	Fugelsang & Dunbar (2003).
In *historico*	Use historical data to reconstruct thought processes that led to a particular discovery.	Data already present.	Time consuming.	Nersessian (1992), Tweney (1992), Tweney et al. (1981), Thagard (1999).
In *silico*	Write computer programs that simulate scientific thinking or discovery.	Specify specific components and how they are related.	May be artificial. Can not be proven necessary, only sufficient.	Langley et al. (1987), Kulkarni & Simon (1988).

scientists *in vivo*, such as Schunn's work on scientists analyzing data (chap. 5, this volume) and Nersessian's work on scientists (chap. 2, this volume). Our *in vitro* research, on the other hand, involves the systematic investigation of people reasoning about scientific phenomena in cognitive laboratory settings. Classic examples of this approach, such as Wason's (1968) 2–4–6 task (Klayman & Ha, 1987), have been the hallmark paradigms used by experimental psychologists since before the turn of the century. Other examples, such as Klahr and Dunbar's (1988) work using a programmable device, Tweney et al.'s (1981) work in an artificial physics universe, and Dunbar's (1993a) work on people discovering a biological mechanism of control in a simulated molecular biology laboratory, have extended this traditional approach to include scientific concepts. Many of these studies have shown that interpreting unexpected findings is a key component of scientific thinking.

In this chapter we broaden our characterization of methods used to investigate scientific thinking by adding four other methods to our taxonomy: (a) *ex vivo*, (b) *in magnetico*, (c) *in historico*, and (d) *in silico*. All of these methods have been used to understand scientific thinking and are commonly used in the biological sciences.

In biology, *ex vivo* refers to a method in which some tissue or an organ is removed from a living organism and is then briefly cultured *in vitro* and mixed with a transfection formulation (cells from another source). The transfected tissue or organ is then reimplanted in the hope that it will reconstitute the normal function. Here we use the term in a similar fashion. Scientists are taken out of their regular scientific context and investigated in the cognitive laboratory but given some of the problems and goals that they normally encounter in their real research. After this, the scientists are returned to their natural *in vivo* context. Researchers using the *ex vivo* method thus take scientists out of their normal scientific context and conduct research on the scientists. A good example of this approach was portrayed by Schunn and Anderson (1999), who asked researchers to design experiments in their field of expertise. Similar work has been conducted with advanced students on experimental design by Schraagen (1993) and by Baker and Dunbar (2000). *Ex vivo* methods allow the researcher to use complex tasks virtually identical to the ones that scientists use in their day-to-day work yet maintain tight control over the parameters of their experiments. Large numbers of scientists can be used, and the work can be replicated. All of this provides many advantages over *in vivo* work but, like *In vitro* research, has the potential of being artificial.

Another method that is becoming increasingly popular in the biological sciences is what we call *in magnetico*. *In magnetico* methods consist of conducting research on thinking using fMRI. Although most research using fMRI has concentrated on fairly simple, lower level cognitive tasks, recent research has begun to examine more complex cognition, such as categorization, inductive thinking, deductive thinking, and scientific thinking (e.g., Fugelsang & Dunbar, 2003; Goel, Gold, Kapur, & Houle,1998; Seger et al., 2000). All of these studies reveal how the human brain performs cognitive processes that are at the heart of the scientific mind. This type of research makes it possible to understand both the brain sites that are involved in specific aspects of scientific thinking and what circuits of brain sites subserve particular aspects of scientific thinking such as induction, deduction, analogy, and causality. Note that there are related methods, such as positron emission tomography and evoked reaction potentials, that also measure levels of energy along an electromagnetic continuum that could also be subsumed under the *in magnetico* category.

Other methods that are important for understanding scientific thinking are *in silico* and *in historico*. These are terms first used by Lindley Darden to describe computational and historical work on scientific thinking (Darden, 1992, 1997). Examples of *in silico* research are found in the groundbreaking work of Langley, Simon, Bradshaw, and Zytkow (1987), which modeled various historically important scientific discoveries. Simulations by Langley et al. (1987), Darden (1997), Thagard (1999), and Nersessian (1992) provide computational models that elucidate the cognitive processes underlying scientific thinking. *In historico* work on scientific thinking has used documents such as diaries and notebooks to propose models of the cognitive processes that scientists use in their research (e.g., Kulkarni & Simon, 1988; chap. 7, this volume). One consistent finding encountered when using both *in historico* and *in silico* work is that the strategies that scientists use when they encounter unexpected findings are a key aspect of the discovery process. Darden (1997), Thagard (2000), and Kulkarni and Simon (1988) have all given unexpected, anomalous, or serendipitous findings a central role in scientific discovery. Thus, *in vivo*, *in vitro*, *ex vivo*, *in silico*, and *in historico* methods have converged on the unexpected as a key factor in science and scientific discovery.

SCIENCE AND THE UNEXPECTED

Not only have most of the methods in our taxonomy given unexpected findings a central role in understanding the scientific mind, but also scientists themselves frequently state that a finding was due to chance or was unexpected. The recent discoveries of reverse evolution (Delneri, Colson, Grammenoudi, Roberts, Louis, & Olivier, 2003), naked DNA, and Buckey balls are among the many significant discoveries that have been attributed to unexpected findings (see also Dunbar, 1999a, for a discussion of some of these discoveries). It is interesting that, other than quoting Pasteur by stating that "Chance favors the prepared mind," scientists have not given many insights into what the prepared mind *is*; neither do they provide a clear account of what strategies they use when they encounter results that were not expected.

In historico analyses of different scientific discoveries have resulted in many models that focus on findings that are anomalous and unexpected. *Anomalies* are findings that are not predicted by a particular framework or theory, and they have been seen to be a key feature in scientific discovery. Theorists such as Thomas Kuhn (1962) have given anomalies a central role in scientific discovery, and theorists such as Darden (1997) have built computational models that use anomalies as a key part of their models. Darden (1997) proposed that scientists engage in a type of fault diagnosis where they try to isolate the portion of their theory that is leading to a faulty prediction. She argued that the scientists iterate through changing the faulty component until they can predict the finding. An alternate approach is that the scientists blame the anomalous finding on a methodological error. Thagard (1999; chap. 8, this volume) has analyzed a number of discoveries and inventions and argued that serendipity is a key component of the discovery process. He sees unexpected or serendipitous findings as providing input to an ongoing problem-solving process that results in the formulation of new concepts and theories. For our purposes, we categorize anomalous, serendipitous, and unexpected findings as equivalent. All three types of findings are unexpected. However, anomalies more explicitly focus on a finding being inconsistent with a prediction of a theory, and serendipitous findings are thought to involve chance events (see also chap. 7, this volume). Overall, these *in historico* and *in silico* analyses reveal that unexpected, serendipitous, or anomalous findings lead to a cascade of cognitive processes that can result in major conceptual shifts and, as a consequence, major scientific discoveries.

AN *IN VIVO* INVESTIGATION OF SCIENTISTS AND THE UNEXPECTED

Given that claims of unexpected findings are such a frequent component of scientists' autobiographies and in historical and computational models of discovery, we decided to investigate the ways that scientists use unexpected findings *in vivo*. Kevin N. Dunbar spent 1 year in three molecular biology laboratories and one immunology laboratory at a prestigious U.S. university. He interviewed the scientists; attended the laboratory meetings, impromptu meetings, and talks by laboratory members; and read grant proposals and drafts of papers. A detailed account of the laboratories he investigated and the precise methodology used can be found in Dunbar (1995, 2001). He found that although science does transpire in the mind of the lone scientist, in hallway conversations, and in one-on-one conversations between scientists, the place where a representative cross-section of the ways scientists think and reason is in the weekly laboratory meeting. It is in these meetings that scientists present their data, reason about findings, and propose new experiments and theories. Thus, by analyzing the reasoning that scientists use at their laboratory meetings it was possible to obtain a representative view into how scientists really reason. Here we focus on laboratory meetings from the three molecular biology laboratories (an HIV laboratory, a bacterial laboratory, and a parasite laboratory).

What Do Scientists Do?

We audiotaped laboratory meetings and followed particular projects in the laboratories that had just started at the time our investigation began. The tapes of laboratory meetings were supplemented by interviews with the scientists both before and after the laboratory meetings. This made it possible to probe deeper into the issues involved in particular research projects and to firmly anchor the scientists' path of knowledge from initial experiment to model building and discovery. In this chapter we focus on 12 meetings: 4 from each of the three molecular biology laboratories. All three laboratories were of similar size and structure and three to four postdoctoral fellows, three to five graduate students, and one or two technicians. We transcribed the audiotapes, identified every experiment conducted, and coded whether the results of the experiment were expected or unexpected and how the scientists reasoned about the results. We were interested in determining how frequent unexpected findings are, what scientists do about such findings, and what reasoning strategies they use once they have obtained unexpected findings.

The analysis of the 12 laboratory meetings yielded 28 research projects, with 165 experiments, and the participants reasoned about 417 results. What do scientists do with their findings? The first step they take is to classify their results. All findings—both expected and unexpected—were classified in one of two main ways. The first was to classify the finding directly. Here, the scientist said immediately what a particular result meant by looking at that result alone. For example, a scientist might look at a newly obtained gel and say, "This is P30." The second way of classifying data is by comparing one result to another result, such as a control condition, and inferring what the result is by making this comparison. For example, the scientists might say that "This band is larger than this [the control one], so this is P30." It is interesting that only 15% of the findings were classified directly, whereas 85% of the findings were classified by comparing results from one condition with results from other conditions. This holds for both expected and unexpected findings. Thus, classification of data is not a simple process. Instead, it usually consists of comparisons between experimental conditions and the numerous controls that scientists use in their experiments. This highlights a very important feature of contemporary biological science: Scientists are aware that other not-yet-understood biological mechanisms may be at work in their experiments and, because of this, they purposely build many experimental and control conditions into their experiments. The use of these multiple conditions is necessary both to interpret results and to expose hitherto-unseen mechanisms. The use of control conditions is usually thought merely to be a check on whether the experiment is working but, in addition, these control conditions serve the vital function of revealing new phenomena. The control conditions do this in two ways. First, they make it possible to determine whether a result is expected or not, and second, the control conditions themselves can produce unexpected findings. By conducting experiments with multiple conditions, scientists are making it decidedly possible for unexpected findings to occur.

How Frequent Is the Unexpected?

Experimental results were coded as unexpected if the scientist stated that the result was not what she or he expected to find. Unexpected results could range from a simple methodological error to a result that was a major anomaly that ultimately resulted in a radical revision of a theory. The reason for treating all unexpected results as equivalent was that initially scientists treat the results as equivalent and only later see them as theoretically important. It was interesting that novice scientists with only 1 or 2 years of postdoctoral experience are more likely to think that a particular unexpected finding is theoretically important than expert scientists. When we divided the scientists' results into expected and unexpected findings, we found that over half of their findings were unexpected (223 out of 417 results). Thus, rather than being a rare event, the unexpected finding was a regular occurrence about which the scientists reasoned. The large number of unexpected findings is important; it is not the case that scientists can take their most recent unexpected finding and make a discovery merely by focusing on it. Rather, the scientists have to evaluate which findings are due to methodological error, faulty assumptions, chance events, or new mechanisms. Two different types of analogical reasoning and two different types of social reasoning contribute to the mental toolkit that scientists use to decipher unexpected findings.

Once a finding was classified, expected and unexpected results were treated in different ways. Expected results usually led to the next step in a sequence of experiments, whereas unexpected findings led to either replication, change in the protocol, or use of an entirely new protocol. It is interesting that scientists initially proposed a methodological explanation for the majority of their unexpected findings. For the 223 unexpected findings, 196 methodological and 27 theoretical explanations were offered. Thus, the first strategy that the scientists used was to blame the method. One reasoning strategy that they used to support their methodological explanations was to draw analogies to other experiments that yielded similar results under similar conditions, usually with the same organism. For example, if one scientist obtained an unexpected finding, another scientist might draw an analogy to another experiment using the same organism in their laboratory and say "Oh. I also incubated the cells at 37 degrees and failed to get digestion of trypsin, but when I incubated the cells at zero degrees I got it; maybe you should do that." The types of analogies that the scientists used at this point in their research were what we term *local*—that is, the analogies were made to very similar experiments, usually with the same organism and within their own laboratory. Using local analogies is a powerful analogical reasoning strategy that scientists use when they obtain unexpected findings, and it is an important part of dealing with such findings. Scientists rarely mention these types of analogies in their autobiographies, as they appear to be uninteresting, yet they constitute one of the first reasoning mechanisms used in dealing with unexpected findings. Another important aspect of scientists' initial reactions to unexpected findings is that the social interactions that occur are circumscribed. Other scientists provide explanations and make suggestions, but the interactions are one on one: One scientist will give a methodological explanation or suggestion to the scientist who obtained the unexpected finding. There is no extensive discussion of the unexpected finding by many members of the laboratory.

Changes in Reasoning With Patterns of Unexpected Findings

The way that the scientists reasoned about their data changed when they obtained a *series* of unexpected findings. This usually occurred when the scientist continued to obtain unexpected findings despite having modified the method, attempted to replicate the finding, or obtained a whole series of unexpected findings in related conditions. It is at this point that a major shift in the reasoning occurred: The scientists began to offer new more general models, hypotheses, or theoretical explanations. Thus, for repeated unexpected findings, 51 theoretical explanations were offered as compared with 33 methodological ones. These models usually covered anywhere from two to five unexpected findings. The types of reasoning that occurred at that point were very different from the reasoning that scientists initially used when they obtained isolated unexpected findings. Three different types of reasoning processes were used to reason about a series of findings that were repeatedly unexpected:

1. Scientists draw analogies to different types of mechanisms and models in other organisms rather than make analogies to the same organism. This also involved making analogies to research conducted outside their laboratory. The scientists switched from using local analogies to more distant analogies. For example, a scientist working on a certain type of bacterium might say that "*IF3* in *E coli* works like this. Maybe your gene is doing the same thing." Thus, the way that analogies are used dramatically changes.

2. Scientists attempt to generalize over a series of findings to propose a new hypothesis that can explain the data. Often the scientists will search for common features of the unexpected finding and attempt to propose a general model that can explain these results. This is a form of inductive reasoning.

3. Social interactions change from one-on-one to group reasoning. It is not one scientist shouting "Eureka!" but a number of scientists building a new model by adding different components to the model. An important feature of this process is that often the members of the laboratory propose alternate models and explanations for the unexpected findings. They assess the plausibility and probability of different causal mechanisms that could have produced their unexpected finding. Group reasoning and the use of more distant analogies are key strategies in dealing with patterns of unexpected findings and greatly facilitate the generation of new models and the evaluation of alternate ones.

These three ways of responding to unexpected findings were consistently used by the scientists, suggesting that the scientists have sets of causal reasoning heuristics for dealing with the unexpected. The use of specialized causal reasoning heuristics is very different from the standard view of scientists having one accidental finding and then making a discovery. Much of their reasoning centers around their use of controls and evaluating the findings relative to the controls that they put into their experiment. Even when the scientists are planning their experiments, they discuss the possibility of error and how they can deal with unexpected findings before they occur. It is interesting that ignored historical data often reveal a similar pattern. For example, one of the most famous examples of an unexpected finding is Fleming's discovery of penicillin. On careful examination of his experimental strategies, one discovers that Fleming had adopted a strategy of following up on unusual and unexpected findings, attempting to replicate anomalous discoveries

(MacFarlane, 1984). It is interesting that it took many follow-up experiments before the importance and relevance of Fleming's unexpected findings led to the discovery of penicillin.

How Scientists Assess Unexpected Findings: The Building of Causal Models

When scientists obtain a set of unexpected findings, what do they do? One activity in which they engage is the building of a *causal model*. They propose a pathway of events that leads them from a cause to an effect. They often assess the plausibility of the causal model by evaluating the links in their causal chain, assessing whether the links fit. Other members of the laboratory engage in adding, deleting, and changing components of the chain. Some propose very different causal models, and members of the laboratory assess the plausibility of the model relative to their existing knowledge and the coherence of the model.

For example, one project in the HIV laboratory was to determine how HIV binds and integrates itself into the host DNA. The laboratory personnel invented a number of novel assays for determining how HIV integrates into the host DNA. HIV secretes a protein called *integrase* that both cuts the host DNA and then sticks the viral DNA into the host DNA. They used these assays to determine what parts of the integrase protein are necessary for binding. At one laboratory meeting, a postdoctoral fellow presented over a dozen experiments in which she had mutated the integrase protein to determine which aspects of integrase are necessary for normal activity. In all of these experiments she obtained results that were unexpected, and much of the laboratory meeting was concerned with interpreting these results, proposing new models, and assessing the plausibility of the various models. In many experiments she unexpectedly found that mutating integrase had a major effect on integration, preventing it from occurring, but little effect on its reverse process, disintegration. These unexpected findings were the impetus for the scientists to propose new mechanisms of integration. They rejected their previous model of the integration process and adopted a different model. One strategy that they used was to focus on the weak link in their model and replace that link with one that was consistent with their new data. This strategy is similar to one that Darden (1997) found in a historical analysis in genetics. Thus, both our *in vivo* and previous *in historico* work points to the building of causal models and replacement of weak links with new links as key reactions to unexpected findings. We wanted to explore the ways these types of causal models are assessed and built in a more controlled experimental environment. To this end, we decided to conduct an *in vitro* investigation of how people reason causally about models of scientific phenomena.

CAUSAL SCIENTIFIC REASONING *IN VITRO*

In this section we outline the findings from two *in vitro* experiments we have conducted that extend our *in vivo* work. In the first experiment we asked participants to generate causal models for phenomena. We asked participants to think out loud while they reasoned causally. We then conducted a verbal protocol analysis of participants' statements, focusing on the components of their causal models. In the second experiment we explored the role of data that were either consistent or inconsistent with participants' prior knowledge. Overall, this new *in vitro* work allowed us to explore the ways that people's underlying theories influence the interpretation of unexpected findings.

Experiment 1: Causal Scientific Thinking and Plausibility

We investigated the ways that people generate causal mechanisms by asking students to generate verbal protocols while they reasoned about causes and effects. Verbal protocols allow us to see the types of causal reasoning that people use and whether they use the same types of strategies that scientists used in our *in vivo* research. We presented undergraduate students with a series of scenarios depicting events that had happened and possible causes for those events. The causal scenarios were designed to vary with respect to the degree to which they were believable precursors to the observed effect. Participants were first given a brief introductory paragraph that depicted an event that had happened and a possible cause for that event. For example, participants were asked to "imagine you are an allergist who is trying to determine the cause of red, swollen eyes in a group of patients. You have a hypothesis that the red, swollen eyes may be due to exposure to tree pollen." Participants were then asked to graphically depict the relationship between the cause and effect by drawing a flow diagram linking the proposed cause and effect including all links between the cause and the effect. In addition, participants were asked to provide three judgments for each cause. Specifically, they were asked to (a) rate their a priori beliefs of how *plausible* the cause was, (b) rate their a priori beliefs of how *probable* the cause was (i.e., the estimated correlation between cause and effect), and (c) rate each of the links in their causal diagrams in terms of *plausibility* and *probability*. Using the techniques of Ericsson and Simon (1980, 1984), we videotaped participants as they reasoned causally and asked them to think aloud as they performed the task. Once participants had completed their causal diagrams and verbal protocols, we entered an explanation phase in which the participants explained why they had assigned links a particular value. Thus, we obtained both verbal protocols and retrospective reports from participants.

All of the protocols were transcribed and coded. Several interesting findings emerged from the analyses of these data. First, although the complexities of the diagrams were similar, the nature of the diagrams differed with respect to the ratings of the *probability* of the links in the diagrams. Specifically, the overall *plausibility* of the cause was directly related to the *probability* of the weakest link in the causal chain. For example, each participant was asked to draw a causal diagram depicting how red pots may cause flowers to bloom. Figure 3.1A illustrates a diagram along with the probability judgments for each link in the causal chain depicted by a prototypical participant.

Here the participant stated that "Sunlight will reflect off, that's a high probability that that'll happen [writes '6' below first link] that sunlight reflects off to, the pot to hit the right spot and that it takes that exact … that's the one, this is the weak link [writes '1' below second link]." Thus, for an implausible mechanism participants isolated the weakest link and gave the weak link a low probability value. We can see this clearly in the explanation phase, where the participant stated, "um … this again [points to the second link] is the weakest link." When participants reasoned about plausible mechanisms, they evaluated the probability of the links very differently. For plausible mechanisms, all the links in a participant's causal diagram received a high value. This can be seen in Fig. 3.1B, which is a causal diagram for the same participant reasoning about a plausible mechanism (flu virus causing a fever). Note that this participant rated each link in the causal diagram as highly plausible (5–7 on the 7-point scale). Most participants reasoned in a similar way. Highly plausible causal diagrams contained links in

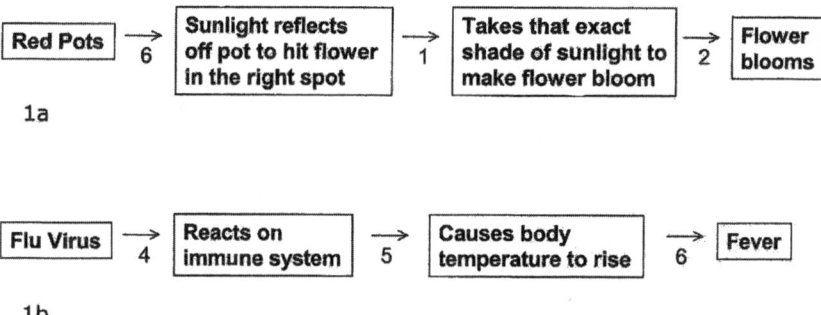

FIG. 3.1. Causal diagrams and probability of links provided by a prototypical participant. Note that Panel 1a contains an implausible mechanism and Panel 1b contains a plausible mechanism.

which the probabilities of each link were high, whereas implausible causal diagrams had at least one link that was rated improbable.

The participants in this experiment were similar to the scientists in that many different causal models could be generated and that the implausible models were the ones with weak links. Scientists in the *in vivo* context rejected explanations for unexpected findings when they thought that one of the steps in the causal explanation was improbable. What is also interesting about the results of this study is that, like the scientists, the participants were able to generate many models, both plausible and implausible, and ultimately reject implausible models due to an improbable link. There were some differences, however. For example, scientists sometimes combined the highly probable links of one implausible model with the highly probable links of another implausible model to produce a plausible model that contains only probable links. We did not observe this strategy in our *in vitro* analyses. This discrepancy is most likely due to the large differences in expertise between the *in vivo* and *in vitro* participants in our studies. Experts, unlike the novice participants in our experiment, both have more knowledge of a domain and are adept at constructing new models based on experience with their paradigms.

Experiment 2: Causal Reasoning About Expected and Unexpected Events

Our *in vivo* work has demonstrated that scientists spend a great deal of time reasoning about unexpected findings. We next wanted to examine the degree to which participants reason about unexpected findings in a more controlled setting. Specifically, we wanted to create a situation where we could present experimental outcomes to people that were either expected or unexpected and (a) measure their reaction times to the data and (b) measure how these unexpected findings contribute to overall causal evaluations. To do this, we created a scientific reasoning simulation. Fifty students participated in a simulation in which their task was to take the role of a research scientist analyzing data. They were asked to test the effectiveness of four pills designed to relieve depressive symptoms in a group of patients. They viewed patient data on a computer screen that were either expected (i.e., consistent with their knowledge) or unexpected (inconsistent with their knowledge). We manipulated the plausibility of the theories by presenting participants with a brief introductory statement of a theory that was plausible or implausible (see Appendix for stimuli). Data were then provided to participants in a trial-by-trial format in which they viewed 20 trials of data for each type of pill. The data were presented in combinations of the cause (*red pill* vs. *blue pill*) and the effect (*happy* versus *neutral* affective outcome) co-occurring. Participants were asked to judge the effectiveness of the *red pill* in causing *happiness* after they viewed each group of 20 patients for each theory. Under some conditions the *red pill* and *happiness* covaried strongly; under other conditions the *red pill* and *happiness* covaried weakly. This was accomplished by varying the number of times the cause (*red pill*) co-occurred with the observed effect (*happiness*). After participants received 20 trials of data, they were asked to judge the effectiveness of the red pill in causing the happiness using a scale that ranged from 1 (*low probability*) to 3 (*high probability*). For example, for one of the plausible conditions the description read:

> You are about to test the causal efficacy of a red pill. Research has demonstrated that happiness is related to the level of serotonin in the brain. The higher the level of serotonin, the more elevated one's mood becomes. The red pill is called a selective serotonin reuptake inhibitor. This pill works by increasing the level of serotonin in the brain. Specifically, this pill results in a blocking of the recycling process for the serotonin that then keeps more of this neurotransmitter in the brain available to communicate with other nerve cells.

Participants controlled the presentation of each trial by pressing the space bar on a computer keyboard to advance to the next trial. This allowed the participants to view each trial for as long as they wished. After the space bar was pressed, an interval occurred, followed by the subsequent trial. Participants were not aware that we were measuring the amount of time that they viewed each of the trials. After they viewed 20 trials of the stimuli, participants were asked to rate how probable they thought it was that the given cause (i.e., red pill) produced the given effect (i.e., happiness) using a 3-point scale that ranged from 1 (*low*) to 3 (*high*). This sequence was repeated once for each of the 4 pills.

We analyzed reaction time at two points in the experiments: (a) amount of time spent viewing each patient datum and (b) amount of time to formulate an overall causal judgment. Figure 3.2 presents the mean reaction times for these two analyses.

Time Viewing Patient Data. As can be seen in Fig. 3.2A, participants spent more time viewing patient data when the pill being tested had an implausible mechanism of action (M = 1,501 msec) than when the pill had a plausible mechanism of action (M = 1,424 msec). In addition, participants spent more time viewing the patient data when the pill being tested covaried weakly with the expected outcome (M = 1,567 msec) than when it covaried strongly with the expected outcome (M = 1,358 msec); that is, participants spent more time analyzing data where a positive outcome (i.e., relieved depression) would be unexpected.

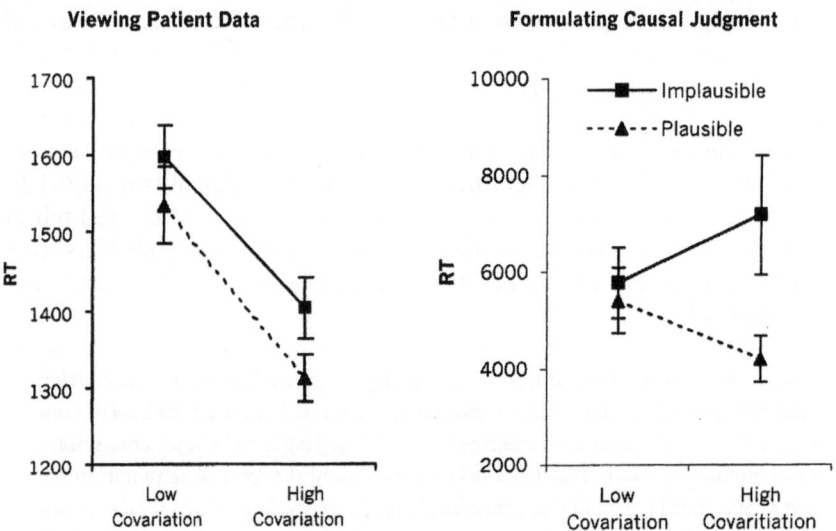

FIG. 3.2. Mean reaction times (RTs) for the analyses of patient data (Panel 2a) and formulation of causal judgments (Panel 2b) as a function of the plausibility of the causal mechanism and the degree to which the cause and effect actually covaried in the simulation. Error bars depict standard error of the mean.

Time to Formulate a Causal Judgment.   As can be seen in Fig. 3.2B, participants spent an equal amount of time determining the causal efficacy for a pill that had plausible (M = 5,417 msec) and implausible (M = 5,785 msec) mechanisms of action when the drug covaried weakly with the expected outcome. When the drug and the expected outcome strongly covaried, however, participants spent significantly more time formulating a causal judgment for the pill that contained an implausible mechanism of action (M = 7,190 msec) compared with the pill that contained a plausible mechanism of action (M = 4,204 msec); that is, people spent the majority of their time formulating a causal judgment when there was strong evidence for an unbelievable candidate. This finding is analogous to that obtained in our *in vivo* analyses: Scientists spent a great deal of time reasoning about the unexpected due to the generation of more general models, new hypotheses, new theoretical explanations, or drawing analogies to different types of mechanisms of action. It is reasonable to conceive that the participants in our simulation were carrying out a similar strategy. Indeed, informal interviews with the participants suggested this to be the case.

Our *in vitro* simulation extended our *in vivo* analyses in two main ways. First, similar to scientists reasoning in their laboratory meetings, the students in our simulation spent considerably more time reasoning about unexpected findings than about expected findings. Second, when formulating a causal judgment, they spent the majority of their time formulating a causal judgment when there was strong evidence in favor of an unbelievable candidate. Based in part on informal interviews with the participants in our simulation, we have found that some of this increased reasoning time was the product of thinking of similar mechanisms of action in other drugs and drawing analogies to these other mechanisms of action. Thus, as with the scientists, a common strategy for dealing with unexpected events was to seek similar mechanisms of action.

Our *in vitro* research on causal thinking has examined the interaction between expectations and the way people interpret data. This research illustrates a general-purpose mechanism by which expectations influence people's interpretation of information that is consistent with other findings in psychological research. For example, the illusory correlation effect (Chapman & Chapman, 1967; Shweder, 1977), the belief-bias effect in causal reasoning (Fugelsang & Thompson, 2000, 2001), the belief-bias effect in deductive reasoning (Klauer, Musch, & Naumer, 2000), and the plausibility of hypotheses in scientific reasoning (Klahr, Fay, & Dunbar, 1993) all have consistently demonstrated that people's knowledge biases the degree to which they reason about data. Our current *in vitro* work demonstrates that prior knowledge and expectations bias not only people's overall judgments but also the immediate interpretation of data as reflected in reaction time. Thus, expectations influence both the immediate and more delayed time course of scientific thinking.

CONCLUSION: BEYOND THE MYTHS OF SCIENTIFIC DISCOVERY

Human beings, in general, and scientists, in particular, appear to have a propensity to attend to the unexpected, have definitive strategies for sifting through these findings, and focus on the potentially interesting. Rather than being the victims of chance, scientists and nonscientists are ready to use the unexpected. How do we understand the role of the unexpected in science? Our *in vivo* studies reveal that scientists have used a number of different strategies for dealing with the unexpected. As we can see, rather than being at the mercy of random events that occur in their laboratories, scientists use the unexpected to foster discovery in the following ways. First, they create multifaceted experiments having many different types of experimental and control conditions (see also Baker & Dunbar, 2000). These controls make it possible for them to determine whether the result is what they expect. The many different conditions in an experiment cast a wide net, allowing the scientists to catch other unpredicted findings. Second, scientists attend to both expected and unexpected findings. They initially give methodological explanations for their findings and support these explanations by making analogies to very similar results obtained in other experiments. Third, if they continue to see the same type of unexpected result despite changes in methodology, they form new models and theories using more distant analogies and group reasoning. Thus, it is only when the scientists pass a threshold in unexpected findings that they propose general theories and discoveries are made.

The *in vivo* analyses paint a different picture of science from the one where the scientist accidentally knocks over a test tube and makes a discovery. What we see is that scientists must deal with unexpected findings virtually all the time. One of the most important places that they anticipate the unexpected is in designing experiments. They build many conditions and controls into their experiments. Thus, rather than being the victims of the unexpected, they create opportunities for unexpected events to occur, and once these events do occur, they have specific reasoning strategies for determining which of these events will be a clue to a new discovery. They focus on the method, using local analogies, and only after repeated demonstration of the unexpected event will they switch to the use of new theoretical explanations using more distant analogies and generalizations. Scientists are not passive recipients of the unexpected; rather, they actively create the conditions for discovering the unexpected and have a robust mental toolkit that makes discovery possible.

Our *in vitro* experiments provide a more detailed account of how unexpected results are responded to and interpreted. Similar to scientists reasoning in their laboratory meetings, the students in our simulation spent a great deal of time reasoning about unexpected findings. First, when viewing patient data, students in our simulation spent more time viewing cases in which a positive finding (i.e., relieved depressive symptoms) would be unexpected. Second, when formulating a causal judgment, the students in our simulation spent the majority of their time formulating a causal judgment when there was strong evidence for an unbelievable candidate. Based in part on informal interviews with the participants in our simulation, we suggest that this increased reasoning time was likely the product of them drawing analogies to different types of mechanisms of action.

The results of our analyses all point to a more complex and nuanced account of the role of the unexpected in science. Both William James (1890) and John Dewey (1896) noted that attention is drawn to unexpected events, and contemporary work on attention has provided both psychological and neurological accounts of the lower level cognitive processes that are involved in this response. What our analyses show is a way of putting together the higher level use of the unexpected in reasoning about complex scientific phenomena with research on the lower level processes that underlie allocation of attention to unexpected perceptual events. The overarching take-home message is that the underlying mental machinery that scientists use when reasoning about unexpected findings is the same as that used by nonscientists when they encounter any type of unexpected event. Dunbar (2002) recently argued that scientific thinking builds on thinking processes that all human beings use.

If this account of the use of the unexpected is correct, then why can't just anyone make a scientific discovery? First, as Dewey (1896) pointed out, what is and is not unexpected is both learned and contextually bound, even for simple perceptual stimuli. What is unexpected to an expert scientist may not be to a novice. Experts understand the deeper underlying principles that novices miss. Second, our *in vivo* analyses show that scientists design experiments that make it possible for unexpected phenomena to occur. We see this as one of the key ways that scientists conduct their research. The construction of experiments that have many controls is crucial for science and allows scientists to obtain unexpected findings. Overall, our *InVivo* analysis reveals that scientists have developed sets of reasoning strategies for dealing with the uncertainties of conducting research that allow them to harness chance events and make scientific discoveries. Our *in vitro* data show a pattern similar to that of the scientists; people spend more time reasoning about unexpected findngs then they do expected findings.

How do we now draw further insights into the role of the unexpected in science? We are currently extending our analyses of scientific reasoning to include a systematic analysis of the brain areas associated with scientific reasoning using fMRI (see Fugelsang & Dunbar, 2003). We are using designs similar to our *in vitro* work on scientific causal thinking. This methodology will allow us to draw deeper inferences regarding the possible underlying cognitive mechanisms that are involved in causal scientific thinking.

ACKNOWLEDGMENTS

The research reported in this chapter was funded by grants from Dartmouth College, The Spencer Foundation, the Natural Sciences and Engineering Research Council of Canada, McGill University, and the Office of Naval Research.

REFERENCES

Bacon, F. (1854). *Novum organum* (B. Montague, Trans.). Philadelphia: Parry & McMillan. (Original work published 1620)

Baker, L. M., & Dunbar, K. (2000). Experimental design heuristics for scientific discovery: The use of baseline and known controls. *International Journal of Human Computer Studies, 53*, 335–349.

Blanchette, I., & Dunbar, K. (2000). How analogies are generated: The roles of structural and superficial similarity. *Memory & Cognition, 28,* 108–124.

Blanchette, I., & Dunbar, K. (2001). Analogy use in naturalistic settings: The influence of audience, emotion and goals. *Memory & Cognition, 29,* 730–735.

Blanchette, I., & Dunbar, K. (2002). Representational change and analogy: How analogical inferences alter target representations. *Journal of Experimental Psychology: Learning, Memory, and Cognition, 28,* 672–685.

Callebaut, W. (1992). *Taking the naturalistic turn, or how real philosophy of science is done.* Chicago: University of Chicago Press.

Chapman, L. J., & Chapman, J. P. (1967). Genesis of popular but erroneous psycho-diagnostic observations. *Journal of Abnormal Psychology, 72,* 193–204.

Darden, L. (1992). Strategies for anomaly resolution. In R. N. Giere (Ed.), *Minnesota Studies in Philosophy of Science: Vol. 15. Cognitive models of science* (pp. 251–273). Minneapolis: University of Minnesota Press.

Darden, L. (1997). *Strategies for discovering mechanisms: Schema instantiation, modular subassembly, forward chaining/backtracking.* In *Proceedings of the 1997 Biennial Meeting of the Philosophy of Science Association.*

Delneri, D., Colson, I., Grammenoudi, S., Roberts, I., Louis, E. J., & Oliver, S. G. (2003). Engineering evolution to study speciation in yeasts. *Nature, 422,* 68–72.

Dewey, J. (1896). The reflex arc concept of psychology. *Psychological Review, 3,* 357–370.

Dunbar, K. (1993a). Concept discovery in a scientific domain. *Cognitive Science, 17,* 397–434.

Dunbar, K. (1993b, March). In Vivo *cognition: Knowledge representation and change in real-world scientific laboratories.* Paper presented at the society for Research in Child Development, New Orleans, LA.

Dunbar, K. (1995). How scientists really reason: Scientific reasoning in real-world laboratories. In R. J. Sternberg, & J. Davidson (Eds.), *Mechanisms of insight* (pp. 365–395). Cambridge MA: MIT Press.

Dunbar, K. (1997). How scientists think: Online creativity and conceptual change in science. In T. B. Ward, S. M. Smith, & S. Vaid (Eds.), *Conceptual structures and processes: Emergence, discovery and change* (pp. 461–493). Washington DC: American Psychological Association.

Dunbar, K. (1999a). Science. In M. Runco & S. Pritzker (Eds.), *The encyclopedia of creativity.* (Vol. 1, pp. 1379–1384). New York: Academic.

Dunbar, K. (1999b). The scientist In Vivo: How scientists think and reason in the laboratory. In L. Magnani, N. Nersessian, & P. Thagard (Eds.), *Model-based reasoning in scientific discovery* (pp. 85–99). New York: Plenum.

Dunbar, K. (2001a). What scientific thinking reveals about the nature of cognition. In K. Crowley, C. D. Schunn, & T. Okada (Eds.), *Designing for science: Implications from everyday, classroom, and professional settings* (pp. 115–140). Hillsdale: NJ: Lawrence Erlbaum Associates.

Dunbar, K. (2001b). The analogical paradox: Why analogy is so easy in naturalistic settings, yet so difficult in the psychology laboratory. In D. Gentner, K. J. Holyoak, & B. Kokinov, (Eds.), *Analogy: Perspectives from cognitive science* (pp. 313–334). Cambridge, MA: MIT Press.

Dunbar, K. (2002). Understanding the role of cognition in science: The science as category framework. In P. Carruthers, S. Stitch, & M. Siegal (Eds.), *The cognitive basis of science* (pp. 154–170). New York: Cambridge University Press.

Dunbar, K., & Blanchette, I. (2001). The *In Vivo/In Vitro* approach to cognition: The case of analogy. *Trends in Cognitive Sciences, 5,* 334–339.

Ericsson, K., & Simon, H. (1980). Verbal reports as data. *Psychological Review, 87,* 215–225.

Ericsson, K., & Simon, H. (1984). *Protocol analysis: Verbal reports as data.* Cambridge, MA: MIT Press.

Feist, G., & Gorman, M. (1998). The psychology of science: Review and integration of a nascent discipline. *Review of General Psychology, 2,* 3–47.

Fugelsang, J., & Dunbar, K. (2003). *How the brain uses theory to interpret data.* Manuscript submitted for publication.

Fugelsang, J., & Thompson, V. (2000). Strategy selection in causal reasoning: When beliefs and covariation collide. *Canadian Journal of Experimental Psychology, 54,* 15–32.

Fugelsang, J., & Thompson, V. (2001). Belief-based and covariation-based cues affect causal discounting. *Canadian Journal of Experimental Psychology, 55,* 70–76.

Galilei, G. (1991). *Dialogues concerning two new sciences* (A. de Salvio and H. Crew, Trans.). New York: Prometheus. (Original work published 1638)

Goel, V., Gold, B., Kapur, S., & Houle, S. (1998). Neuroanatomical correlates of human reasoning. *Journal of Cognitive Neuroscience, 10,* 293–302.

James, W. (1890). *The principles of psychology.* New York: Holt.

Klahr, D. (2000). *Exploring science: The cognition and development of discovery processes.* Cambridge, MA: MIT Press.

Klahr, D., & Dunbar, K. (1988). Dual space search during scientific reasoning. *Cognitive Science, 12,* 1–48.

Klahr, D., Fay, A., & Dunbar, K. (1993). Heuristics for scientific experimentation: A developmental study. *Cognitive Psychology, 25,* 111–146.

Klahr, D., & Simon, H. (1999). Studies of scientific discovery: Complementary approaches and convergent findings. *Psychological Bulletin, 125,* 524–543.

Klauer, K., Musch, J., & Naumer, B. (2000). On belief bias in syllogistic reasoning. *Psychological Review, 107,* 852–884.

Klayman, J., & Ha, Y. (1987). Confirmation, disconfirmation, and information in hypothesis testing. *Psychological Review, 94,* 211–228.

Kuhn, T. (1962). *The structure of scientific revolutions.* Chicago: University of Chicago Press.

Kulkarni, D., & Simon, H. (1988). The processes of scientific discovery: The strategy of experimentation. *Cognitive Science, 12,* 139–175.

Langley, P., Simon, H., Bradshaw, G., & Zytkow, J. (1987). *Scientific discovery: Computational explorations of the creative processes.* Cambridge, MA: MIT Press.

Lave, J. (1998). *Cognition in practice.* Cambridge, England: Cambridge University Press.

MacFarlane, G. (1984). *Alexander Fleming: The man and the myth.* Oxford, England: Oxford University Press.

Nersessian, N. (1992). How do scientists think? Capturing the dynamics of conceptual change in science. In R. N. Giere (Ed.), *Minnesota studies in the philosophy of Science: Vol. 15. Cognitive models of science* (pp. 3–45). Minneapolis: University of Minnesota Press.

Schraagen, J. (1993). How experts solve a novel problem in experimental design. *Cognitive Science, 17,* 285–309.

Schunn, C., & Anderson, J. (1999). The generality/specificity of expertise in scientific reasoning. *Cognitive Science, 23,* 337–370.

Seger, C., Poldrack, R., Prabhakaran, V., Zhao, M., Glover, G., & Gabrieli, J. (2000). Hemispheric asymmetries and individual differences in visual concept learning as measured by functional MRI. *Neuropsychologia, 38,* 1316–1324.

Shweder, R. (1977). Illusory correlation and the MMPI controversy. *Journal of Consulting and Clinical Psychology, 45,* 917–924.

Suchman, L. (1987). *Plans and situated action: The problems of human–machine communication.* Cambridge, England: Cambridge University Press.

Thagard, P. (1999). *How scientists explain disease.* Princeton, NJ: Princeton University Press.

Tweney, P. (2000). *Coherence in thought and action.* Cambridge, MA: MIT Press.

Tweney, R. (1992). Inventing the field: Michael Faraday and the creative "engineering" of electromagnetic field theory. In R. J. Weber & D. N. Perkins (Eds.), *Inventive minds: Creativity in technology* (pp. 31–47). London: Oxford University Press.

Tweney, R., Doherty, M., & Mynatt, C. (Eds.). (1981). *On scientific thinking.* New York: Columbia University Press.

Wason, P. (1968). Reasoning about a rule. *Quarterly Journal of Experimental Psychology, 20,* 273–281.

Appendix

Stimuli used in the *In Vitro* study

IMPLAUSIBLE MECHANISM

Past research has demonstrated that the growth of small amounts of the bacteria *staphylococcus* in the body has <u>no</u> direct link to people's feelings of happiness. The <u>red pill</u> is a "topoisomerase inhibitor." Topoisomerase is an enzyme that is necessary for the reproduction of *staphylococcus* in the body. "Topoisomerase inhibitors" inhibit this enzyme, thus restricting the ability of *staphylococcus* to replicate.

Past research has demonstrated that the growth of small amounts of the bacteria *clostridium* in the body has <u>no</u> direct link to people's feelings of happiness. The <u>red pill</u> is a "protein binder." The cell walls of bacteria are continuously expanding through the synthesis of proteins and amino acids. In order for a bacteria cell to flourish and reproduce, the cell wall must be able to expand with the growing interior. "Protein binders" bind to specific amino acids and proteins, thus inhibiting the cell wall of *clostridium* to synthesize.

PLAUSIBLE MECHANISM

Past research has demonstrated that people's feelings of happiness <u>are</u> directly related to the level of *serotonin* in the brain. The <u>red pill</u> is a "selective *serotonin* reuptake inhibitor." This pill blocks the recycling process for the *serotonin*, which then keeps more of this neurotransmitter in the brain available to communicate with other nerve cells.

Past research has demonstrated that people's feelings of happiness <u>are</u> directly related to the level of *norepinephrine* in the brain. The <u>red pill</u> is a "monoamine oxidase inhibitor." Monoamine oxidase is an enzyme that breaks down *norepinephrine* in the brain. Monoamine oxidase inhibitors inhibit this enzyme, thus allowing a greater supply of this neurotransmitter to remain available in the brain.

4

A Framework
for Cognitive Studies
of Science
and Technology

David Klahr
Carnegie Mellon University

Over the past 40 years or so, an extensive body of knowledge about the "cognition of discovery" has emerged. For most of this period, the work was based in the psychology laboratory, using ordinary participants working on simplified (but not always simple) problems that represented one or another important aspects of the discovery process. Nevertheless, as these types of studies have become increasingly complicated—progressing from simple concept formation (Bruner, Goodnow, & Austin, 1956) and rule-induction studies (Wason, 1960) to difficult and complex "discovery microworlds" (Dunbar, 1993; Klahr, 2000; Mynatt, Doherty, & Tweney, 1978; Schunn & Klahr, 1992)—they have provided important insights, findings, and conceptual frameworks for better understanding the cognitive processes that support discovery and invention.

Although many of the chapters in this volume push the envelope of complexity and reality of cognitive studies of science and technology, a structured review of the psychology laboratory based studies (which Bruner et al., 1956, called *in vitro studies,* by analogy to common practice in the biological sciences) may help to explicate the distinct components of human cognition involved in scientific discovery. My aim in this chapter is to offer such a framework, based in part on my own laboratory research but also incorporating a wide range of studies of the process of scientific reasoning in both adults and children.

Laboratory investigations of scientific reasoning can be classified along two dimensions: one representing the degree of domain specificity or domain generality, and the other representing the type of processes involved. Table 4.1 depicts this characterization of the field.[1] The two rows correspond to the difference between domain-general knowledge and domain-specific knowledge, and the three columns correspond to the major components of the overall discovery process: generating hypotheses, designing experiments, and evaluating evidence.

DISTINCTIONS BETWEEN DOMAIN-GENERAL AND DOMAIN-SPECIFIC KNOWLEDGE

Both domain-specific and domain-general knowledge influence the scientific reasoning process, but they can be considered partially independent entities. On the one hand, acquisition of domain-specific knowledge not only changes the substantive structural knowledge in the domain (by definition) but also influences the processes used to generate and evaluate new hypotheses in that domain (Carey, 1985; Keil, 1981; Wiser, 1989). Thus, it is not surprising that, after years of study in a specific domain, scientists exhibit reasoning processes that are purported to be characteristic of the field (e.g., "she thinks like a physicist") and very different from those used by experts in other fields, or by novices or children. On the other hand, in simple contexts that are nearly devoid of domain-specific knowledge, professional scientists are often indistinguishable from laypersons (Mahoney & DeMonbreun, 1978), and even first graders can reason correctly about hypotheses and select appropriate experiments to evaluate them (Sodian, Zaitchik & Carey, 1991).

TABLE 4.1
Types of Foci in Psychological Studies of Scientific Reasoning

Type of Knowledge	Hypothesis Generation	Experiment Design	Evidence Evaluation
Domain specific	A	B	C
Domain general	D	E	F

---

[1]This taxonomy, first proposed by Klahr and Carver (1995), was used by Zimmerman (2000) to organize her extensive review of the developmental literature on scientific reasoning.

Psychologists' attempts to disentangle the relative influence of general versus specific knowledge have produced two distinct literatures: one on domain-specific knowledge and the other on domain-general reasoning processes. This distinction, corresponding to the two rows in Table 4.1, will be explained later, after I have described the other dimension in the table.

PROCESSES IN SCIENTIFIC REASONING

The three columns in Table 4.1 reflect a view of scientific discovery as a problem-solving process involving search in a problem space (Newell & Simon, 1972). In the case of scientific discovery, there are two primary spaces to be searched: (a) a space of hypotheses and (b) a space of experiments (Klahr & Dunbar, 1988).[2] These spaces are sufficiently different that they require different representations, different operators for moving about in the space, and different criteria for what constitutes progress in the space. The distinction between searching the hypothesis space and searching the experiment space is sufficiently important that in the natural sciences people are often trained to be experts in one, but not the other, aspect of their discipline: that is, as theorists or as experimentalists. It is clear that the problems to be solved in each space are different, even though they have obvious and necessary mutual influences.[3]

Search in the two spaces requires three major interdependent processes: (a) *hypothesis space search*, (b) *experiment space search*, and (c) *evidence evaluation*. In searching the hypothesis space, the initial state consists of some knowledge about a domain, and the goal state is a hypothesis that can account for some or all of that knowledge. When one or more hypotheses are under consideration, it is not immediately obvious what constitutes a "good" experiment. In constructing experiments, scientists are faced with a problem-solving task paralleling their search for hypotheses; that is, they must search in the experiment space for an informative experiment. The third process—evidence evaluation—involves a comparison of the predictions derived from the current hypothesis with the results obtained from experimentation.

These three processes, shown as the column headings in Table 4.1, correspond to the conventional view of science as involving the generation of hypotheses, the formulation and execution of experiments, and the revision of

---

[2]A detailed account of problem space search and these two spaces in particular can be found in Klahr (2000). In this chapter I provide only a brief sketch.

[3]To make things even more complicated, in many cases there are other problem spaces involved in invention and discovery, including, among others, a space of representations, a space of instrumentation, and a space of paradigms, to name just a few. This issue is addressed further in Schunn and Klahr (1995, 1996) and Klahr and Simon (2001).

hypotheses on the basis of empirical evidence. During the course of normal scientific discovery, the various cells in Table 4.1 are traversed repeatedly. However, it is very difficult to study thinking processes that involve all of them simultaneously. Consequently, the early research in the field started out with investigations designed to constrain the topic of interest to just one or two cells. As the field has matured, more complex contexts involving multiple cells have been used. In the following brief review, I describe a few investigations that exemplify various cells and cell combinations from Table 4.1.

STUDYING VARIOUS CELLS AND CELL COMBINATIONS

Cell A: Domain-Specific Hypothesis Space Search

In a series of pioneering studies, McCloskey (1983) investigated people's naive theories of motion. His basic paradigm was to present participants with depictions of simple physical situations (e.g., an object being dropped from an airplane, or a ball exiting a curved tube) and ask them to predict the subsequent motion of the object. McCloskey found that many college students held naive impetus theories (e.g., the belief that the curved tube imparted a curvilinear impetus to the ball, which dissipated slowly and made the ball continue in a curved trajectory that eventually straightened out). Note that in this kind of study, participants are asked about their knowledge about a specific domain. They are not allowed to run experiments, and they are not presented with any evidence to evaluate. Nor is there an attempt to assess any domain-general skills, such as designing unconfounded experiments, making valid inferences, and so on. Thus McCloskey's investigations—as well as similar studies on children's understanding of the distinction between heat and temperature (Carey, 1985)—are classified as examples of research in Cell A of the taxonomy.

Cell B: Domain-Specific Experiment Space Search

In some investigations, participants are asked to decide which of a set of pre-specified experiments will provide the most informative test of a prespecified hypothesis. There is no search of the hypothesis space, and the experiment space search is limited to choosing from among the given experiments, rather than generating them. The goal of such investigations is to find out how participants map the features of the given hypothesis onto the features of the given set of experimental alternatives.

Tschirgi (1980) investigated the ability of children and adults to make valid inferences in familiar, knowledge-rich contexts. Participants were presented with an everyday scenario—for example, a description of a successful (or unsuccessful) cake baking situation in which some novel ingredients were used—and then provided with a hypothesis about a specific factor (e.g., honey rather than sugar made a great cake). Next, they were asked to decide which one of three possible variant recipes would best determine whether the hypothesized factor (honey) was the important one for a great cake. Tschirgi found that, at all ages, the extent to which people chose an unconfounded experiment depended on whether the outcome of the initial experiment was good or bad; that is, if the initial outcome was bad, the participants tended to vary the hypothesized culprit factor while holding all other factors constant. (Tschirgi called this VOTAT, or "vary one thing at a time.") In contrast, if the initial outcome was good, then participants tended to keep the hypothesized single cause of the good outcome and vary many other things simultaneously (HOTAT: "hold one thing at a time"). Note that this study required neither hypothesis space search (the sole hypothesis to be evaluated was provided by the investigator) nor the actual evaluation of evidence, because participants did not get to see the outcome of the experiment that they chose.

VOTAT is a domain-general strategy for producing unconfounded experimental contrasts. Thus, Tschirgi's (1980) study could be placed in Cell E. However, because its main purpose was to demonstrate the influence of domain knowledge on people's ability to design an experiment from which they could make a valid inference, it exemplifies the type of studies in Cell B.

Cell E: Domain-General Experiment Space Search

Studies of people's ability to design factorial experiments (e.g., Case, 1974; Siegler & Liebert, 1975) focus almost entirely on effective search of the experiment space. The use of domain-specific knowledge is minimized, as is search in the hypothesis space and the evidence evaluation process. For example, Kuhn and Angelev (1976) explored children's ability to design unconfounded experiments to determine which combination of chemicals would produce a liquid of a specific color. Children had no particular reason to favor one chemical over another, nor were they supposed to propose any causal mechanisms for the color change. The goal of the study was focused on the development of children's domain-general skill at experimental design.

Cells C and F: Domain-Specific
and Domain-General Evidence Evaluation

Studies in this category focus on people's ability to decide which of several hypotheses is supported by evidence. Such studies typically present tables of covariation data and ask participants to decide which of several hypotheses is supported or refuted by the data in the tables. In some cases, the factors are abstract and arbitrary (e.g., Shaklee & Paszek, 1985)—in which case the studies are classified in Cell F—and in others they refer to real world factors (e.g., studies that present data on plant growth in the context of different amounts of sunlight and water; Amsel & Brock, 1996; Bullock, Ziegler, & Martin, 1992). In such cases participants have to coordinate their prior domain knowledge with the covariation data in the tables (e.g., Ruffman, Perner, Olson, & Doherty, 1993). These studies involve both Cell C and Cell F.

Cells A and C: Domain-Specific Hypothesis Space Search and Evidence Evaluation

In investigations that combine Cells A and C, children are asked to integrate a variety of forms of existing evidence in order to produce a theory that is consistent with that evidence. They do not have the opportunity to generate new evidence via experimentation, and the context of their search in the hypothesis space is highly domain specific. This type of investigation is exemplified by Vosniadou and Brewer's (1992) analysis of elementary school children's mental models of the earth. The challenge faced by children is how to reconcile adults' claims that the earth is round with their firsthand sense experience that it is flat, which, in turn, is contradicted by their secondhand sense experience of photographs of the round earth. Based on an analysis of children's responses to an extensive clinical interview about features and aspects of the earth's shape, Vosniadou and Brewer were able to describe a progression of increasingly mature mental models of the earth, including a rectangular earth; a disc earth; a dual earth; a hollow sphere; a flattened sphere; and, finally, a sphere.

The work of Koslowski and her colleagues (Koslowski, 1996; Koslowski & Okagaki, 1986; Koslowski, Okagaki, Lorenz, & Umbach, 1989) also combines Cells A and C as well as F. In these studies, participants are presented not only with covariation data but also with suggestions about possible causal mechanisms, analogous effects, sampling procedures, and alternative hypotheses. For example, in one study, participants were first presented with covariation information about a gasoline additive and car performance, and they (correctly) rated perfect correlation as more likely to indicate causation than partial correlation. However, when presented with additional information about particular mechanisms that had been ruled out, participants revised these estimates downward, even though the covariation data itself remained the same. The results of these studies demonstrate that abstract consideration of the strength of covariation data is not the only process that determines people's judgments about causality. Instead, both adults and children draw on many other aspects of both their specific knowledge about the domain as well as domain-general features—such as sample size or the elimination of plausible mechanisms—in making causal inferences.

Cells D, E, and F: Domain-General Hypothesis Space Search, Experiment Space Search, and Evidence Evaluation

Bruner et al. (1956) created their classic concept-learning task in order to better understand people's appreciation of the logic of experimentation and their strategies for discovering regularities. Their participants' task was to discover an arbitrary rule (e.g., "large & red") that was being used to classify a set of predefined instances consisting of all possible combinations of shape, color, number, and so on. Bruner et al. ran several different procedures that collectively span the bottom row of Table 4.1: Participants had to generate hypotheses, choose among "experiments" (i.e., select different cards that displayed specific combinations of attributes), and evaluate the evidence provided by the yes–no feedback that they received. Because the task is abstract and arbitrary, none of the domain-specific cells are involved in Bruner et al.'s studies. This type of study is mainly classified as involving Cells D and F. Cell E is implicated only marginally, because participants did not have to generate instances, only select them.

Another venerable task that spans Cells D, E, and F is Wason's (1960) 2–4–6 task, in which participants are asked to discover a rule (predetermined by the investigator) that could classify sets of numerical triads. They are told that "2–4–6" is an example of a triad that conforms to the rule, and they are instructed to generate their own triads—which the investigator then classifies as an instance or noninstance of the rule. This simple task has been used in hundreds of investigations of so-called "confirmation bias" (see Klayman & Ha, 1987, for a review and theoretical integration of the extensive literature on this task, and Samuels & McDonald, 2002, for an recent example of how it has been used to investigate children's scientific reasoning skills.) This task does involve Cell E (as well as Cells D and F), because participants have to generate their own "experiments" (test triads) in their attempt to discover the classification rule. Like Bruner et al.'s (1956) paradigm, the 2–4–6 task was designed to study people's reasoning in a context that "simulates a miniature scientific problem" (Wason, 1960, p. 139).

A more complex example of the a study that spans Cells D, E, and F is the pioneering investigation conducted by Mynatt, Doherty, and Tweney (1977, 1978) in which people were asked to determine the laws governing the motion of "particles" in an arbitrary (and non-Newtonian) universe consisting of objects displayed on a computer screen. The participants' task was to infer the laws of motion by running experiments in which they fired particles at objects.

INTEGRATIVE INVESTIGATIONS OF SCIENTIFIC REASONING

Although the earliest laboratory studies of scientific thinking tended to focus on highly selected aspects of the discovery process, it is now possible to do empirical studies of people performing tasks that simultaneously evoke—in abstracted and idealized form—several of the essential components of the discovery process (Dunbar, 1993; Koslowski, 1996; Krems & Johnson, 1995; Kuhn, 1989; Kuhn, Garcia-Mila, Zohar, & Anderson, 1995; Schauble, 1990, 1996; Schauble, Glaser, Raghavan, & Reiner, 1991). For example, Mynatt et al.'s (1977, 1978) studies are clear precursors of the kinds of microworld tasks that my colleagues and I have used to study the development of scientific reasoning processes in children and adults (Klahr, Fay, & Dunbar, 1993; Klahr & Dunbar, 1988; Schunn & Klahr, 1992).

We started these investigations because we felt that although research focusing on either domain-specific or domain-general knowledge had yielded much useful information about scientific discovery, such efforts are, perforce, unable to assess an important aspect of this kind of problem solving: the interaction between the two types of knowledge. Similarly, the isolation of hypothesis search, experimentation strategies, and evidence evaluation begs a fundamental question: How are the three main processes integrated, and how do they mutually influence one another? The goal of our work has been to integrate the six different aspects of the scientific discovery process represented in Table 4.1 while still being cast at a sufficiently fine grain so as not to lose relevant detail about the discovery process.

Several of our studies have used tasks in which domain-general problem-solving heuristics play a central role in constraining search while at the same time participants' domain-specific knowledge biases them to view some hypotheses as more plausible than others. Furthermore, in these tasks, both domain-specific knowledge and domain-general heuristics guide participants in designing experiments and evaluating their outcomes. With respect to the integration of the three processes, such tasks require coordinated search in *both* the experiment space and the hypothesis space, as well as the evaluation of evidence produced by participant-generated experiments. A summary of this work can be found in Klahr (2000).

EXTENSIONS TO THE TAXONOMY

Like any taxonomy, the one presented here is necessarily oversimplified. I have already mentioned that there are often more than just two problem spaces involved in the discovery process, so it is clear that more columns are required to fully capture the different types of search. An equally important extension would recognize the fact that science and invention are often highly collaborative activities. Thus, the simple two-dimensional table that I have been using should be extended as suggested in Fig. 4.1 to include dyads, groups, or much large social collectives.

Of course, as we move along this dimension we move from traditional cognitive psychology to the social psychological study of dyads, groups, and ultimately to sociological perspectives on scientific discovery. Social psychologists have been studying the process of science, but without explicit attention to the aspects of the framework presented here (e.g., Fuller & Shadish, 1994), and psychologists have approached pieces of the overall framework from the perspective of small groups (e.g., Azmitia, 1988; Castellan, 1993, Resnick, Levine, & Teasley, 1991). The types of microworld studies described earlier have been easily extended to investigations of collaborative discovery (e.g., Okada & Simon, 1997; Teasley, 1995).

A personal example of the type of study that might fill some of the cells in the dyadic "slice" of this three-dimensional taxonomy is provided in the introduction to my own monograph (Klahr, 2000) in which I offer a brief account of how I first came to use microworlds to study scientific reasoning. The account begins with a quote from B. F. Skinner (1956): "Here was a first principle not formally recognized by scientific methodologists: When you run onto something interesting, drop everything else and study it" (p. 223).

I then go on to reveal a bit of the personal history of the interplay of ideas between two scientists struggling to work out a productive research paradigm and how the goal of the research itself evolved during that interplay. The account continues, as follows:

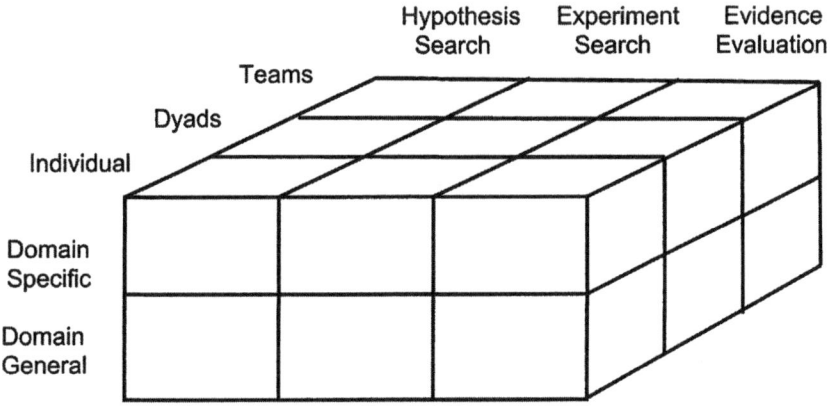

FIG. 4.1. Knowledge Type × Number of Minds × Phase.

About 15 years ago I did run into something interesting and, as Skinner rec-ommended, for quite a while I dropped nearly everything else to study it. Ac-tually, the thing of interest ran into me when one of my graduate students—Jeff Shrager—programmed a toy robotic tank—called "BigTrak"—to travel down the corridors of Baker Hall, turn into my office, and startle me with several shots from its laser cannon. "Now this," I thought, "is very interesting!" because I realized that Jeff had produced a solution to several problems that he and I had been struggling with.

One such problem was how to investigate they way in which people construct mental models of moderately complex devices that they encounter in their everyday lives, such as calculators, VCRs and ATM machines. We speculated that by observing people learning to use such devices—without instruc-tion—we might gain some insight into aspects of both device design and com-plex problem solving. At the time, we were motivated by a mixture of interests in human–computer interaction and complex problem solving. In addition, we were influenced by the work of Herbert Simon and his colleagues on the modeling the inductive processes underlying several important scientific dis-coveries (Simon, Langley, & Bradshaw, 1981). We felt that this work was rele-vant to our interests because we had a vague notion that what people do as they discover how such devices work is not so different from what scientists do when they attempt to discover how the world works. Our problem was to devise a paradigm with which to study these issues.

In addition, I was seeking a context in which to extend my research on how children solve formal problems, such as the Tower of Hanoi (Klahr & Robin-son, 1981), to less well-structured domains. Thus, I wanted to include a de-velopmental perspective in any study of learning to use complex devices, but I felt that calculators and ATM machines would be too difficult to use with young children. We were familiar with the "simulated research envi-ronments" used by Mynatt, Doherty, and Tweney (1977) to study hypothe-sis-testing strategies in adults, and we had been considering the idea of creating our own computer games as a context, but we had not found a satis-factory solution. Then Jeff had the brilliant idea of using a toy he had ac-quired as an undergraduate: BigTrak. As soon as I saw it come lumbering into my office, I knew that the BigTrak was indeed "something interesting" that would enable us to study the processes of scientific discovery in chil-dren and adults.

Having selected the BigTrak as a suitable context for our work, we went on to use it to study complex cognition in adults. In the investigations of "instruc-tionless learning," adult subjects were simply presented with the BigTrak, with no instruction whatsoever, and asked to figure out how it worked. Shrager (1985) argued that in this kind of situation initial hypotheses were generated through an analogy-like process he called "view application," in which previously stored knowledge structures are mapped to specific BigTrak elements. For example, if the "calculator view" is activated, then a mapping is made between Big Trak's keys and calculator keys, and the associated knowledge that calculators have memories is used to hypothesize that BigTrak has one also. Shrager's model focused almost entirely on this first phase of the discovery process (Shrager, 1987).

It soon became clear, however, that with a device as complex as BigTrak, the instructionless learning approach was too open-ended to provide sufficiently tractable data to support the formulation of a clear theoretical account of the discovery process in adults (see Shrager & Klahr, 1986), and too difficult for children. At this point we made a small but important change in our paradigm. We decided that rather than give people no information about how BigTrak worked, we would provide instruction on its basic commands, and then ask people to discover only one additional complex function. We felt that by providing the initial training we would have a much stronger analogy to the real world of scientific discovery in which scientists are always working in a context of a body of knowledge that they then attempt to extend. Moreover, because this initial training was mastery-based it allowed us to reduce some of the variance in the initial domain-specific knowledge with which our subjects started their discovery efforts. From this evolved the program of research on the developmental psychology of scientific discovery described in this book. (Klahr, 2000, pp. xi–xii)

Of course, this account has several severe shortcomings. For one thing, it was written long after the fact, and without any reference to diaries, logs, or any other objective and verifiable record. For another, it is not framed in the context of any theory about dyadic interaction. However, it does give the flavor of what the addition of a social dimension to the study of collaboration might entail. A more extended, but highly personal account of the interplay of personal, social, and scientific issues can be found in Gorman (1992). It is clear that much more could be done. Consider, for example, the recent controversial analyses and reanalysis of the evidence about the effects of day care on children (NICHD Early Child Care Research Network, 2002), or the efficacy of postmenopausal hormone replacement therapy on heart disease (Enserink, 2002), or the efficacy of periodic mammograms (Costanza et al., 2000). The focus of research in each of these cases involves highly charged issues with important policy implications, and investigations of the complex interactions among individual scientists, teams of researchers, and groups of policy makers in these domains—corresponding to the Domain Specific–Evidence Evaluation–Teams cell in Fig. 4.1—would break new ground in advancing our understanding of the interactions among cognitive, social, sociological, economic, and political processes in the pursuit of scientific discovery.

CONCLUSION

Science is a fundamental feature of human activity, and many of its aspects make it enormously appealing as an object of study. To some people, science exemplifies the impenetrable mystery of creativity, whereas for others it epitomizes the best that systematic thought can produce. It has been an object of study by just about every one of the humanistic and social science disciplines, from anthropology to economics, from history to sociology, from education to philosophy, and each approach has its inherent strengths and weaknesses (see Klahr & Simon, 1999, for an extensive discussion). The taxonomy used to review the laboratory studies described here may also be useful for scientists studying the nature of discovery and invention in the "real world," which may in turn suggest new questions that can be addressed in the cognitive psychology laboratory. Given the important role of collaboration in science, I hope that others interested in the problems addressed in this volume will be inspired to join with its contributors in a distributed collaboration as we continue to search for ways to advance our knowledge about the processes of discovery and invention.

REFERENCES

Amsel, E., & Brock, S. (1996). The development of evidence evaluation skills. *Cognitive Development, 11*, 523–550.

Azmitia, M. (1988). Peer interaction and problem solving: When are two heads better than one? *Child Development, 59*, 87–96.

Bruner, J. S., Goodnow, J. J., & Austin, G. A. (1956). *A study of thinking*. New York: New York Science Editions.

Bullock, M., Ziegler, A., & Martin, S. (1992). Scientific thinking. In F. E. Weinert & W. Schneider (Eds.), *LOGIC Report 9: Assessment procedures and results of Wave 6*. Munich: Max Plank Institute for Psychological Research.

Carey, S. (1985). *Conceptual change in childhood*. Cambridge, MA: Bradford Books/ MIT Press.

Case, R. (1974). Structures and strictures: Some functional limitations on the course of cognitive growth. *Cognitive Psychology, 6*, 544–573.

Castellan, J. J. (Ed.). (1993). *Current issues in individual and group decision making*. Hillsdale, NJ: Lawrence Erlbaum Associates, Inc.

Costanza, M. E., Stoddard, A. M., Luckmann, R., White, M. J., Spitz, A. J., & Clemow, L. (2000). Promoting mammography: Results of a randomized trial of telephone counseling and a medical practice intervention. *American Journal of Preventive Medicine, 19*, 39–46.

Dunbar, K. (1993). Concept discovery in a scientific domain. *Cognitive Science, 17*, 397–434.

Enserink, M. (2002, August 2). Women's health: U.K. hormone trial to pause for review. *Science, 297*, 755–757.

Fuller, S., & Shadish, W. R. (Eds.). (1994). *The social psychology of science.* New York: Guildford.

Gorman, M. E. (1992). *Simulating science: Heuristics, mental models and technoscientific thinking.* Bloomington: Indiana University Press.

Keil, F. C. (1981). Constraints on knowledge and cognitive development. *Psychological Review, 88,* 197–227.

Klahr, D. (2000). *Exploring science: The cognition and development of discovery processes.* Cambridge, MA: MIT Press.

Klahr, D., & Carver, S. M. (1995) Scientific Thinking about Scientific Thinking. *Monographs of the Society for Research in Child Development,* #245, Vol. 60, No. 4, 137–151.

Klahr, D., & Dunbar, K. (1988). Dual space search during scientific reasoning. *Cognitive Science, 12,* 1–48.

Klahr, D., Fay, A. L., & Dunbar, K. (1993). Heuristics for scientific experimentation: A developmental study. *Cognitive Psychology, 24,* 111–146.

Klahr, D., & Robinson, M. (1981). Formal assessment of problem solving and planning processes in preschool children. *Cognitive Psychology, 13,* 113–148.

Klahr, D., & Simon, H. A. (1999). Studies of scientific discovery: Complementary approaches and convergent findings. *Psychological Bulletin, 125,* 524–543.

Klahr, D., & Simon, H. A. (2001). What have psychologists (and others) discovered about the psychology of scientific discovery. *Current Directions in Psychological Science, 10*(3), 75–83.

Klayman, J., & Ha, Y. (1987). Confirmation, dis-confirmation and information in hypothesis testing. *Psychological Review, 94,* 211–228.

Koslowski, B. (1996). *Theory and evidence: The development of scientific reasoning.* Cambridge, MA: MIT Press.

Koslowski, B., & Okagaki, L. (1986). Non-Human indices of causation in problem-solving situations: Causal mechanisms, analogous effects, and the status of rival alternative accounts. *Child Development, 57,* 1100–1108.

Koslowski, B., Okagaki, L., Lorenz, C., & Umbach, D. (1989). When covariation is not enough: The role of causal mechanism, sampling method, and sample size in causal reasoning. *Child Development, 60,* 1316–1327.

Krems, J., & Johnson, T. R. (1995). Integration of anomalous data in multicausal explanations. In J. D. Moore & J. F. Lehman (Eds.), *Proceedings of the 17th Annual Conference of the Cognitive Science Society* (pp. 277–282).

Kuhn, D. (1989). Children and adults as intuitive scientists. *Psychological Review, 96,* 674–689.

Kuhn, D., & Angelev, J. (1976). An experimental study of the development of formal operational thought. *Child Development, 47,* 697–706.

Kuhn, D., Garcia-Mila, M., Zohar, A., & Andersen, C. (1995). Strategies of knowledge acquisition. *Monographs of the Society for Research in Child Development, 60*(4, Serial No. 245).

Mahoney, M. J., & DeMonbreun, B. G. (1978). Psychology of the scientist: An analysis of problem-solving bias. *Cognitive Therapy and Research, 1,* 229–238.

McCloskey, M. (1983). Naive theories of motion. In D. Gentner & A. L. Stevens (Eds.), *Mental models* (pp. 299–324). Hillsdale, NJ: Lawrence Erlbaum Associates.

Mynatt, C. R., Doherty, M. E., & Tweney, R. D. (1977). Confirmation bias in a simulated research environment: An experimental study of scientific inference. *Quarterly Journal of Experimental Psychology, 29,* 85–95.

Mynatt, C. R., Doherty, M. E., & Tweney, R. D. (1978). Consequences of confirmation and disconfirmation in a simulated research environment. *Quarterly Journal of Experimental Psychology, 30*, 395–406.

Newell, A., & Simon, H. A. (1972). *Human problem solving*. Englewood Cliffs, NJ: Prentice Hall.

NICHD Early Child Care Research Network. (2002). Child-care structure → process → outcome: Direct and indirect effects of child-care quality on young children's development. *Psychological Science, 13*, 199–206.

Okada, T., & Simon, H. A. (1997). Collaborative discovery in a scientific domain. *Cognitive Science, 21*, 109–146.

Resnick, L. B., Levine, J. M., & Teasley, S. D. (1991). *Perspectives on socially shared cognition*. Washington, DC: American Psychological Association.

Ruffman, T., Perner, J., Olson, D. R., & Doherty, M. (1993). Reflecting on scientific thinking: Children's understanding of the hypothesis–evidence relation. *Child Development, 64*, 1617–1636.

Samuels, M. C., & McDonald, J. (2002). Elementary school-age children's capacity to choose positive diagnostic and negative diagnostic tests. *Child Development, 73*, 857–866.

Schauble, L. (1990). Belief revision in children: The role of prior knowledge and strategies for generating evidence. *Journal of Experimental Child Psychology, 49*, 31–57.

Schauble, L. (1996). The development of scientific reasoning in knowledge-rich contexts. *Developmental Psychology, 32*, 102–119.

Schauble, L., Glaser, R., Raghavan, K., & Reiner, M. (1991). Causal models and experimentation strategies in scientific reasoning. *Journal of the Learning Sciences, 1*, 201–238.

Schunn, C. D., & Klahr, D. (1992). Complexity management in a discovery task. In the *Proceedings of the 14th Annual Conference of the Cognitive Science Society* (pp. 177–182). Hillsdale, NJ: Lawrence Erlbaum Associates.

Schunn, C. D., & Klahr, D. (1995). A 4-space model of scientific discovery. In J. D. Moore & J. F. Lehman (Eds.), *Proceedings of the 17th Annual Conference of the Cognitive Science Society* (pp. 106–111). Mahwah, NJ: Lawrence Erlbaum Associates

Schunn, C. D., & Klahr, D. (1996). Integrated yet different: Logical, empirical, and implementational arguments for a 4-space model of inductive problem solving. In G. W. Cottrell (Ed.), *Proceedings of the 18th Annual Meetings of the Cognitive Science Society* (pp. 25–26). Mahwah, NJ: Lawrence Erlbaum Associates.

Shaklee, H., & Paszek, D. (1985). Covariation judgment: Systematic rule use in middle childhood. *Child Development, 56*, 1229–1240.

Shrager, J. (1985). *Instructionless learning: Discovery of the mental model of a complex device*. Unpublished doctoral dissertation, Carnegie Mellon University.

Shrager, J. (1987). Theory change via view application in instructionless learning. *Machine Learning, 2*, 247–276.

Shrager, J., & Klahr, D. (1986). Instructionless learning about a complex device: The paradigm and observations. *International Journal of Man–Machine Studies, 25*, 153–189.

Siegler, R. S., & Liebert, R. M. (1975). Acquisition of formal scientific reasoning by 10- and 13-year-olds: Designing a factorial experiment. *Developmental Psychology, 10*, 401–402.

Simon, H. A., Langley, P., & Bradshaw, G. L. (1981). Scientific discovery as problem solving. *Synthese, 47*, 1–27.

Skinner, B. F. (1956). A case history in scientific method. *American Psychologist, 11,* 221–233.

Sodian, B., Zaitchik, D., & Carey, S. (1991). Young children's differentiation of hypothetical beliefs from evidence. *Child Development, 62,* 753–766.

Teasley, S. D. (1995). The role of talk in children's peer collaborations. *Developmental Psychology, 31,* 207–220.

Tschirgi, J. E. (1980). Sensible reasoning: A hypothesis about hypotheses. *Child Development, 51,* 1–10.

Vosniadou, S., & Brewer, W. F. (1992). Mental models of the earth: A study of conceptual change in childhood. *Cognitive Psychology, 24,* 535–585.

Wason, P. C. (1960). On the failure to eliminate hypotheses in a conceptual task. *Quarterly Journal of Experimental Psychology, 12,* 129–140.

Wiser, M. (1989, April). *Does learning science involve theory change?* Paper presented at the biannual meeting of the Society for Research in Child Development, Kansas City, MO.

Zimmerman, C. (2000). The development of scientific reasoning. *Developmental Review, 20,* 99–149.

5

Puzzles and Peculiarities: How Scientists Attend to and Process Anomalies During Data Analysis

Susan Bell Trickett
George Mason University

Christian D. Schunn
University of Pittsburgh

J. Gregory Trafton
Naval Research Laboratory, Code 5513

Carl Linnaeus published the first edition of his classification of living things, the *Systema Naturae*, in 1735. Shortly thereafter, while having lunch with a colleague at the University of Leiden, he was in the middle of explaining the nature of his classification system when the colleague stopped him in mid-explanation. A beetle had crawled onto the table, and the colleague wanted to know where this particular type of beetle fit into the classification system. Linnaeus examined the bug carefully and frowned. Then he squished the bug with a thumb, flicked it from the table, and asked "What beetle?"

This amusing anecdote, although very unlikely to have actually happened, illustrates one view of scientists: that they are very fond of their own theories and frameworks and not particularly tolerant of data that contradict the theory. By contrast, philosophers of science have argued that anomalies are crucial to the advancement of science, both in terms of refining existing theories and in terms of developing whole new theories and frameworks (Kuhn, 1970; Lakatos, 1976). Moreover, when asked, scientists themselves believe anomalies are important (Knorr, 1980).

However, the general question for the psychology of science on this topic is how scientists *actually* deal with anomalies, rather than how science *ought* to proceed or how scientists perceive themselves as operating. Psychology has had two general approaches to investigating how scientists deal with anomalies. Researchers using the first approach asked how people respond to negative evidence in a concept identification task. In particular, they asked what sort of evidence they seek for their hypotheses and what they do when the evidence contradicts a working hypothesis (Mynatt, Doherty, & Tweney, 1977; Wason, 1960). Results generally showed (or were interpreted as showing) that scientists are susceptible to confirmation bias, in that they appear to seek confirming evidence and ignore negative "anomalous" evidence (Mahoney & DeMonbreun, 1977). Additional evidence for this apparent confirmation bias came from sociological studies based on interviews of practicing scientists (Mitroff, 1974).

Researchers using the second approach have focused on practicing scientists performing analyses of authentic scientific data (as opposed to abstract hypothesis-testing tasks), using either historical case studies or *in vivo* observations of practicing scientists as they work. The historical case studies have found mixed results (Chinn & Brewer, 1992; Gorman, 1995; Kulkarni & Simon, 1988; Nersessian, 1999; Tweney, 1989; Tweney & Yachanin, 1985). Kulkarni and Simon (1988, 1990) have found that some famous scientists used an "attend to surprising results" heuristic as the impetus for important discoveries, but Chinn and Brewer (1992) suggested there is a range of responses that scientists use, from focusing on the anomaly and changing one's theory all the way to ignoring the anomaly entirely. Tweney (1989) argued that Faraday profitably used a "confirm early, disconfirm late" strategy that might also relate to how scientists respond to anomalous data when they occur. Kevin Dunbar (1995, 1997) pioneered the *in vivo* approach to analyzing the cognitive activities of practicing scientists as they work. To date, his emphasis has been on molecular biology laboratories and the activities that take place in regular laboratory group meetings. In an analysis of how scientists respond to anomalous data, Dunbar (1995, 1997) found that scientists do indeed discard hypotheses that are inconsistent with evidence and that they devote considerable attention to anomalies when they occur.

It is important to note that Dunbar's (1995, 1997) studies deal with reasoning in larger groups working in semiformal settings.[1] The question still remains how individual scientists or pairs of scientists in less formal settings

---

[1] Dunbar also examined laboratory notebooks and grant proposals and interviewed scientists in more individual settings to make sure that focusing on laboratory groups missed no important activities.

(e.g., as they analyze data in their offices or laboratories) might respond to anomalous data. Indeed, Dama and Dunbar (1996) showed that important reasoning occurs across several individuals in these laboratory group settings, such that inductions, for example, are not generated entirely by just a single individual but rather collaboratively across individuals.

Alberdi, Sleeman, and Korpi (2000) combined the *in vitro* and *in vivo* approaches and conducted a laboratory study of botanists performing a categorization task. In particular, they presented the botanists with "rogue" items (anomalies to an apparent categorization scheme) and recorded the scientists' responses. They found that scientists did indeed attend to anomalies. Alberdi et al. also identified several strategies by which scientists tried to resolve anomalous data. *Instantiate* was a key strategy. In using the instantiate strategy, participants activated a schema they had not hitherto considered, which directed them to investigate new features in the data. Thus, they were essentially exploring their theoretical domain knowledge in an attempt to fit the rogue and non-rogue items into a category. They generated new hypotheses about the rogue items that would assimilate the anomalies into their current theoretical understanding of the domain.

The goal of this chapter is to build on this past work by developing a methodology by which expert scientists can be studied as they conduct their own research in more individual settings. We then investigate scientists' responses to anomalies, both immediately and over time, and determine whether their response to anomalies is different from their response to expected phenomena. This will enable us to ask (and answer) three specific questions: (a) Do scientists commonly notice anomalies during data analysis? (b) Do scientists focus on anomalies when they are noticed (as opposed to attending to them but not substantially diverting their activities in response to them)? and (c) When scientists examine anomalies, by what processes do they do so?

METHOD

To investigate the issues we have just discussed, we have adapted Dunbar's (1997; Trickett, Fu, Schunn, & Trafton, 2000) *in vivo* methodology. This approach offers several advantages. It allows observation of experts, who can use their domain knowledge to guide their strategy selection. It also allows the collection of "on-line" measures of thinking, so that the scientists' thought processes can be examined as they occur. Finally, the tasks the scientists perform are fully authentic.

Participants

Our participants were 1 individual and one pair of scientists, working on data analysis in their own offices. These scientists were videotaped while conducting their own research. All the scientists were experts, having earned their PhDs more than 6 years previously. In the first data set, two astronomers—one a tenured professor at a university, the other a fellow at a research institute—worked collaboratively to investigate computer-generated visual representations of a new set of observational data. At the time of this study, one astronomer had approximately 20 publications in this general area, and the other had approximately 10. The astronomers had been collaborating for some years, although they did not frequently work at the same computer screen and at the same time to examine data.

In the second data set, a physicist with expertise in computational fluid dynamics worked alone to inspect the results of a computational model he had built and run. He was working as a research scientist at a major U.S. scientific research facility and had earned his PhD 23 years beforehand. He had inspected the data earlier but had made some adjustments to the physics parameters underlying the model and was therefore revisiting the data.

All participants were instructed to carry out their work as though no camera were present and without explanation to the experimenter. The individual scientist was trained to give a talk-aloud verbal protocol (Ericsson & Simon, 1993). We recorded the two astronomers' conversation as they engaged in scientific discussion about their data. It is important to emphasize that all participants were performing their usual tasks in the manner in which they typically did so. At the beginning of the session, the participants gave the experimenter an explanatory overview of the data and the questions to be resolved, and after the session the experimenter interviewed the participants to gain clarification about any uncertainties. During the analysis session, however, the experimenter did not interrupt the participants, and the interactions between participant and experimenter were not included in our own analyses. The relevant part of the astronomy session lasted approximately 53 minutes, and the physics session lasted approximately 15 minutes.

All utterances were later transcribed and segmented. A segment consisted of either a short sentence or a clause, if the sentence was complex. All segments were coded by two coders as on task (pertaining to data analysis) or off task (e.g., jokes, phone interruptions, etc.). Interrater reliability for this coding was more than 95%. Off-task segments were excluded from further analysis. On-task segments ($N = 649$ for astronomy and $N = 173$ for physics) were then grouped into episodes on the basis of visualizations ($N = 11$ for astronomy and $N = 8$ for physics). A new episode began when the scientists opened a new data visualization. This grouping of the protocol into episodes allowed us to focus on the more immediate reaction to anomalies.

The Tasks and the Data

Astronomy. The data under analysis were optical and radio data of a ring galaxy. The astronomers' high-level goal was to understand the galaxy's evolution and structure by understanding the flow of gas in the galaxy. To understand the gas flow, the astronomers had to make inferences about the velocity field, represented by contour lines on the two-dimensional display.

The astronomers' task was made difficult by two characteristics of their data. First, the data were one- or, at best, two-dimensional, whereas the structure they were attempting to understand was three-dimensional. Second, the data were noisy, with no easy way to separate noise from real phenomena. Figure 5.1 shows a screen snapshot of the type of data the astronomers were examining. To make their inferences, the astronomers used different types of images, representing different phenomena (e.g., different forms of gas), which contain different information about the structure and dynamics of the galaxy. In addition, they could choose from images created by different processing algorithms, each with advantages and disadvantages (e.g., more or less resolution). Finally, they could adjust some features of the display, such as contrast or false color.

Physics. The physicist was working to evaluate how deep into a pellet a laser light will go before being reflected. His high-level goal was to understand the fundamental physics underlying the reaction, an understanding that hinged on comprehending the relative importance and growth rates of different parts of the pellet and the relationship of that growth to the location of the laser. The physicist had built a computer model of the reaction; other scientists had independently conducted experiments in which lasers were fired at pellets and the reactions recorded. A close match between model and empirical data would indicate a good understanding of the underlying phenomenon. Although the physicist had been in conversation

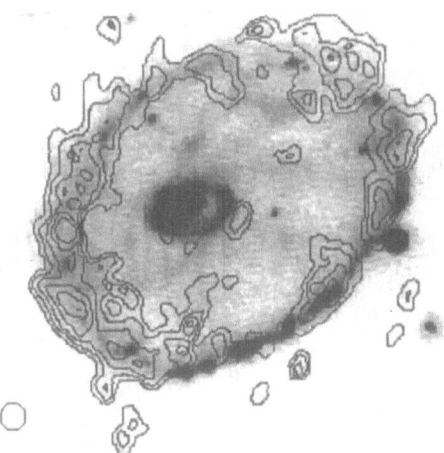

FIG. 5.1. Example of data examined by astronomers. Radio data (contour lines) are laid over optical data.

with the experimentalist, he had not viewed the empirical data, and in this session he was investigating only the results of his computational model. However, he believed the model to be correct (i.e., he had strong expectations about what he would see), and in this sense this session may be considered confirmatory.

The physicist's data consisted of two different kinds of representation of the different modes, shown over time (in nanoseconds). The physicist was able to view either a Fourier decomposition of the modes or a representation of the "raw" data. Figure 5.2 shows an example of the physicist's data. He could choose from black-and-white or a variety of color representations and could adjust the scales of the displayed image as well as some other features. He was able to open numerous views simultaneously. A large part of his task was comparing images, both different types of representation of the same data and different time slices represented in the same way.

Coding Scheme

Our goal in this research was to investigate scientists' responses to anomalous data. First, we wanted to establish whether and to what extent the scientists noticed and attended to anomalies. Second, we wanted to investigate the processes by which they responded to anomalous data. Both

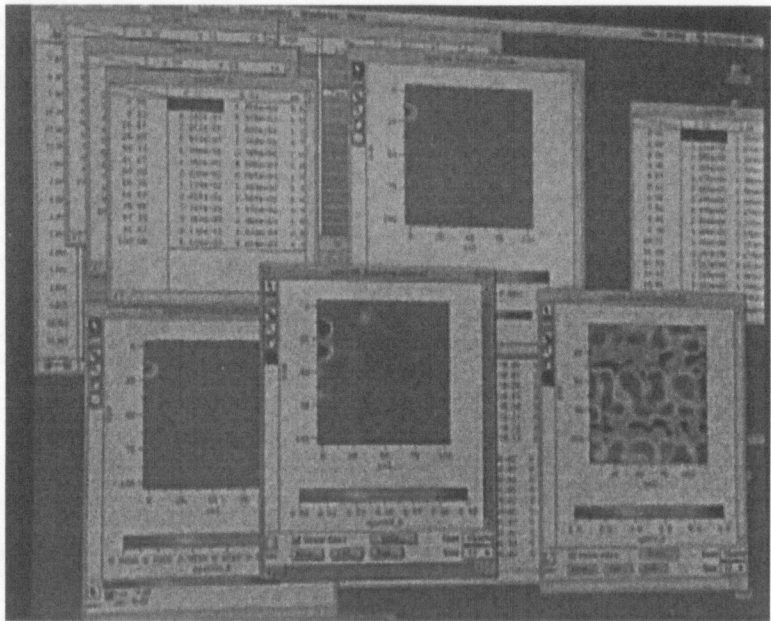

FIG. 5.2.   Example of data examined by physicist: Fourier modes (left) and raw data (right).

protocols were coded independently by two different coders. Initial inter-rater reliability for each code was greater than 85%. Disagreements were resolved by discussion. Any coding disagreements that could not be resolved were excluded from further analysis.

Noticings.   In order to establish which phenomena—anomalous or not—the scientists attended to, we first coded for the scientists' *noticing* phenomena in the data. A noticing could involve merely some surface feature of the display, such as a line, shape, or color, or it could involve some interpretation, for example, identifying an area of star formation or the implied presence of a mode. Only the first reference to a phenomenon was coded as a noticing; coding of subsequent references to the same phenomenon is discussed later.

Because our investigation focused on the extent to which the scientists attended to anomalies in the data, we further coded these noticings as either *anomalous* or *expected,* according to one or more of the following criteria: (a) in some cases, the scientist made explicit verbal reference to the fact that something was anomalous or expected; (b) if there was no explicit reference, domain knowledge was used to determine whether a noticing was anomalous;[2] (c) a phenomenon might be associated with (i.e., identified as like) another phenomenon that had already been established as anomalous or not; (d) a phenomenon might be contrasted with (i.e., identified as unlike) a phenomenon that had already been established as anomalous or not; (e) a scientist might question a feature, thus implying that it is unexpected. Table 5.1 illustrates these codes.

Subsequent References.   One of our questions was the extent to which the scientists attended to anomalies. The coding of noticings captured only the first reference to a phenomenon of interest; we needed to establish how frequently they made subsequent reference to each noticing. Consequently, all subsequent references were also identified and coded.[3] Not all subsequent references immediately followed a noticing; frequently, the scientists returned to a phenomenon after investigating other features of the data. Subsequent references were identified both within the episode in which the noticing had occurred and across later episodes.

The rest of the coding scheme addresses *how* the scientists responded to the anomalies, in particular, immediately after they noticed the anomalies. To investigate the scientists' immediate response to their anomalous find-

---

[2]The coders' domain knowledge came from textbooks and interviews with the scientists.

[3]In the astronomy data set, frequently the first interaction between them after a noticing was to make sure the scientists were both looking at the same thing. Subsequent references that served purely to establish identity were *not* included in the analyses.

TABLE 5.1
Noticings (in Italics): Anomalous or Expected

Criterion	Code	Example
Explicit	Anomalous	What's *that funky thing* ... That's odd.
		Look at that! Ooh! That's interesting. That's very interesting. Okay, what's interesting about this is that *this mode here* ...
Domain Knowledge	Expected	You can see that *all the H1* is concentrated in the ring ...
		OK, that was the *initial 2D spike*, all right, the 2D perturbation that we put on the surface.
Association	Anomalous	You see *similar kinds of intrusions* along here.
Contrast	Expected	That's odd ... As opposed to *these things*, which are just the lower contours down here.
Question	Anomalous	I still wonder why *we don't see any H1* up here in this sort of northern ring segment?
		So, I should be seeing *another peak* somewhere in here. And I don't see that there.

ings, we coded 10 utterances following each noticing, whether anomalous or expected (excluding utterances in the astronomy data set that merely established which phenomenon was under discussion). We coded the presence or absence of each type of utterance identified in the next section. We anticipated that scientists would attempt to produce hypotheses for the anomalies and that some of these hypotheses would be discussed further. On the basis of the results reported by Alberdi et al. (2000), we investigated the extent to which elaboration of hypotheses was grounded in theory or in the visual display of the data. We also anticipated the use of additional strategies and inspected the data to identify strategies that emerged, as discussed later.

Identify Features.   Perhaps because the data the scientists were analyzing were visual, we found that when the scientists noticed a phenomenon, they often elaborated on specific features of the phenomenon that either had attracted their attention or were characteristic of the phenomenon.

Hypotheses.   Statements that attempted to provide a possible explanation for the data were coded as *hypotheses*. All hypotheses were further coded as *elaborated* or *unelaborated*. Elaboration consisted of one or more statements that either supported or opposed the hypothesis. Hypotheses that were not discussed further after they were proposed were coded as unelaborated.

When a hypothesis was elaborated, we coded whether the elaboration was *theoretical* or *visual*. When evidence for or against a hypothesis was grounded in theoretical domain knowledge, elaboration was coded as theoretical; when evidence came from the display, it was coded as visual.

Place in Context. A strategy that emerged from our examination of the data was considering the noticed phenomenon in relation to other data. Thus, we coded whether the scientist placed the noticing in context and whether that context was another part of the data set (*local*) or the scientist's own theoretical knowledge (*global*).

RESULTS

Noticing Anomalies

Our first question was whether the scientists commonly noticed anomalies. Recall that a "noticing" is a first-time reference to a phenomenon. Table 5.2 presents the total number of noticings for each data set and the percentages of anomalous and expected phenomena ("not coded" refers to noticings on which the coders did not agree). As Table 5.2 shows, at least one third of the phenomena on which the astronomers commented, and almost one half the phenomena physicist noticed, were coded as anomalous. Thus, out of all the phenomena that caught the scientists' attention, a notable proportion was unusual in some way.

Another measure of how commonly the scientists noticed anomalies in the data is the rate at which they commented on an aspect of the data that was anomalous. In the astronomy data set, the 9 anomalies were identified over the course of 53 minutes, or 1 anomaly for approximately every 6 minutes of data analysis. In the physics data set, the 4 anomalies were identified over the course of 15 minutes, or 1 anomaly for approximately every 3.75

TABLE 5.2
Frequency of Anomalous and Expected Noticings

Domain	Total Noticings	Anomalous	Expected	Not Coded
Astronomy	27	33%	48%	19%
Physics	9	44%	44%	12%

minutes of analysis. It appears, then, that noticing anomalies in the data was a relatively common occurrence for these scientists.

Attending to Anomalies

Once the scientists had identified something unusual in the data, what did they do with this observation? There are several possible reactions, including immediately attempting to account for the anomaly; temporarily disregarding it and returning to it later; or moving on to investigate another, perhaps better understood aspect of the data. One way to investigate this issue is to determine whether the scientists' response to anomalies was different from their response to expected phenomena.

We investigated this by counting how often the scientists made subsequent references to a noticing immediately on identifying it. If anomalies and expected phenomena are of equal interest, we would expect the scientists to make a similar number of references to both the anomalous and the expected patterns. However, if anomalies play a more important role in scientists' efforts to understand the data, we would expect them to pay more attention (measured by the number of subsequent references) to anomalies than to expected observations.

As Table 5.3 shows, for both the astronomy and physics data sets, scientists paid more attention to anomalies than to expected phenomena, $t(28) = 3.22, p < .01$. In the case of astronomy, the anomalies received more than four times as many subsequent references within the same episode as the expected phenomena. The physics data set follows a similar pattern, with more than three times as many references to anomalies as to expected phenomena. The results indicate that the scientists were more interested in phenomena that did not match their expectations and are thus in stark contrast to the findings of the confirmation bias literature.

Recall that the studies of confirmation bias, even those that involved scientists, involved abstract tasks that required no theoretical or domain knowledge. These scientists we observed, however, were working in their

TABLE 5.3
Mean Numbers of Subsequent References Per Noticed Object for Anomalies and Expected Phenomena *Within* the Same Visualization Episode

Domain	Anomalies	Expected
Astronomy	7.4	1.7
Physics	5.0	1.2

own domain and, more important, on their own data. Thus, they were likely to have strong expectations grounded in deeply held beliefs that had been built up over years of practice. Insofar as anomalies had the potential to challenge those beliefs, the scientists were likely to have a real-life invest-ment in understanding and eventually resolving anomalous data.

Another measure of the attention the scientists paid to anomalies is the number of subsequent references they made to phenomena across episodes. Recall that a new episode was coded when the scientists switched to a new visualization. A new visualization presented the scientists with the oppor-tunity to direct their attention to new phenomena that may have been newly visible (or more salient) in the new visualization. Thus, refocusing their attention on a phenomenon they had already investigated indicates a relatively high measure of interest on the scientists' part. In both data sets, the scientists continued to pay more attention to the anomalies than to the expected phenomena, $t(27) = 2.08, p < .05$. As Table 5.4 shows, both sets of scientists increased their attention to the anomalies over visualizations, more so than to the expected phenomena. This pattern was especially strong in the astronomy data set.

Examples of Attending to Anomalies

In both data sets, some anomalies in particular seemed to take hold of and keep the scientists' attention. In these cases, the scientists revisited the anomalies multiple times and over many different intervening visualiza-tions and noticings. For example, one of the astronomers noticed an area in the galaxy with "a substantial amount of gas" but where no star formation was apparently occurring. The other astronomer proposed two hypotheses that might account for the lack of star formation and then adjusted the con-trast of the image in order to allow a better view of the phenomenon. This view allowed the astronomers to confirm the lack of star formation in the relevant area. They then created a new visualization, specifically to allow a closer inspection of the phenomenon. Unsatisfied with what they learned from this visualization, they decided to create yet another visualization

TABLE 5.4

Mean Numbers of Subsequent References Per Noticed Object for Anomalies and Expected Phenomena Across Visualization Episodes

Domain	Anomalies	Expected Phenomena
Astronomy	8.3	0.6
Physics	7.7	5.2

("Let's return to good old black and white") and compared the anomalous area with surrounding phenomena in the galaxy. This comparison led to one astronomer's proposing a new hypothesis, which was rejected by the other. The astronomers then seemed to concede that this anomaly, although potentially important, could not be accounted for at this time:

> *Astronomer 2:*   I mean, it's, it's a strong feature. Can't ignore it.
> *Astronomer 1:*   Can't ignore it. Yeah, well … Gloss over it.

The astronomers then directed their attention to other phenomena in the galaxy. However, despite both their apparent agreement to table further discussion of this anomaly and their intervening focus on several other phenomena, the astronomers continued to return to this anomaly several more times ("Getting back to the blob …" ; "Note that this, that's the only sort of major blob of H1 that we don't see associated with, you know, really dense pile of H1 we don't see associated with star formation").[4] Furthermore, they proposed four additional hypotheses about this particular anomaly, considered whether it would be cost effective to collect more data to investigate it further, and proposed it as the focus of some computational modeling in which one astronomer was to engage. The anomaly, originally noticed in the fifth visualization episode, continued to hold their attention into the ninth visualization episode—approximately 20 minutes later. A similar situation occurred in the physics data set. The physicist identified an anomalous growth pattern among several modes: "Okay, what's interesting about this is that this mode here, while being very large on this plot, at roughly a nanosecond later, is not very big and that while that original seed dominates, we have to go two modes up to actually see a result." He determined that this would be of interest to the experimentalist (recall that the physicist's data derived from computational models, which he hoped would predict the experimental data collected by another scientist). He then opened a new visualization and noted that the same pattern occurred in this new visualization and proposed an explanation. He concluded, "That's basically what the lesson is here," suggesting that the explanation satisfactorily accounted for the phenomenon. He viewed one more visualization in order to examine a different phenomenon before apparently concluding that the session was complete ("All right, at this point I got to, ah, show these things to our experimentalist and get his experimental data and see how well we agree.") At this point, the scientist began to discuss procedural issues with the experimenter, explaining, for example, what his next steps would be. However, despite having acknowledged that the data exploration session was over ("Yeah, I think I'm pretty much done"), the physicist abruptly in-

---

[4]*H1* refers to hydrogen gas.

terrupted himself to consider in more detail the anomalous growth pattern he had identified two visualizations earlier. Although he appeared earlier to have been satisfied with his explanation for the anomaly, he expressed renewed interest and continued perplexity: "Was outrun by the next one down. And I don't know, just don't know. I'll have to get someone else's interpretation of that. I don't understand that. The high modes are supposed to take off, they're supposed to run faster." He then proposed and explored a new hypothesis for this discrepant growth rate, but was, in the end, unable to conclusively account for it: " It'll be nice when I, um, show this to the actual experimentalist because they have a better feel for this stuff than I do." Thus, the anomaly continued to hold the physicist's attention even after he had proposed an apparently satisfactory explanation for it. He continued to devote attention to this anomaly across two visualization episodes beyond his original noticing of it and after attending to other phenomena in the data, as well as discussing his future intentions for this data set.

Immediate Response to Anomalies

In addition to this prolonged attention to several anomalies, we have shown that in general, whenever the scientists noticed an anomaly, they paid immediate attention to it. However, we have not analyzed the content of that immediate attention to anomalies. To understand how the scientists dealt with the anomalies on noticing them, we now turn to the results of the second part of our coding scheme, which was applied to the 10 utterances that immediately followed the initial noticing of anomalies and expected phenomena.

Identify Features. In approximately half the instances when they noticed a phenomenon, the scientists elaborated on that phenomenon by identifying one or more of its features. However, as Fig. 5.3 shows, the scientists were only slightly (and nonsignificantly) more likely to identify specific features of the anomalies as the expected noticings, and this pattern held for both domains.

Although identifying the features of a phenomenon occurred equally for both types of noticing, it seems likely that this strategy served a different purpose for the expected noticings than for the anomalies. For the expected noticings, the scientists seemed to elaborate on the details almost as a kind of mental checklist of the expected characteristics, as though affirming the match between what would be expected and the phenomenon itself. For example, the physicist began his session by plotting a visualization of something he acknowledged that he understood well (a "baseline") and confirmed, "OK, that was the initial 2D spike." He then identified more specific features of this spike, by noting its origin location ("All right, the 2D

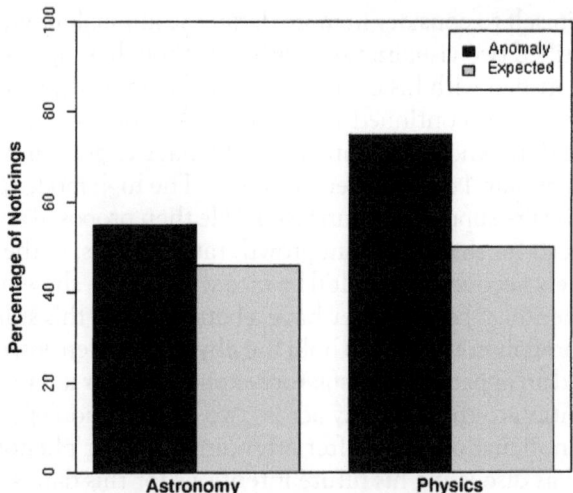

FIG. 5.3.   Percentages of noticings for which scientists identified features.

perturbation that we put on the surface") and finally its nature ("That's basically the sixty micron, uh, the sixty micron perturbation"). This additional information provided a more complete, confirmatory description of the phenomenon itself.

In contrast, when the scientists identified features of anomalous noticings, they seemed to focus on aspects of the phenomenon that were unusual. Early in the session, for example, one of the astronomers noticed an anomalous pattern in the velocity contours ("Look at the little sort of, er, sort of intrusion of the velocity field here"). He then identified specific features of the "intrusion" by describing it further: "The velocity, the velocity contour goes 'woop.' It sort of dips under, sort of does a very noncircular motion thing. I mean, notice that the, the contours don't sort of cross the ring, they sort of go...." These utterances all elaborate on specific features of the phenomenon that specify how it *deviates* from what the scientist expected.

Propose Hypothesis.   When the scientists encountered an anomaly, in most cases they proposed a hypothesis or explanation that would account for it, whereas they rarely did so for the expected phenomena (see Fig. 5.4). This difference was significant, $\chi^2(1) = 7.5$, $p < .05$, and the pattern was very strong in both domains.

As Fig. 5.4 shows, most anomalies were followed within the next 10 utterances (i.e., almost immediately) by some effort to explain what might have caused them, whether based in the underlying theories of the relevant science or some characteristic of the data collection process or the visualization itself. For example, for the anomalous "intrusion of the velocity field"

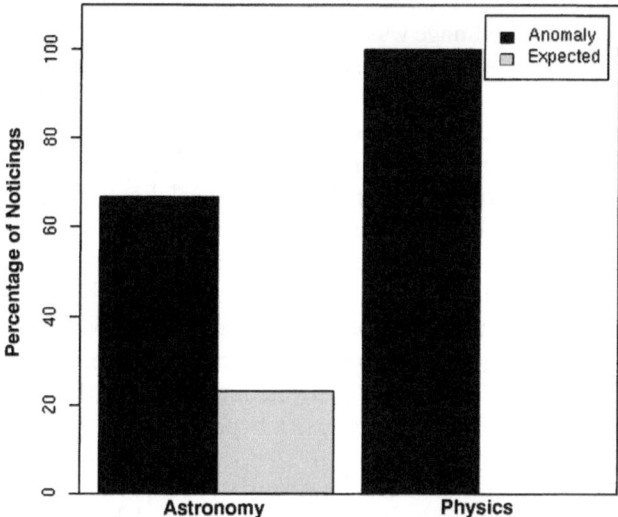

FIG. 5.4.   Percentages of noticings for which hypotheses were proposed.

mentioned earlier, one of the astronomers asked if this pattern were "significant." His colleague immediately proposed a theoretically based cause for the "dip" in the velocity contour: "Oh yeah, that's sort of like, um, what can it mean? A streaming motion." Somewhat later, one of the astronomers noticed an anomalous lack of hydrogen gas, where he expected to find it: "I still wonder why we don't see any H1 up here in this sort of northern ring segment." He then suggested, "Ahh. I'm beginning to wonder if we didn't have enough velocity range all of a sudden." In other words, he proposed that the observation itself was not sensitive enough to detect the hydrogen gas, a hypothesis grounded in the data collection process that would explain why the gas did not appear on the image.

The physicist was also likely to propose hypotheses about anomalous phenomena. After he noticed the discrepant growth pattern among the modes discussed earlier, he asked "So what happened? What happened?" He then suggested "That's the nonlinear feature of this, because I added this mode with that mode and got him. The higher coupling coefficient, and as a result he peaks up sooner. And once this guy starts off, he goes very quickly." He thus proposed a hypothesis that pertained to the physics underlying the reaction. In another instance, he failed to detect an expected peak: "So I should be seeing another peak somewhere in here. And I don't see that here." As with the preceding astronomy example, he proposed an explanation for this that was grounded in a characteristic of the display, that is, that the scale of the image was simply too small for the expected peak to be visible: "So let's re-scale. Let's see if I can see it. Because this amplitude may not be very big over here."

The scientists' treatment of the expected phenomena was quite different. For example, one astronomer noted, "Star formation is concentrated in the ring as well" (an expected phenomenon). He elaborated slightly on this observation by indicating the grounds for his comment ("Which is what this color map shows, you've got all the star formation is really concentrated inside the ring") but made no further comment about this phenomenon. The physicist showed the same tendency not to account for expected phenomena: "Now, clearly we got a fundamental growth right here. Which is the ... uh, second harmonic." Again, he identified a feature of the growth ("the second harmonic") but did not propose a hypothesis to explain it.

Hypothesis Elaboration.   Once the scientists had proposed a hypothesis (primarily about the anomalies), in most cases they elaborated on that hypothesis. Elaboration might consist of more detailed specification of the hypothesis itself or of evidence for or against the hypothesis. Figure 5.5 presents the proportion of hypotheses that were elaborated within each domain for expected and anomalous noticings. In most cases, scientists attempted to elaborate the hypotheses, for both expected and anomalous noticings (note that there were no hypotheses to elaborate in the expected physics case). As Fig. 5.5 shows, the scientists rarely proposed a hypothesis without considering it further; however, as our next analysis shows, the source of the elaboration was quite different for anomalies than for expected phenomena.

Source of Elaboration.   For the physics data set there were not enough elaborated hypotheses to analyze further. However, the astronomy data set showed a strong contrast in the type of elaboration the scientists used to further consider their hypotheses, and this difference is illustrated in Fig. 5.6. For the anomalies, evidence pertaining to four of the five hypotheses came from the visual display, whereas the two hypotheses about expected noticings were developed theoretically.

To illustrate the difference in the type of elaboration, consider the following exchange as the astronomers attempted to understand an anomalous pattern of hydrogen gas in the galaxy:

Astronomer 2:	I mean, OK, let's go with er, cosmic conspiracy, think there's another galaxy in the background?
Astronomer 1:	No.
Astronomer 2:	OK. The velocity contours, would they be any different there?

Astronomer 1:   No. Remember the, um, let's see, that corresponds to right about—here.

Astronomer 2:   Well, the velocity contours are doing something there.

Astronomer 2 proposes the hypothesis that another (unobserved) galaxy could account for the pattern of gas and considers whether, if this were the case, the representation of velocity contours in the display would appear

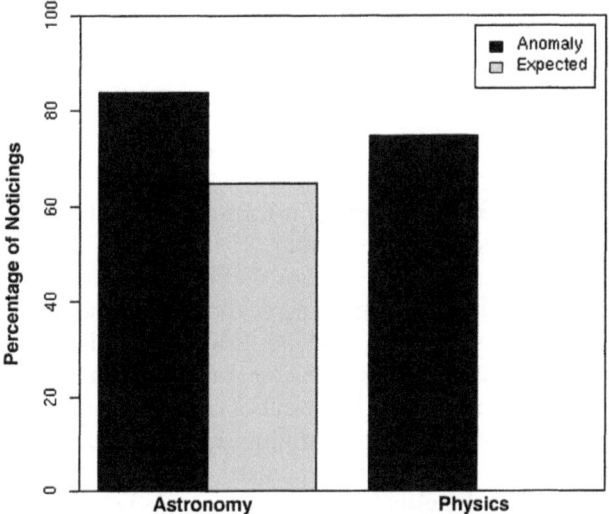

FIG. 5.5.   Percentage of hypotheses that were elaborated.

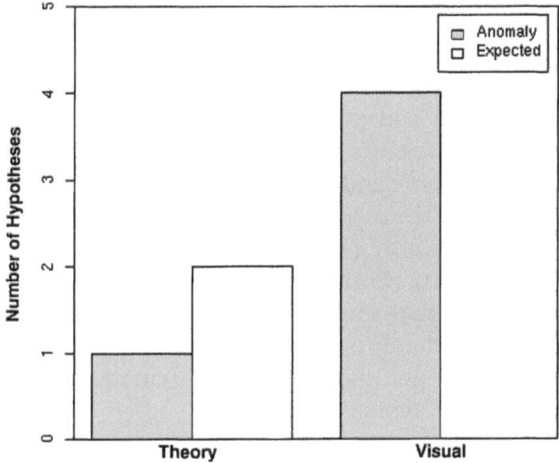

FIG. 5.6.   Elaboration type for hypotheses (astronomy data set only).

different in a specific location (*there*). He looks strictly to the visual pattern to investigate the implications of the hidden galaxy hypothesis. Astronomer 1 rejects the hypothesis (*No*) and similarly focuses on the visual display (*that corresponds to right about—here*). Astronomer 2 continues to focus on the unusual pattern of movement shown by the displayed contours (*the velocity contours are doing something there*). The focus of the whole exchange is the visual display itself rather than any theoretical issues that underlie the representation.

In contrast, the following exchange occurs as the astronomers discuss an area of the galaxy about which they have agreed: "Yeah, exactly, I mean that's not too surprising,"—in other words, the phenomenon is expected.

Astronomer 1:   That might just be gas blowing out from the star-forming regions.
Astronomer 2:   But that's not a star-forming region, though … that one.
Astronomer 1:   No, absolutely not.

In this case, first the astronomers agree that the phenomenon represents nothing unusual. Astronomer 1 proposes a hypothesis to account for it (*That might just be gas blowing out from the star-forming regions*), an explanation that Astronomer 2 rejects. He rejects it on the theoretical grounds that the area in question is not a star-forming region, to which Astronomer 1 readily agrees. The astronomers are focused on theoretical domain knowledge rather than on the visual patterns represented on the display.

Place in Context.   In addition to (or instead of) developing hypotheses about the noticings, the scientists also might consider the noticing in relation to other information, either theoretical information in memory (global context) or information about the current data set (local context), or they might not place it in either context. In fact, none of the noticings was considered directly in the context of the scientists' theoretical knowledge (global). However, the scientists considered the noticings in the context of the current data set (local), and this sequence occurred more frequently for the anomalies than for the expected phenomena, especially in the astronomy data set (see Fig. 5.7), $\chi^2(1) = 9.21, p < .01$.

How did the scientists consider phenomena in the context of information about the current data set? Again, referring to the anomalous "intrusion of the velocity field" noticed by one of the astronomers, after identifying the features and proposing the "streaming motion" hypothesis discussed earlier, the discussion continued:

Astronomer 1:   You sort of see that all through this region. You see velocity contours kind of, sort of cutting across the outer ring.

Astronomer 2:   Yeah, it's kinda odd, it seems like the outer ring's …
Astronomer 1:   And it's only in this region.

Thus, the astronomers made an explicit comparison between the anomaly and the immediately surrounding area of the galaxy, apparently in an effort to fit the phenomenon into its surrounding context and further understand its possible significance.

Although there was less contrast in his response to the anomalous and expected noticings, the physicist also considered the noticings in the context of the surrounding data between the phenomenon and its context. For example, the physicist noticed that the data implied the existence of an unexpected mode: "That means there's somebody in between him. Oh man, oh man." He then considered this implication in the context of the visible modes: "Does that mean that guy's supposed to be …?" before rebuilding the entire visualization and then carefully counting the placement of these contextual modes: "Okay, now. Where is that? [Counting along the visualization axis] That's roughly about thirty. Thirty, right. And this one is roughly about …" In both cases, the scientists appear to be trying to understand the anomaly in terms of the larger local context, that is, the area immediately surrounding it.

GENERAL DISCUSSION AND CONCLUSIONS

We have examined the behavior of scientists at work, analyzing their own data. Our results show that these scientists not only commonly notice

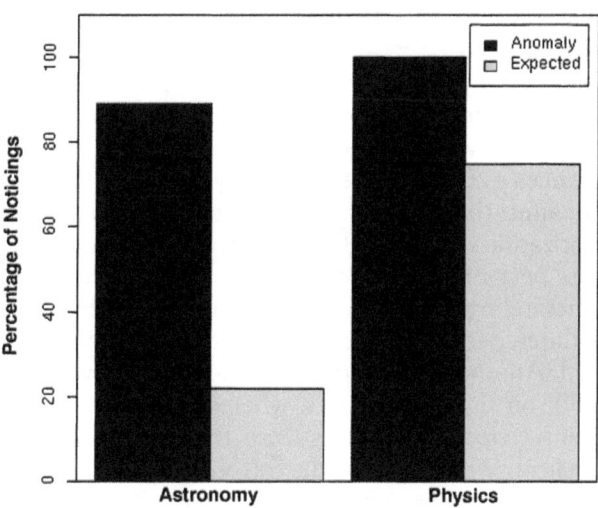

FIG. 5.7.   Percentage of noticings put in local context.

anomalies in the data but also attend to them, both immediately and over multiple visualization episodes. These results are in keeping with the findings of Dunbar (1997) and Alberdi et al. (2000), although they contrast with the results of the confirmation bias literature (e.g., Mahoney & DeMonbreun, 1977).

Beyond corroborating the finding that scientists notice and attend to anomalies, our results also show that these scientists' response followed a particular pattern. Furthermore, their investigation of anomalies was quite different from their treatment of expected phenomena. When they noticed an expected phenomenon, after identifying or describing its features, the scientists were likely to engage in no further elaboration of the phenomenon. On the rare occasions when they did attempt to account for it by proposing a hypothesis, they sought evidence in their own theoretical knowledge rather than in the visually displayed data. By contrast, however, for anomalous noticings the scientists attempted to account for the anomaly by proposing a hypothesis. They then elaborated the hypothesis, primarily by seeking evidence in the visual display, and finally considered how the anomaly related to neighboring phenomena within the same display.

Our results mesh in part with those of other researchers in that they provide further evidence for the important role played by anomalies as scientists analyze and reason about data. However, our results differ from those of Alberdi et al. (2000) in some significant ways. When the botanists in their study encountered an anomaly, they were most likely to use a strategy of theory-driven search for an explanation. The scientists in our study, however, sought support for hypotheses in the visually displayed data and attempted to place the anomaly in the local context of neighboring phenomena. Only hypotheses about expected phenomena were developed at a theoretical level.

Thus, we found an anomaly with respect to previous work in our own data, and here we attend to it, providing some theoretical discussion of its origins. There are several possible explanations for this difference. Situational differences in the tasks performed by the participants in these two studies might affect their strategy. For the botanists in Alberdi et al.'s (2000) study, categorization was the goal *per se*. Although the astronomers and the physicist were performing some categorization tasks, this was done in service of understanding the data as a whole, in order to build a mechanistic theory. The difference in their goals might account for the different strategies they used. Another possibility is that the botanists were getting immediate feedback on their hypotheses, whereas the other scientists had to generate their own feedback. In this sense, the botanists' task is similar to a supervised learning task, whereas the astronomers and physicist were in a situation where learning was unsupervised (Hertz, Krogh, & Palmer, 1991). It is plausible that the uncertainties inherent in this situation can account for the fact that these scientists sought feedback in the empirical data in the display rather than jumping immediately to their theoretical domain knowledge. One might expect that later in the research process the scientists would shift to more theoretical explorations of the anomalies.

Future work—historical, *in vivo*, or *in vitro*—should explore this theoretical explanation in further detail. We have elaborated further the account of how scientists attend to anomalous data, providing some insights into the strategies that scientists use for elaborating and resolving these anomalies. However, much work remains, and we expect that the *in vivo* method will continue to be fruitful in continuing this work.

ACKNOWLEDGMENTS

This research was supported in part by Grant 55-7850-00 to J. Gregory Trafton from the Office of Naval Research, Grant DASWO1-00-K-0017 to Christian D. Schunn from the Army Research Institute, and Grant NOOO14-01-0-0321 to Christian D. Schunn from the Office of Naval Research. The views and conclusions contained in this chapter are those of the authors and should not be interpreted as necessarily representing the official policies, either expressed or implied, of the U.S. Navy.

REFERENCES

Alberdi, E., Sleeman, D. H., & Korpi, M. (2000). Accommodating surprise in taxonomic tasks: The role of expertise. *Cognitive Science, 24,* 53–91.

Chinn, C. A., & Brewer, W. F. (1992, July–August). *Psychological responses to anomalous data.* Paper presented at the 14th annual meeting of the Cognitive Science Society, Bloomington, IN.

Dama, M., & Dunbar, K. (1996). Distributed reasoning: An analysis of where social and cognitive worlds fuse. In G. W. Cottrell (Ed.), *Proceedings of the 18th annual conference of the Cognitive Science Society* (pp. 166–170). Mahwah, NJ: Lawrence Erlbaum Associates.

Dunbar, K. (1995). How scientists really reason: Scientific reasoning in real-world laboratories. In R. J. Sternberg & J. E. Davidson (Eds.), *The nature of insight* (pp. 365–395). Cambridge, MA: MIT Press.

Dunbar, K. (1997). How scientists think: On-line creativity and conceptual change in science. In T. B. Ward & S. M. Smith (Eds.), *Creative thought: An investigation of conceptual structures and processes* (pp. 461–493). Washington, DC: American Psychological Association.

Ericsson, K. A., & Simon, H. A. (1993). *Protocol analysis: Verbal reports as data* (2nd ed.). Cambridge, MA: MIT Press.

Gorman, M. E. (1995). Hypothesis testing. In S. E. Newstead & J. S. B. T. Evans (Eds.), *Perspectives on thinking and reasoning: Essays in honour of Peter Wason* (pp. 217–240). Mahwah, NJ: Lawrence Erlbaum Associates.

Hertz, J., Krogh, A., & Palmer, R. G. (1991). *Introduction to the theory of neural computation*. Reading, MA: Addison-Wesley.

Knorr, K. D. (1980). *Manufacture of knowledge: An essay on the constructivist and contextual nature of science*. New York: Pergamon.

Kuhn, T. S. (1970). *The structure of scientific revolutions* (2nd ed.). Chicago: University of Chicago Press.

Kulkarni, D., & Simon, H. A. (1988). The process of scientific discovery: The strategy of experimentation. *Cognitive Science, 12*, 139–176.

Kulkarni, D., & Simon, H. A. (1990). Experimentation in machine discovery. In J. Shrager & P. Langley (Eds.), *Computational models of scientific discovery and theory formation* (pp. 255–276). San Mateo, CA: Morgan Kaufmann.

Lakatos, I. (1976). *Proofs and refutations*. Cambridge, England: Cambridge University Press.

Linnaeus, C. (1964/1735) *Systema Natura* (M. Sara, J. Engel-Ledeboer, & H. Engel, Trans.). Nieuwkoop, The Netherlands: B. de Graaf.

Mahoney, M. J., & DeMonbreun, B. G. (1977). Psychology of the scientist: An analysis of problem-solving bias. *Cognitive Therapy and Research, 1*, 229–238.

Mitroff, I. I. (1974). *The subjective side of science*. New York: Elsevier.

Mynatt, C. R., Doherty, M. E., & Tweney, R. D. (1977). Confirmational bias in a simulated research environment: An experimental study of scientific inference. *Quarterly Journal of Experimental Psychology, 29*, 85–95.

Nersessian, N. J. (1999). Model-based reasoning in conceptual change. In L. Magnani, N. J. Nersessian, & P. Thagard (Eds.), *Model-based reasoning in scientific discovery* (pp. 5–22). New York: Kluwer Academic/Plenum.

Trickett, S. B., Fu, W.-T., Schunn, C. D., & Trafton, J. G. (2000). From dipsy-doodles to streaming motions: Changes in representation in the analysis of visual scientific data. In L. R. Gleitman & A. K. Joshi (Eds.), *Proceedings of the 22nd annual conference of the Cognitive Science Society* (pp. 959–964). Mahwah, NJ: Lawrence Erlbaum Associates.

Tweney, R. D. (1989). A framework for the cognitive psychology of science. In B. Gholson, W. R. Shadish Jr., R. A. Neimeyer, & A. C. Houts (Eds.), *Psychology of science: Contributions to metascience* (pp. 342–366). Cambridge, England: Cambridge University Press.

Tweney, R. D., & Yachanin, S. A. (1985). Can scientists rationally assess conditional inferences? *Social Studies of Science, 15*, 155–173.

Wason, P. C. (1960). On the failure to eliminate hypotheses in a conceptual task. *Quarterly Journal of Experimental Psychology, 12*, 129–140.

6

On Being and Becoming a Molecular Biologist: Notes From the Diary of an Insane Cell Mechanic

Jeff Shrager
Carnegie Institute of Washington

What is it like to be a bat?
—Thomas Nagel (1974)

Over a decade ago, Pat Langley and I predicted that an important source for cognitive models of science would come from "the developmental psychology of socialization," which studies the way that children become part of their culture (Shrager & Langley, 1990). We suggested that "insights into this process may provide hypotheses about the paths through which graduate students and junior scientists become members of their scientific community—mastering the ways of thinking, operating, and communicating that constitute the institution of science" (pp. 19–20). An interesting methodological refinement of this proposal can be found in Tanya Luhrmann's fascinating participant-observer study of modern witchcraft in England (Luhrmann, 1989). Luhrmann joined a coven, participating in the rituals, and reported on the process of *becoming* a neo-pagan witch. Others have engaged in participant–observer studies of science, and even specifically of molecular biology (e.g., Latour & Woolgar, 1979/1986), but they have not, to my knowledge, done so with the goal of observing the process of *becoming* molecular biologists. In this chapter I present preliminary results from the first study of this kind.

DIARY OF AN INSANE CELL MECHANIC

In 1996, I set out to become an environmental phytoplankton molecular biologist. I began with informal study, and in the summer of 2000 I joined a laboratory full time. My only prior formal training, aside from a typical precollege K–12 education, was in computer science (BSE, 1980, MSE, 1981) and cognitive psychology (MS, 1982, PhD, 1985). I had only a few classes in chemistry, biology, and physics in high school and college. Beginning around 1997, while commuting by train to work at a startup company building drug discovery software, I studied organic chemistry, biochemistry, and molecular biology from textbooks. In 1999, I took the Cell, Molecular Biology, and Biochemistry Graduate Record Examinations, and in the fall of 1999 I volunteered in the molecular microbiology laboratory of Dr. Arthur Grossman at the Carnegie Institution of Washington, Department of Plant Biology, located at Stanford University. I worked half-time in Grossman's laboratory for about a year, finally joining full time in the summer of 2000. At that time, I began to keep a "cognitive diary" (Shrager, 2000) in which I recorded my insights about my activities and thinking in the laboratory. Between May and December of 2000 I logged a total of 75 entries.[1]

It is important to keep in mind that my primary motivation has been to contribute to biology, *not* to the cognitive studies of science. I wrote the diary out of personal interest, not for the purpose of this chapter, so it is quirky, subjective, and incomplete. In this chapter I have selected specific entries to make my specific points, rather than trying to demonstrate the breadth or depth of the diary, as it is neither broad nor deep. We shall see that it is of somewhat limited utility for cognitive analysis, and in the last section I propose a new study that might provide much more careful and useful data. For now, however, this is what we have to work with. Let us see what we can wring out of it.

ON "SHALLOW" REASON

One way to ask whether a study such as this is useful is to compare my observations with those of others who have studied molecular biology laboratories. The activity of a molecular biology laboratory consists primarily of pipetting minute quantities of clear liquids into other clear liquids, heating

---

[1]The full text of the diary is on-line and can be easily found either through your favorite search engine (try, e.g., "Shrager" AND "insane"), or by writing to me at jshrager@stanford.edu

or centrifuging them, recording numbers and letters in lab notebooks, producing printouts from arcane machines and computers, and occasionally producing a paper telling fantastical stories about invisible entities. The naive observer may be misled by the apparent mindlessness of most of laboratory practice to conclude that molecular biology is merely the manipulation of water and inscriptions, and the execution of "shallow" heuristics—rules not shaped or underlain by understanding, explanation, or analysis. Indeed, this misapprehension appears to be a majority opinion from both the anthropological left and the cognitive right of science studies.

Representing the left, please consider Latour and Woolgar's (1979/1986) famous study, "Laboratory Life." These anthropologists undertook an extended participant–observer study of a molecular biology laboratory; one of them worked as a technician in the laboratory for 21 months. They became fascinated by the process of *inscription*, noting that the members of the laboratory are "compulsive and almost manic writers" (p. 48), including the writing of scientific papers, filling out of tracking forms, labeling of tubes and materials, and entering information into notebooks of various sorts. Latour and Woolgar understand the laboratory as "a factory where facts [are] produced on an assembly line" (p. 236), implying that science involves the production of "facts" through the manipulation of inscriptions—scientific facts and theories being thus reduced to a kind of inscription.

Representing the right, please consider the work of Kevin Dunbar and his coworkers (e.g., Dunbar, 2001). For Dunbar, it is lab meetings, not the laboratory itself, that offers an "accurate picture of the conceptual life of a laboratory" (p. 120). In some cases, Dunbar finds that scientists use subtle skill, for example, using surface versus distant analogy in appropriate settings. In other cases, he describes scientists using cursory heuristics, for example, claiming that "causal reasoning strategy number one for an unexpected finding is to 'blame the method' " (p. 131).

In my experience, both the analyses from the left and the right fall short in important ways.

Let me begin with some points of agreement. It is certainly correct that molecular biologists rely heavily on various sorts of inscription. Indeed, no truer words have been written than Latour and Woolgar's (1979/1986): "Between scientists and chaos, there is nothing but a wall of archives, labels, protocol books, figures, and papers" (p. 236). The following excerpt is from my diary:[2]

---

[2]Dates are represented as YYYYMMDD; for example: the 14th day of July, 2000, becomes 20000714. People's names have been replaced by initials, for example, JS. I have taken some liberty in cleaning up these quotations by silently deleting irrelevant material before or after the quotation and by removing the none-too-rare expletive. However, I have not *added* anything to the content of the diary entries, or rewritten them, except as indicated by brackets.

20000714: Molecular biology is both work intensive and record intensive. I've come to rely heavily, as do all molecular biologists [...] upon three sources: My protocol book, my lab notebook, and a [...] set of protocols called "Molecular Cloning" by Sambrook, Fristch, and Maniatas. [...] If you lose your protocols, you might be okay, because they are mostly collected from other people. If you lose Maniatis, you just buy another, or borrow it from the lab next door. [...] The protocols in [Maniatis] are so complete that people usually just read them straight out of the book. It's sort of The Joy of Cooking for molecular biology [...]. But if you lose your lab notebook, you're hosed, mainly because you'll never figure out what [...] is in the hundreds of obscurely-labelled tubes in the various freezers in the various boxes with the various obscure markings on them. I haven't come upon a perfect scheme for organizing all this yet. [...] [E]very tube has a date on it, and these refer back to my lab notebook [....] In my lab notebook, I'll have a page with that date [...] and it'll have some notes telling me what I was up to, so that I can tell what I'm working on. Also on the tube there's usually, some sort of sequence number [...] scribbled in very small, and hopefully permanent penmanship [...]. Then these get put into a set of boxes [...]. [T]here are 4°C [...] boxes, –20°C boxes, and –80°C boxes, as well as room-temperature boxes. It's an organizational nightmare held together by a very thin thread of paperwork.

However, it is an unfounded leap to conclude from this that the creation and manipulation of inscription is paramount in science. Both inscription and understanding are profoundly important to molecular biologists, especially because it is very difficult to keep track of complex experiments, and, like quantum physics, it is absolutely invisible.

20000724: I've written before about the weirdness of pipetting clear liquids into clear liquids and ending up with something wonderful, for example, DNA. Well, today I pipetted water into *air*, with the firm belief that I'd end up with something similarly wonderful: The tools needed to clone a gene. Specifically, the other day I ordered two "primers," which are custom synthetic oligonucleotides that match the ends of the gene I'm after. You order them on the web, and they show up a couple of days later in a FedEx package that contains two apparently empty vials, which, so I was told in the enclosed materials, contains my custom primers. To make them useful, you just add water and stir. Literally! So, I did, and tonight I'll try to use them to amplify my gene.

Moreover, like physics, much of molecular biology operates by carrying out huge numbers of "micro-experiments" in order to capture and amplify low-probability events, such as the hybridization (joining) of two specific DNA sequences among thousands:

20000516: The scale of these reactions—both large and small at the same time—is hard to comprehend. I am constantly pipetting single microliter quantities, and sometimes a quarter of that! You can hardly see a microliter, much less 0.25 microliters! It amazes me that the pipette can actually do it. At the same time, a single microliter of dATP, for example, at 100mM/liter is, let's see, .1 mole, which is 6e23 * .1 or 6e22. That's, let's see 22 orders of magnitude or about a million million billion ATP molecules.[3]

This is not to say that the procedures are poorly thought out. Quite the contrary, molecular techniques are extremely clever—a mix of combinatorics and filtering; operating on many objects at once and then picking out the few that one wants from among the many. However, it is a kind of cleverness that relies on very large numbers, and careful record keeping, to work.

Precisely because of the invisibility of the materials involved, one has to learn to keep track of what it is that one is doing, either by keeping them all in mind, or, more reliably, by keeping careful notes. Latour and Woolgar's (1979/1986) observations of the obsessiveness with which molecular biologists write is a response to this problem; *Our notes and data are not the objects of our work, but the objects that we are concerned with, proteins, DNA, etc., can only be tracked through notes and data.*

20000902: What is most interesting [...] is not the skill itself, but how it relates to what the protocol is doing. [...] In molecular biology, it's largely gene manipulation, carried out by fairly simple (from my point of view) chemical reactions, mostly involving nano-machines—enzymes. [One] thing that is hard, and this is in my experience the most difficult thing of all, is to keep track of what-all is going on when you do this. Here's where it's not like car mechanics! In car mechanics, the car has a few parts, and the parts are different enough from one another that you don't get them confused. Moreover, the car stays put (usually) when you stop working on it, and you can look it over, choose your next tool, unbolt something, set it aside, take a break for a week, and the state of the world as you left it tells you where you were. In molecular biology, there are a million (more!) almost-the-same pieces of material floating around invisibly. If you lose track of what you were doing, you can't just look at the state of things and tell where you were and what you have to do next. Even if you could SEE DNA, there'd be too much of it in too many [possible] configurations to be able to figure out by looking at it where you were. So there's a huge cognitive load imposed upon the experimentalist to keep track of what's going on; what state things are left in, and where things

---

[3]This calculation turns out to be wrong; I left out the 1ml part. The correct version is: 0.1 × 0.001 Moles, which is 6e23 × 1e − 4 or 6e19, which is about six billion billion ATP molecules—still pretty big by any measure.

are going. Moreover, unlike the car, DNA (and especially living organisms) will take off and do whatever they do if you just leave them to themselves—hybridise or degenerate or overgrow or die. So not only do you have to know what [is] going on, you have to be able to make decisions on a schedule. Lab notebooks help a lot in keeping things straight, and being able to freeze away things helps a lot with this, but you can't freeze, for example, organisms, and when unfrozen, reactions will continue in perhaps undesired ways, so lots of things are done on ice, and still more things just have to be done in a timely way.

Because Latour and Woolgar's (1979/1986) observer did not have the goal to actually become a molecular biologist, he did not find himself compelled to reason about the relevant objects: proteins, DNA, an so on. Their observer's goal was literally to observe, and one cannot observe reason. To the contrary, *my* goal was to *become* a molecular biologist, and one cannot be a molecular biologist without coming to reason quite deeply about molecular biological objects. Indeed, it seems that only a beginning molecular biologist (like myself) does not have very deep and specific understanding of every step of the process and cannot operate without such understanding:

20000727: My understanding has been so blocked by the difficulties of the DOING part, that until very recently—like this week!—I have not really felt like I had any understanding at all of what is going on. Was it that I didn't have the handles or hooks onto which to hang pieces of content, or maybe I was just in such a fog that I didn't have the mental space, time, or energy to make the connections, or maybe I just hadn't done each procedures [sic] more than once, and so really wasn't thinking about them, or maybe there were little pieces of understanding that had to coalesce, just as little pieces of the protocols themselves coalesce into the whole in my head. Probably a little of all of these, and other factors. But today, although things are going no better than average, I think that I can actually begin to THINK about what I'm doing at a meaningful level, instead of merely thinking about how to do the next step— or even just how to figure out what the next step should be. [...] I'm still nearly exhausted all the time, but I don't feel both exhausted and lost—just exhausted.

20000731: I'm doing the latter steps of the inactivation protocol again, this time with my own gene: slr1176, and I'm not in too much of a fog about it either! I'm actually planning the next step based upon the LOGIC of the operation, rather than just reading LZ (or AK's) protocols and doing what they say, and I can reason about things like the blunting reaction and the ligation reactions. [...] and although I shouldn't count my clones before they hatch, I'm feeling quite confident about the steps I've taken so far, and I'm expecting some success this time through. All of this apparent understanding and positive feeling about what I'm up to has come upon me rather rapidly; in the past week or so [....] I've been full time in the lab since mid June, so only about a month and a half, and I've only really been concentrating on techniques for that period of time. Although I had done once or twice pretty much all of these things during my part-time period, I was definitely not focusing at that time, and, as I've written, I was fogged in until just recently. What has happened is complicated. I've proceduralized the procedures well enough that I can think about them, and I can tune them slightly, and debug complete errors successfully [...]. I've mentally cross-indexed the procedures in many ways: what reagents they use, what they do, how long they take, what larger-scale procedures they appear in. Also, I've gotten myself organized to the point that I'm not spending a lot of time searching for things any longer. Also, I've become less timid regarding taking action—trying things out.

20000815: When I have a protocol well in hand [...] I can start to think about its function instead of its form. In fact, I can dwell upon its function, which is often what I find myself doing when things don't work—a very common occurrence for me! In that case, I dwell upon the functions of procedures, and in so doing take small steps toward a full understanding of what's going on. For example, what could be wrong with my ligations? If the LB is wrong, it could be that the pH is wrong; what does that do? I don't know, but I could ask. [...] Each of these small analyses only leads a couple of inferential steps, and these are mostly mundane inferences and guesses, but taken together, they form a rich understanding of the function of what I'm doing [...].

Let us now consider Dunbar's heuristic claims, for example that when something goes wrong, scientists first "blame the method" (Dunbar, 2001, p. 131).

Once again, let me begin by agreeing with Dunbar that indeed biologists often use cursory heuristics in place of deeply thinking about every detail of every step of every protocol.

20000803: Things went fairly well today [....] I learned some new tricks from DB. Mainly that you shouldn't pay much attention to the kit manual's precise directions, but do things a little more aggressively to get better results.

20000811: I've produced many lab rituals of my own. The ones that I'm most aware of [...] have to do with this [...] gel extraction process that I can't ever get good yields from. So I've taken to heating the elution buffer, running the materials through the column multiple times, and leaving the various steps sit for ten minutes each [...]. Are these functional? Who knows. I can make up ways in which they are, but I'm not sure. Taken together, I seem to be getting somewhat better yields, but still not great.

It is important to understand, however, the very limited sense in which these are unreasoned heuristics and the highly circumscribed settings in which such heuristics are used. We very often use protocols blindly, but we do this in areas that would be considered unsurprising and generally unproblematic, akin to cooking: Few cooks think about why you add a pinch of salt to bread—you just do, and it works. However, experienced molecular biologists are much better off than cooks, because when things do *not* work, we *can* and we *do* engage in analysis. Indeed, quite to the contrary of Dunbar's (2001) claim, it has been my experience that we engage in very complex reasoning when something surprising happens even in the simplest of processes!

> 20000720: When we discovered that glycogen pathway enzymes were apparently involved in high light adaptive response, we said to ourselves: Hmmm, well, that's weird, we'll see if it's confirmed, but in the meantime, let's think about what could be happening. In so doing, we produced an explanation that turns out to be the one that we have stuck to, about glycogen production being an electron valve. It turned out that our discovery was confirmed—that is, we found the same genes in the same phenotype over and over, and different genes in the same pathway, which is possibly the best possible confirmation! In this case we didn't blame the data, nor just believe that data. Instead, we did a very reasonable thing: We thought about it, [....]

The result may be the same: To end up blaming the method, but, far from being a blind heuristic, this is the result of subtle understanding. If we did not engage in this sort of reason, we would never get anywhere, rechecking every single result all the time.

An interesting example of the relationship between understanding and executing protocols comes from my struggle with the simplest of techniques, PCR:

> 20000713: My PCRs were only sort of working, even though I was meticulously following LZ's protocols, and LZ had even checked my materials prep and PCR programming. [...] The PCR process, which involves cycling the temperature of the reaction mixture repeatedly, was causing the liquids in the reaction vessel to evaporate! [...] LZ had never seen this before, nor had anyone else I spoke to. The reaction was sort of working, but not very well, as one might expect if the thing had cycled a few times before evaporating the water. In fact, the effect was, at first, not a major anomaly to me, and so I didn't think to really push it among my lab-mates. [...] Was it right or wrong? A clue came

from someplace [....] It used to be [...] that, presumably in order to avoid this exact problem, you put a drop of mineral oil on top of the PCR reaction, but no one has done that for years, as, somehow, the technology had gone beyond that. I assumed that something about the physical structure of the cycler had made this advance, [....] I had taken to ice cooling my reactions after the run so that I'd get the liquid back—a private procedure that made up for an error. [...] I was walking with DB one day, to a lab meeting, bemoaning the problem to her, which she had never heard of either, when she, sort of mused about the heated lid. Heated lid? Yeah, in fact, I think that that was turned off. Aha! She was now adamant that that was the problem, and, in fact, she was right. You have to enable the heated lid. I asked LZ, and she, and everyone else said, "Of course! Of course you have to heat the lid. That's the whole principle!" So, what's with the heated lid? Why wasn't I using it? What could it have to do "in principle" with PCR? When you start the PCR program, which is the only part of the process that LZ had NOT watched me through, it asks: "Enable heated lid?," where the default it "Yes" I had been consistently saying "No." Why? I don't know. No one told me to use the heated lid, and the reaction wasn't in the lid. I guess I just assumed that the heated lid was for something special and that it was just the way the machine was left, or something, so I'd been saying "No," don't heat the lid, thank you. [...] You figure out little things that no one told you about, that you were doing wrong. Another is that once I've done it a few times, I've got most of the "unit tasks" down pretty well, and the little mistakes that one makes get worked out.

It is difficult but important to understand the subtle differences among theoretical thought, shallow heuristic application, and what might be called "promixal" understanding and reason. Consider:

20001019: One of the things that my work in the lab has led me to understand in a way that I don't think I could have otherwise, is just how little theory there is in molecular biology, and how hard it is to get anything resembling facts. If you were to read the literature that has been produced by sitting in on lab meetings, or from the literature, or in the popular press, it sounds like biologists sit around all day and think, and then do some experiments to confirm their thoughts, or get more data, or to generate new hypotheses or test existing ones. But in reality, we spend 99% of our time at the bench tinkering with protocols and genes and such, and only a tiny bit of time—probably less than an hour a week—actually thinking about the implications, or theory, or whatever of it all.

I present this entry not to buy into Dunbar's (2001) heuristic assertions but to complexify the subtleties of the differences at hand. A linguistic analogy may help make this clearer. There are three ways in which one could use language: (a) the typical way that one uses language, to operate in daily life, with understanding but not much philosophical depth; (b) as a form without content, for example, in speaking canned phrases in a foreign language that one does not understand; and (c) to speak about deep philosophical issues, presumably with understanding. What my representatives on the left and right, Latour and Woolgar (1979/1986), and at times Dunbar, appear to claim is that scientists often operate in Mode (b), without clear conceptualization, whereas I claim that skilled molecular biologists nearly always operate in Mode (a), action with understanding—what I call *proximal reason*. What rarely goes on, however, in accord with the above diary entry, is Mode (c): beyond proximal reason, into theoretical understanding and analysis. This distinction, evident throughout the diary, leads one naturally to ask the question: How does one learn to think about what one is doing? Here is where my diary experiment may hold the most promise: To understand how one *becomes* a molecular biologist. I deal with this issue in the remainder of this chapter.

ON COMMONSENSE PERCEPTION

The real voyage of discovery consists not in seeking new landscapes but in having new eyes.

—Marcel Proust

Thus far I have drawn on my experiences of *being* a molecular biologist to address some existing claims about how molecular biology works, but my diary really is a record of *becoming* a molecular biologist, and it is here that it offers the most interesting and novel resource for analyzing how one masters "the ways of thinking, operating, and communicating that constitute the institution of science" (Shrager & Langley, 1990, p. 20). There is much of possible interest here, but I focus on just one topic: the role of Commonsense Perception in learning to become a molecular biologist.

What is fundamental about most sciences, and that which differentiates them from other day-to-day cognition, is perfectly captured in Proust's quote at the beginning of this section, which I shall render less eloquently to the following: Scientists "see" (i.e., interpret) the world in terms of theories—generally in terms of process theories. Although this is true of every science, it is most apparent in what might be called the "sciences of the invisible": chemistry, high-energy physics, astronomy, molecular biology, and other sciences in which the objects of research are literally invisible. Every assertion in these sciences is an act of interpretation, but near the beginning of one's experience in science one does not have the eyes through which to clearly see:

20000902: [Imagine] trying to drive through a thick fog in an unknown city, at or near rush hour, and get someplace on time that you've never been before. Until you've done it a hundred times—until you've lived there!—you just don't know the physical layout of the city, and the temporal structure of rush hour. And since you're in a fog, you can't look around you and see what's coming; where the curves are, when the traffic is stopped, and for how long. Imagine the frustration of doing this the first few times, until you know some workarounds, and know what's up ahead, and moreover, know where to expect traffic at what times. Now imagine that you not only need to know all that, but your task is actually to use your knowledge of that to manage the streetlights for the city! That's more like the job of the molecular biologist, to thread long strings of cars (DNA) through the city efficiently, using gross manipulations like changing the traffic lights. Imagine someone who does not know the city well trying to do that job for the first time—in a thick fog!, or without the aid of cameras or other mechanisms of telling the state of things other than, every once in awhile being able to call some of your friends who are in cars in various parts of the city and ask them if they are in traffic, and what kind of cars are around them, and occasionally, with great effort, being able to ask for a very high level traffic report (running a gel).

I have previously argued the case for the importance of a specific interpretive process, called common-sense perception (CSP, a.k.a. *view application*; Shrager, 1987, 1990), especially in combinatorial chemistry (Shrager, 2001). Here I shall take the liberty to further my agenda to infuse CSP into molecular biology. Put too briefly, CSP is the cognitive process that imparts conceptual content to perceptions. The claim is that scientists (and ordinary people as well) construct their understandings of things through a process of interpretation and reinterpretation, where at each such stage new abstractions are layered into one's current understanding of what is going on. (See Shrager, 1987, 1990, for detailed accounts of the process.)

Molecular biology is a clear case where, in order to understand what is going on requires extensive interpretive work, if for no other reason than, as we have seen, the objects of analysis are invisible. Clear cases of interpretation take place where actual inscriptions are involved, for example, in gel electrophoresis, a very common process used to separate DNA fragments of different sizes. The complexity and importance of interpretation is well exemplified in this interesting, complex interpretive "scream":

20000724: This is a complete information nightmare. I'm only working with a few genes—in fact, only ONE gene, a couple of very common and well-documented vectors, and well known and well understood restriction enzymes, yet I'm already overloaded with information to the point that I can't compute what's going on, so none of my debugging strategies work! There are LOTS of reasons for this confusion. First off, I don't know the EXACT sequence of the gene, since I'm actually using a slightly larger form of it [...]. Second, the information you get from the gel is approximate at best; the ladder (the size standard) is all bunched up where I want to read it, and the legend doesn't seem to match the ladder quite right, so I can't really tell what the sizes of my fragments are, except approximately. And, to make matters worse, if two fragments are the same size they end up in the same place. And to make matters worse yet, there is a ghost on the gel from uncut DNA, which either runs faster, or slower, or the same—[no] one seems to know (although the consensus seems to be that it runs slower!), and to make matters even more worse yet, I'm getting a LARGER fragment when I cut it than I get when I think that I'm NOT cutting it! Ugh. So all my debugging procedures have gone to hell.

One may ask how expert molecular biologists (which I am not, as is clearly demonstrated in the preceding diary excerpt!) learn to form these interpretations. This is, I believe, the most important contribution that may come from study of the sort that I have conducted. The concept from the literature that rings most clearly to me in understanding how scientists learn to form interpretations is the concept of *interpretive drift*, which derives from a source that is methodologically very close to mine: Luhrmann's (1989) studies of becoming a witch, which I mentioned in the opening section of this chapter. Luhrmann focused on the slowly changing conceptualizations that magicians experience by engaging in ritual—or, to put it in molecular biology terms, the slowly changing conceptualizations that molecular biologists experience by engaging in laboratory work. Luhrmann put it this way: "Magical ideas begin to seem normal in the process of becoming a magician [...]. I call this 'interpretive drift'—the slow, often unacknowledged shift in someone's manner of interpreting events as they become involved with a particular activity." (p. 312) She goes on:

> There are systematic changes in the very structure of interpretation, in the sort of analysis which the magician brings to bear in making sense of any particular event. The new magician learns to identify evidence for magic's power, to see patterns in events, to explain the success or failure of the rites. She acquires a host of knowledge which enables her to distinguish between events and to associate them to other events in ways not done earlier. She becomes accustomed to particular assumptions about the constitution of her world. These are changes which affect the structure of her intellectual analysis, the form of the arguments she finds appealing. They are learnt, often informally, through conversations or books. [...] The magician slowly gains a sense of mastery over the material, and with that sense comes a sense of naturalness and pleasure in using the ideas to interpret one's world. Use is more important

than testability. As this material makes clear, confirmation and disconfirmation are fairly unimportant parts of most people's lives. (p. 313)

Can cases of interpretive drift be identified in my diaries? An excellent
example of this comes from the transformation protocol. I had studied
plasmid vectors, and separately studied about restriction, but in performing
a plasmid transformation one puts these two together in a specific way, and
in order to make the decisions that are required in this protocol, one needs
to understand both of these moderately deeply and realize that there are
issues in their interaction.

20000709: When I make a mistake, getting myself out the problem, or, as is
usually the case, LZ or someone getting me out of the problem, helps my understanding greatly. There are many examples of this. In one recent one, I
was running a protocol from a kit, whose purpose is to clone genes. LZ said to
select a restriction enzyme that will cut the vector, but not my gene. She suggested EcoRI. [...] [S]o I did the whole thing, and then was supposed to
check the results by running a gel. Now, although I can run gels now easily, I
am not yet sure what in all, or even in most cases I'm supposed to get as a result. So I asked LZ, and she showed me what to expect. HOWEVER, in
showing me how to do this, she asked me which vector I was using [...]. Is
there more than one?! I had only the vaguest idea of what she was talking
about. So she showed me in the manual that there are two vectors, one of
which is cut by EcoRI and one of which isn't. Which one had I used? Now, I
had studied the manual for this kit VERY VERY CAREFULLY before beginning into this process. But it was suddenly clear to me that I hadn't understood a word that it was saying! As it turns out, I was lucky and had used the
vector that is, in fact, cut by EcoRI, [and the restriction worked.] [...] So
this time it happened to work out, and I learned a small but important fact
about the use of this kit, which is that you have to keep track of which vector
you're using. [...] [N]early at the moment at which LZ explained to me
about the two vectors, something much larger clicked into place for me.
[...] [S]omehow I had all the pieces of the puzzle [...] in hand and identified, but hadn't put them into the frame. When LZ showed me the picture in
the manual of the two vectors, with their various restriction sites, that was
the frame for the whole procedure, and all the pieces fell right into it, and I
very suddenly—literally in a matter of a few seconds—"saw" what I had
been doing for the past day: I could see why we were cutting the vector and
amplifying the gene, and ligating them together and why I had to use EcoRI.
And then I understood, all in that same perceptual unit, how to figure out
what to expect from the gel. Maybe this was just the first time I had actually
had time to think, as opposed to feverishly cooking and being lost, but it
doesn't feel like that.

Indeed, in addition to learning the methodological details of protocols, learning to understand what is going on inside one's reactions may be among the principal hurdles in learning to be a molecular biologist; at least, this seems to have been my experience:

20000821: On the T in Boston—the Red line—traveling from South Station to Alewife [...], I "know" the places that we are traveling [...] underground as we go—diving underground after Charles, Kendall at MIT, Central—a church, I think—Lori used to live near there. And then Harvard Sq.—Steve used to live near there—and then out to places unknown. The feeling is very subtly different where I know what is above my head, even vaguely—at least I think I know—from where I do not know, have not been walking or driving aboveground. As a result, I know the lay of things, the direction with respect to the city that we have come, and where we are going, how long it will take to get there, whether it makes sense to take a cab or bus or walk instead of taking the T. However, from Harvard onward and out to Alewife, I do NOT know what's above my head. As a result, the feeling of taking the T in these circumstances is very slightly different. [...] That is a little of how it is with the molecular biology protocols: I can do them, but I don't know what's going on "above ground," so to speak—inside the test tube or cells, and I can't make decisions about whether to walk or drive or heat or cool or add more buffer, or take the T.

20000902: It is these sorts of Knowing: Knowing what the protocols do, Knowing how to put them together to do larger tasks, Knowing how to make the whole thing work together smoothly, Knowing where you can start and stop and pause things, Knowing how to debug problems, Knowing what's going on all the time in all of your hundred variously labeled tubes in six different freezers at six different temperatures, Knowing [what] the sequences are of all your genes and how to manipulate them and reason about them, and Knowing in the end what the results mean in the larger conceptual space of biological science. It is these sorts of Knowing that make this an interesting and difficult thing to do.

There is much more that could be said about this, but aside from these insights, the evidence in my diaries is sparse; perhaps I was in such a deep fog for so long that I was unable to pick out these instances. Indeed, nearly all of my dairy deals with the coevolution of my understanding and my skill. (There is much to say about that as well, but little room in this chapter. Sahdra and Thagard [2003] have used my diary, among other sources, to analyze the complex relationship between "knowing how" and "knowing what," sometimes called the *procedural–declarative* distinction in cognitive science, although, as they argued, these are far from distinct.)

What might help us to delve more deeply into Commonsense Perception and interpretive drift, beyond my rambling introspections, is evidence that is closer to a verbal protocol collected from scientists learning to form interpretations in their fields. In the next section of this paper I conclude with some thoughts about how one could gather the sort of data that might be more useful toward this and other examinations of the cognitive aspects of becoming a scientist.

WHAT IS IT LIKE TO BE AND BECOME A MOLECULAR BIOLOGIST?

It will not help to try to imagine that one has webbing on one's arms, which enables one to fly around at dusk and dawn catching insects in one's mouth; that one has very poor vision, and perceives the surrounding world by a system of reflected high-frequency sound signals; and that one spends the day hanging upside down by one's feet in an attic. In so far as I can imagine this (which is not very far), it tells me only what it would be like for *me* to behave as a bat behaves. But that is not the question. I want to know what it is like for a *bat* to be a bat.

—Thomas Nagel (1974, p. 439)

On the one hand, psychologists of science might do well to embrace Nagel's (1974) critique of the difficulty of understanding another's subjective experience: How could one know what it is like to be, for example, a molecular biologist, without actually being one? But elsewhere in his article, Nagel himself, *en passant,* offers us a program for resolving this problem: "Even if I could by gradual degrees be transformed into a bat, nothing in my present constitution enables me to imagine what the experiences of such a future stage of myself thus metamorphosed would be like" (p. 439). Although one cannot in fact transform oneself into a bat (unless one happens to be a vampire), one *can* transform oneself into a molecular biologist. Thus, in order to know what it is like to be a molecular biologist, one needs merely to *become* one.

I don't know that I have shed very much light on what it is like to become a molecular biologist, but all that I have presented in this chapter is only a very small part of my diary, and that is from a very few months in the very beginning of the career of a molecular biologist. I hope that others will be able to mine my notes for other interesting analyses (e.g., Sahdra & Thagard, 2003). However, in the same way that these excerpts cannot represent the whole of the diary, the diary cannot represent the whole of becoming a molecular biologist. What can be done to perfect, or at least greatly improve this project to the level of a significant contribution? I have one suggestion.

In 1964, Michael Apted, a British documentarian, produced a now-famous TV program called "7up," which documented the lives of 14 London children at age 7. Each 7 years thereafter, Apted returned to as many of those children as he could find and reexamined their lives at 14, 21, 28, 35, and, most recently 42 years of age. The interviews dealt with mere mundanity, but it is in mere mundanity that life most often progresses, so these films are fascinating studies in many aspects of everyday life, including friendship, aging, happiness, and sorrow. What I propose is a longer and broader version of my diary study, one that might be conducted approximately as follows: In the early college years, when students are just at the point of declaring a major, we post a call for volunteer "diarists" who think that they are likely to declare a major in the sciences. (There is no reason to restrict this project to science, but I shall retain that focus for the nonce.)

We offer our diarists courses in participant–observer methodology and pay them a small stipend (as one would any participant) to keep a diary about their entry into, and progress in, science. Each quarter, we meet, either as a group or individually, and discuss observations, technique, and so on, and each year we offer them somewhat more advanced courses in cognition of science, anthropology of science, philosophy of science, and so forth. By the end of their undergraduate careers, only a few of the original students will probably still be participating, and we would discuss with them continuing to follow them through their future careers, including graduate school and perhaps their "real world" careers.

Many details of this proposal need to be worked out, but my commitment to biology does not allow me the time to follow up this idea by myself. Perhaps these few notes will inspire the next generation of psychologists, anthropologists, and sociologists of science to take up this work and give it the attention it deserves.

ACKNOWLEDGMENTS

My thanks go especially to the organizers of the Workshop on Cognitive Studies of Science and Technology meeting at the University of Virginia in 2001 and the editors of this volume—Mike, Ryan, David, and Alexandra— for the opportunity to share this exercise with my colleagues, and to David Klahr, Kevin Dunbar, Bruce Seely, and Paul Thagard for encouraging me to actually go for it. Rea Freeland and Carrie Armel helped me think about some of the difficult problems in trying to write a paper of this sort. I am indebted to Arthur Grossman for allowing me to work in his laboratory, to the members of the laboratory, especially Ling Zhang, who helped me overcome uncountable stupid mistakes, and to Lisa Moore, who suggested that I contact Arthur in the first place, one day when I was simply musing aloud about becoming a phytoplankton biologist. I especially thank Laura McIntosh, who helped me understand some of the arcane ways that biologists think, who reviewed a late draft of this chapter, and who taught me my very first laboratory skills, and Betsy Dumont, who got me interested in biology and the importance of the biological environment in the first place.

REFERENCES

Dunbar, K. (2001). What scientific thinking reveals about the nature of cognition. In K. Crowley, C. D. Schunn, & T. Okada (Eds.), *Designing for science* (pp. 115–140). Mahwah, NJ: Lawrence Erlbaum Associates.

Latour, B., & Woolgar, S. (1986). *Laboratory life: The construction of scientific facts.* Princeton, NJ: Princeton University Press. (Original work published 1979)

Luhrmann, T. M. (1989). *Persuasions of the witches craft.* Cambridge, MA: Harvard University Press.

Nagel, T. (1974). What is it like to be a bat? *Philosophical review, 83,* 435–450.

Sahdra, B., & Thagard, P. (2003). Procedural knowledge in molecular biology. *Philosophical Psychology, 16*(4), 477–498.

Shrager, J. (1987). Theory change via view application in instructionless learning. *Machine Learning, 2,* 247–276.

Shrager, J. (1990). Commonsense perception and the psychology of theory formation. In J. Shrager & P. Langley (Eds.), *Computational models of scientific discovery and theory formation* (pp. 437–470). San Mateo, CA: Morgan Kaufmann.

Shrager, J. (2000). *Diary of an insane cell mechanic.* Retrieved May 15, 2002, from http://aracyc.stanford.edu/~jshrager/personal/diary/diary.html

Shrager, J. (2001). High throughput discovery: Search and interpretation on the path to new drugs. In K. Crowley, C. D. Schunn, & T. Okada (Eds.), *Designing for science* (pp. 325–348). Mahwah, NJ: Lawrence Erlbaum Associates.

Shrager, J., & Langley, P. (1990). Computational approaches to scientific discovery. In J. Shrager & P. Langley (Eds.), *Computational models of scientific discovery and theory formation* (pp. 1–25). San Mateo, CA: Morgan Kaufmann.

7

Replicating the Practices of Discovery: Michael Faraday and the Interaction of Gold and Light

Ryan D. Tweney
Ryan P. Mears
Bowling Green State University

Christiane Spitzmüller
University of Houston

The late historian of science Frederic Lawrence Holmes began a recent book on the early researches of Antoine Lavoisier (Holmes, 1998) by noting the extreme difference between the public face of the great 18th-century scientist and the image suggested by his laboratory notebooks: "We ... see a public Lavoisier announcing for the first time an 'epochal' theory with a grand gesture and an aura of self-assurance that contrasts strikingly with the trepidations of the private investigator" (p. 4). For Holmes, this discrepancy was important, and his search, as a historian, for the roots of scientific creativity led him to the intensive examination of notebooks and laboratory diaries, not just of Lavoisier but of Hans Krebs and of Claude Bernard as well.

Understanding the roots of creativity in science and technology might seem—obviously—to demand close attention to such sources, but it is also true that notebooks and diaries cannot be studied in isolation. Scientific discovery depends on a context far larger than that of the single investigator alone in a laboratory, and even the fullest of laboratory records must acknowledge that discovery always begins with a context external to the discoverer and ends (with text or demonstration or built artifact) in an external—public—context. It is thus no surprise that, complementing the work of historians such as Holmes, whose focus has been on "private" science, there have been richly revealing accounts of the "public" side of science. For example, Golinski (1992) analyzed the debate between Lavoisier and Priestley, and Rudwick (1985) examined the emergence of the Devonian concept in geology by tracking the enterprise of a network of amateur and professional scientists.

Part of the context of scientific discovery resides in its material entities; thus, historians and philosophers of science are paying more attention to the artifacts of science (instruments, graphs, mathematical representations, etc.), just as cognitive scientists are beginning to analyze cognitive artifacts, those external objects and representations that serve as computational substitutes for "in the head" cognition (Hutchins, 1995; Zhang & Norman, 1994). It has long been recognized that the analysis of cognitive behavior is incomplete unless the complexity of the environments of cognition is accounted for (e.g., Simon, 1967/1996). What seems new, however, is the growing attention given to the need for an interactive play among both internal and external aspects of cognition (e.g., Gooding, 1992). Such blurring of the boundaries of cognition is most essential in the attempt to understand the formation of representations in the first place.

In this chapter we describe aspects of our investigation of the surviving experimental artifacts used by Michael Faraday in his 1856 research on the properties of gold films and his resulting discovery of metallic colloids and the associated optical phenomenon now known as the *Faraday–Tyndall effect*.[1] These are important findings, although, as we shall see, Faraday felt frustrated with the research program, which aimed at larger questions. Even so, our study can add to knowledge of the personal side of Faraday's endeavors, especially because we focus on what can be learned from the material objects that accompany Faraday's very complete diary records. Previous readings of Faraday's 1856 diary records have sometimes argued that the record was merely a confusing hodgepodge of unconnected pieces (e.g., Williams, 1965, and perhaps also Zsigmondy, 1909, who referred to Faraday's

---

[1] A *colloid* is formed when finely divided particles of a substance are held in suspension in a fluid. Colloids differ from *solutions* in that solutions represent ionized particles of atomic size, carrying an electrical charge. Ions, the particles that form a solution, are much smaller than colloidal particles. Faraday's discovery that metals could form colloids was important, as was his demonstration that the metallic colloids were particulate and possessed optical properties—the Faraday–Tyndall effect —that are very different from solutions (as we explain later).

methods as "crude," p. 75). We instead make the case that the apparent gaps between Faraday's experiments and his finished (and unfinished) mental models must be filled with additional information to understand his work. The discovery of Faraday's slides allowed a new and crucial visual domain to aid in interpretation of the diary, and our replications, as we show, extend this even further. Far from a muddle, the combination of diary *plus* specimens, *plus* replications, allows a deep understanding of the progression from confusion to knowledge.

By replicating some of Faraday's experiments, we show that the understanding of his scientific practices requires more than what a purely textual analysis could achieve and that Simon's (1967/1996) point about the analysis of the environments of cognition constitutes a necessary imperative for the understanding of discovery. The experiments carried out in our laboratory were formulated with the intent of understanding how Faraday combined his beliefs about light, matter, electricity, and magnetism; wove these beliefs into a creative and productive series of laboratory experiments; and used his experiences in the laboratory to formulate new ideas and explanations. Our experiments resemble his in one sense: Like Faraday, we began with only a series of broad beliefs about the nature of scientific cognition, beliefs that have been productive in other domains (including other aspects of Faraday's research). Like Faraday, we sought to use our experiences to refine and sharpen our own beliefs about scientific discovery.

Our results, including the early phases in which we "stumbled around," trying to get it right, allowed us to account for an important conceptual change that occurred in Faraday's thinking early in the project he conducted in 1856. Thus, later in the chapter we argue for a constructive role of such "confusion" in the process by which Faraday reconceptualized his ideas. Furthermore, our replications led us to a new description of the differences between the discovery processes early in a research project versus those that occur later, when the research is more developed.[2] As we argue, the value of programmatic replication is most evident in this psychological domain in which the cognitive processes of an investigator are under examination; such processes otherwise leave no direct evidentiary trace, and only in their reconstruction can their dynamics be made visible.

To begin our account, we first describe an unexpected opportunity (a historical "discovery") that inspired the replications. We then describe two series of our replication experiments, one on colloids and precipitates of gold

---

[2]Others have used replication as an aid in understanding specific experiments in the history of science. For example, Heering (1994) reconstructed the electrostatic balance by which Coulomb established the inverse square law of electrostatic attraction and repulsion. The results suggested that Coulomb's study could not have been done exactly as he described it in his formal report.

and the other on "deflagrations," that is, on gold wires that are exploded with rapid bursts of electric current. After each replication series, we place our results in context with Faraday's results, as reported in his diary and as manifest in the surviving specimens of his work. Finally, we consider the implications of the project for the understanding of scientific thinking.

FARADAY AND THE PROBLEM OF GOLD

In 1856, Michael Faraday (1791–1867) was nearing the end of a very productive scientific career. He was, of course, best known for his discoveries in the domains of electricity and magnetism, but his activities as a chemist were no less consequential, particularly his 1832 demonstration of the electrochemical equivalent of electric current, a discovery that quantified the close relation between electricity and the compositional forces of matter. Throughout his career, a constant theme was the attempt to unify the seemingly separate domains of force, and this had led to the precise articulation of his comprehensive (if not yet complete) field theory (Gooding, 1981; Nersessian, 1985; Tweney, 1992a; Williams, 1966). By 1850, Faraday had shown that the electric and magnetic fields could be regarded as the manifestation of physically real (but immaterial) lines of force that extended throughout space, and his conviction that light must be related to these lines of force was partly responsible for his discovery (in 1846) of the rotation of a plane polarized beam of light passing through a dense optical medium to which a magnetic field was applied. If not yet a complete "unification," the phenomenon was an important clue that light was somehow implicated in the nature of electromagnetic fields (James, 1985; Tweney, 2002).

Thus, when Faraday carried out nearly a year's worth of research in 1856 on the optical properties of thin gold films (our focus in this chapter), he was seeking more than just the explanation of certain optically pretty effects. The program was also intended to explore some of the deepest questions of physics: Were particles of matter best considered as "force centers"? Was there an ether that served as the medium of light transmission? Could this speculation about force centers be integrated with the demonstration that field forces (electricity and magnetism) consisted of nonmaterial lines of force? James (1985; see also Chen, 2000) has argued that Faraday was committed in many ways to an "optical mode of investigation" and that he frequently used optical criteria, for example, in his development of the distinction between diamagnetic and paramagnetic substances, and his long-standing concern with whether there was an ether. In the gold research, Faraday sought to extend his optical research to an account of light itself.

Gold has richly varied optical properties and was thus a prime suspect in the search for deeper understanding of the interaction of light and matter. Gold is so malleable that it can be hammered out into extremely thin transparent films, and it was known by 1856 that such films were much thinner than the wavelength of light. Furthermore, these films interested Faraday because thin transparent gold films are a different color by transmitted light than by reflected light. When a beam of light is passed through such a film, it manifests a green, blue, or purple color rather than the familiar shiny yellow–gold seen by reflected light. For Faraday, the initial question was simple: How could such very thin (and apparently continuous) films alter light so dramatically? In the end, Faraday failed to answer this question, but the research had important consequences—not only the discovery of metallic colloids and the Faraday–Tyndall effect, as we have noted, but also the first clear indication that all of the color effects of gold were the product of the interaction of *particles* of gold and light. That is, contrary to his initial expectation, Faraday had to confess at the end that he had no evidence for the existence of *continuous* matter, not even for the shiny gold films!

In this chapter we argue that replications of some of Faraday's earliest experiments leads to a perspective on how the research was launched, how Faraday resolved some of his initial "confusions," and how these were important to his final conclusions.

There have been many studies of Faraday's research conducted by historians, philosophers of science, and psychologists, most of them based on his surviving laboratory diary. This extraordinary set of documents, a large part of which has been published (e.g., Martin, 1936), covers nearly his entire career, and the completeness of the record has caused it to be compared to a "think aloud" protocol (Tweney, 1991). Some have used it as the basis for a reconstruction of his research practices (e.g., Steinle, 1996). In some cases, however, as in the case of his research on gold, much of the diary is hard to interpret by itself: He makes reference to specimens and results that are only partially described, and the text, by itself, leaves many questions unanswered.

Faraday wrote 1,160 numbered entries on his gold research, roughly 250 printed pages in the transcribed version (Martin, 1936). These are dated from February 2, 1856, to December 20, 1856. In an earlier account (Tweney, Mears, Gibby, Spitzmüller, & Sun, 2002), we noted that the distribution of entries was roughly bimodal, with the greatest density of entries occurring at the beginning of the series and toward the end. The very first entries appear to be summaries of previous notes. These first entries also include several dozen entries in which Faraday outlined possible experiments, much as he had earlier kept an "idea book" in which to record possible studies (Tweney & Gooding, 1991). Faraday's (1857) published article on gold was submitted to the Royal Society on November 15, 1856, and read before the Society on February 15, 1857, just before it appeared in *Philosophical Transactions,* the society's official journal and one of the premier scientific journals in the world at that time. Not surprisingly, the character of the entries in the second peak suggests that Faraday was "mopping up" prior to ending the research: conducting some necessary control experiments, trying again to resolve some inconsistencies, replicating key preparations, and so on.

As noted at the beginning of the chapter, our program of research was initiated by an unexpected discovery: more than 600 surviving microscope slides and other specimens made by Faraday as part of the 1856 research. The specimens, mostly gold films mounted directly on ordinary 1" × 3" glass microscope slides, were "hidden in plain sight," on display in the museum area at the Royal Institution in London (Tweney, 2002). Examination of the slides revealed that each slide was numbered by Faraday, and each was indexed and referenced in Faraday's laboratory diary covering this work (Martin, 1936). Thus, we have nearly the complete set of metallic film specimens used by Faraday in 1856, as well as a few of his colloidal specimens.

Faraday's surviving colloidal specimens have long been noted, and there are at least five bottles of these on display in the museum area of the Royal Institution. One especially notable one has still (a century and a half later!) the characteristic pink color described by Faraday. The others are pale and nearly colorless. Four of the five bottles show a clear Faraday–Tyndall effect; that is, when a narrow beam of light is passed through the colloid, the light is scattered sideways, rather like a sunbeam through smoky air. This characteristic property marks each as a true colloid, and it is this property that suggested to Faraday their particulate nature.[3]

The surviving specimens fill in the missing dimension of Faraday's diary; those specimens and results that are not adequately described in the text itself can now be examined. Furthermore, by replicating his preparations and comparing them to the originals, we can gain even more insight into the processes Faraday used to create the specimens, and we can reproduce Faraday's often-destructive manipulations (heating the slides, burnishing them, treating them with corrosive substances, etc.). As one of our first efforts, we prepared our own colloids (using modern methods, but obtaining results identical to Faraday's). In this chapter we focus on our efforts to re-create two kinds of experiments carried out near the beginning of Faraday's work on gold: (a) the precipitation of gold from solution and, (b) the "deflagration" of gold wire, that is, exploding a gold wire using sudden surges of electric current. In a later publication (Tweney, in preparation), we plan to discuss our efforts to produce gold colloids and thin metallic films of gold using Faraday's favored technique, reduction by phosphorous.

REPLICATING PRECIPITATES

Faraday's first diary entry on gold (February 2, 1856) describes a visit the week prior to Warren De la Rue's home, to examine gold leaf through his

---

[3]It is interesting that an examination using a parallel beam of light revealed that one of the "colloids" is actually a solution of an unknown substance. Although it has a color nearly that of a gold colloid, it does not manifest a Faraday–Tyndall effect, unlike the true colloids.

friend's better microscope. By the following week, De la Rue had made some especially thin films using phosphorous reduction, and on February 2, Faraday examined these in his laboratory. Three days later, he began his first active work on gold in his own laboratory. It is surprising that, at first sight, he began by making some precipitated gold. Because the precipitation reaction of gold was long familiar by 1856, Faraday could learn nothing new here, and the text of the diary alone does not indicate why he initiated his gold research with such a common procedure. In fact, the experiments with precipitates were far from trivial, as we learned when we replicated his procedure. We were thereby able to detect a "confusion" that served a heuristic role in the important step of arguing that the colors of gold are due to particles interacting with light.[4]

When a reducing agent is added to a solution of a gold salt, metallic gold (Au) is precipitated as a solid; in modern terms, the positively charged gold ions combine with electrons from the reducing agent, forming uncharged particles of elemental gold. These aggregate together, forming larger particles, which then are prone to settle out of the fluid medium. Such simple chemical reactions are familiar to every beginning chemistry student, and we accordingly thought that replicating these first experiments of Faraday would be a simple exercise for our laboratory group, if not a particularly revealing one. However, the chemistry of gold in solution is more complex than we had anticipated, and we thereby experienced our first "confusion"!

Faraday's diary entry stated only that he "prepared a standard weak solution of Gold" and a "standard solution of proto sulphate of Iron … consist-[ing] of 1 vol. saturated solution at 54° F. plus 2 vols. Water, and a little sulphuric acid to keep all in solution during the changes" (Entries 14291 and 14292, February 15, 1856). In modern terms, "proto sulphate of iron" is *ferrous sulfate*, and the fact that it was saturated allowed us to reproduce the exact substance used by Faraday. However, gold salts are complex,[5] and Faraday stated only that he used a "standard weak solution of Gold." What was this? In the end, we used pure gold wire (0.025 in. [0.06 cm] diameter,

---

[4]Williams (1965) suggested that Faraday's work on gold in 1856 was an indication of his "declining powers," perhaps due to aging or to the many toxic exposures he experienced over the years. At first, the seeming aimlessness of the precipitation experiments appears to support the claim, but our analysis suggests that the experiments were not at all aimless. Our results have failed to reveal anything deficient about the mental powers displayed by Faraday during this research.

[5]Gold chlorides exist in solution as $[AuCl_4]^-$ ions and various hydrolyzed ions as well. These more complex species and reactions were not known to Faraday. As we discovered, however, and as Faraday must have known, the complexity of the reactions is reflected in a very complex phenomenology: Gold salts are unstable, dissolve in water to varying degrees, leave varying undissolved residues, and manifest a variety of colors—all of which was extremely confusing in our first efforts. See Puddephatt (1978) for further details on the chemistry, and Tweney et al. (2002) for a more detailed account of our procedures.

99.99%), which we dissolved completely in Aqua Regia, a 3:1 combination of hydrochloric acid and nitric acid. A saturated solution of the reducing agent was then prepared by dissolving crystalline ferrous sulfate ($FeSO_4$) in heated water. When cooled, three drops of ferrous sulfate solution were added to 5 ml of the dissolved gold solution. At first, no reaction was apparent, but on the following day a yellow–orange residue of metallic gold had settled at the bottom of the experiment tube. This could be redispersed by shaking, although it would gradually settle again over the course of half an hour or so. After shaking, the fluid had a muddy, brownish-yellow appearance, in which individual particles could be seen moving about, some glinting with the familiar bright metallic color of gold.

With the advantage of modern knowledge, we of course knew that the precipitate was physically similar to the colloid, except that the particles in the precipitate were much larger than those in the colloid. Yet, when placed side by side, the precipitate looked very different than the colloid we had prepared earlier; the colloid was a clear fluid, red in color but transparent, and very unlike the nearly opaque precipitate. In fact, except for the fact that the overall color was red, rather than the yellow–gold of the dissolved gold wire solution, the colloid more nearly resembled the solution than it did the precipitate.

The relative appearance of these three changed, however, when directional lighting was passed through the fluids, as shown in Fig. 7.1. In the figure, a parallel beam of light produced by a fiberoptic illuminator (entering from the left) is being directed through our prepared gold colloid, a solution of gold chloride, and the precipitated gold preparation, respectively. (Color images of these three can be accessed at http://personal.bgsu.edu/~tweney.) The precipitate was shaken just before the photograph was taken. Note that the colloid scatters light to the side, illuminating the path of the beam. This is the Faraday–Tyndall effect, the effect that first suggested to Faraday that colloids were particulate. By contrast, the solution does not scatter light, except for some small reflections from the sides of the glass test tube being visible in the photograph, and the precipitate scatters light like the colloid.

The important point here is that the colloid and the precipitate resemble each other most closely under *transmitted* light conditions, whereas the colloid and the solution most resemble each other under *reflected* (ambient) light conditions. There is no record in the diary of Faraday placing these three in one context (as we did in Fig. 7.1), but we now believe that he was attending these differences very carefully; they later constituted part of the basis for his conclusion that the colloids were in fact metallic particles of gold. In the light of our own experiments, we believe that the sequence of Faraday's entries in his diary can be reconstructed in the following way.

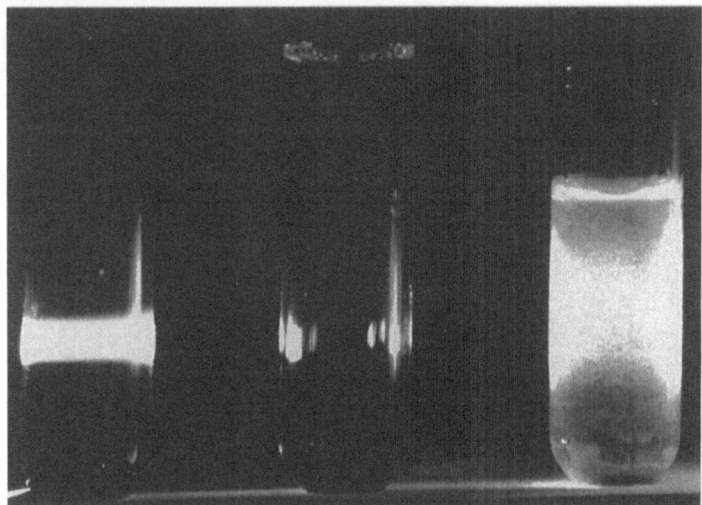

FIG. 7.1. Colloid, solution and precipitate.

Recall that Faraday had visited his friend Warren De la Rue the week prior to February 2 (the date of his first diary entry), to examine some gold leaf through the microscope. Faraday recorded this (Entry #14243, February 2, 1856), indicating that, also on February 2, he received the thin gold films prepared by De la Rue, who had used phosphorous to reduce the gold (Faraday later used this technique himself). On February 6 (1 day after preparing the precipitates), Faraday used a careful optical method to examine the precipitates and recorded that, in the evening, he went to De la Rue's again and observed how the thin gold films were made. In his description of De la Rue's method, Faraday recorded an apparently incidental observation, made during the cleaning up of the glassware used to make the films; "A very fine red fluid is obtained [from] the mere washing" (Entry #14321). With hindsight, we know that this was a colloid, but it is significant that Faraday noticed it in this context.[6] In fact, Faraday saved the fluid, returning to it 2 weeks later, on February 18 (Entry #14437), after his experiments with precipitates and his first examinations of thin films. At that point, he was able to ask of this red fluid: "The question is, is it [i.e., the gold] in the same state as whilst apparently *dissolved* in the fluid" (Entry #14437). During the intervening period, Faraday had referred to the red fluid using two terms interchangeably: *fluid* and *solution*. Only later was he sure that the red fluid was not a solution. It is clear, however, that the possibility had suggested itself very early.

---

[6]Parkes (1822, p. 500) described a similar experiment that may have resulted in a gold colloid. He dissolved gold in Aqua Regia, evaporated it to dryness, and then dissolved the crystals in water. By adding a salt, he observed a faint violet hue that changed to a deep purple. This may have been colloidal gold, although he did not provide an account of the nature of the reaction or the resulting fluid.

The sequence of Faraday's ideas must then have been the following. He first compared thin films (which he suspected to be gold in a *continuous* state) to the precipitates, which he knew to be made of *discrete particles*. Because gold film (i.e., gold in a continuous state) changes appearance in transmitted light and reflected light, he developed an "optical method" for examining precipitates under the same two conditions, namely, "reflected" light (ambient) and "transmitted" light (passing a beam through the substance). We enclose these terms in quotation marks to suggest that each is slightly different from the analogous procedure with thin films. To prepare the precipitates, he must have had before him the clear yellow–gold solution of gold chloride, and this could have suggested, while he was at De la Rue's, a question about why the washing fluids were clear. The substances used to produce that clear red solution (phosphorous, carbon disulfide, and a gold chloride solution) could only have produced metallic gold. But why did it look like a solution? Resolving this confusion is then the reason why he examined the red fluids more closely—using both "transmitted" and "reflected" light, just as he had done with the precipitates.

According to Hanne Andersen (2002), taxonomic change in scientific concepts is best construed as model-based change, because existing "family resemblance" and structural accounts are too limited when seen in the context of change in actual scientific concepts. In the present case, Faraday eventually saw the "red fluid" as gold in a "divisible state," like the precipitates. Note that stating the reorganization in this manner is, in fact, model based, but it describes the end-product of his thinking and experimentation. At the earliest stages, the ones we are concerned with, Faraday was not in a position to make a model-based claim, because the "model" was still too vague; it was really based only on a set of unusual appearances. The text of the diary alone does not, of course, reveal these, because they were obvious, visually, to Faraday—and they became obvious to us only when present as the result of our own "makings." At the phenomenological level, the term *confusions* is thus a better description of what he was faced with (see Cavicchi, 1997). Reorganization of the appearances, not reorganization of the taxonomy of model-based classification, is the issue.

The red fluids provided Faraday with a first important clue that his inquiry into the color of gold in its various states was going to have to focus on the influence of particulate gold on light. Earlier (Entry 14279, February 2, 1856), he had speculated that the color of metallic gold could perhaps be a manifestation of particulate effects, but the discovery of the colloidal state of gold was a strong clue that perhaps size of particle was an important variable. Because the colloids could be seen to be particulate only under certain lighting conditions, were there other conditions that would suggest that even apparently continuous gold (e.g., as in mounted gold leaf) was also particulate? This suggests why Faraday took up the determination of the optical properties of gold in a state in which it was clearly particulate: "Would a metallic surface made up of particles, like de la Rue's films, reflect light so as to give colours of thin plates? Perhaps may find here a test for continuity" (Entry 14407, February 12, 1856).

REPLICATING DEFLAGRATIONS

Faraday did not immediately take up the question of continuity that he had posed on February 12. Instead, his work with the thin films produced chemically using De la Rue's method occupied him for almost 2 months. Then, quite without warning in the ongoing text, he suddenly recorded using a Grove's battery (Grove, 1839) as a source of current to explode gold wire.

Like all metals, gold has a melting point (1,064°C) and a boiling point (2,856°C). However, the liquid state that stands between these two values can be bypassed: If gold is heated quickly enough, it can be vaporized directly, a process known as *deflagration*. One method of deflagration involves vaporizing material with heat energy generated by a rapid current of electricity. The experiment is seemingly simple; in Faraday's words:

> Provided some gold terminals and a voltaic battery of Grove's plates, then brought the terminals suddenly together and separate, so as to have a momentary deflagration. Did this upon and between glass plates, so that the deflagrated gold should be deposited on the glass in films more or less graduated. (Entry 14664, April 9, 1856)

Some of the results (Slides 236, 238, and 239, noted in the same diary entry) are shown in Fig. 7.2. Faraday exposed each slide to multiple deflagrations, each of which has produced a spot or arc of metallic gold deposit.

Similar means of deflagrating gold were well known by 1856; Charles Wilkinson (1804), for example, described a series of experiments in which Leyden jars were charged by a frictional machine and then discharged across a strand of metallic wire (see Fig. 7.3).

Others had conducted similar experiments (e.g., Bostock, 1818), and there was a consensus that the films of metal deposited on a card or a glass plate held near the deflagration represented particulate matter. Wilkinson (1804) believed that the particles consisted of oxides of the metal, but Faraday was suspicious of this claim in the case of gold. Because gold will not burn in air or oxygen, it is likely that only pure gold particles are deposited when gold wire is deflagrated. Later, Faraday was able to verify this belief; when gold is deflagrated, only gold is deposited on a nearby slide. In any case, he did accept that the deposits were particles of gold.

At first, it was not clear what motivated Faraday's choice of a battery of Grove's cells. In fact, this appears to be a rather odd choice; a Grove's cell is made using *platinum* plates in a zinc cell. The special virtue of such a cell, which would have been extremely expensive in Faraday's time, as in ours, is

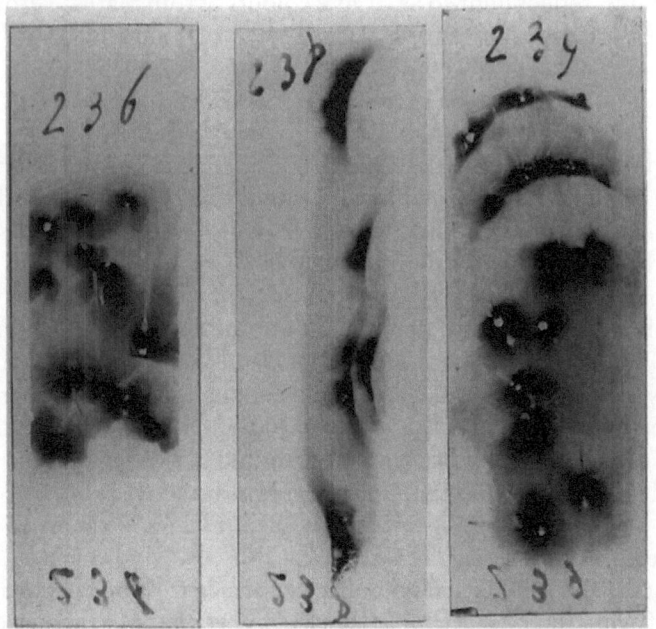

FIG. 7.2.   Three deflagrated slides (exploded gold wire) made by Faraday.

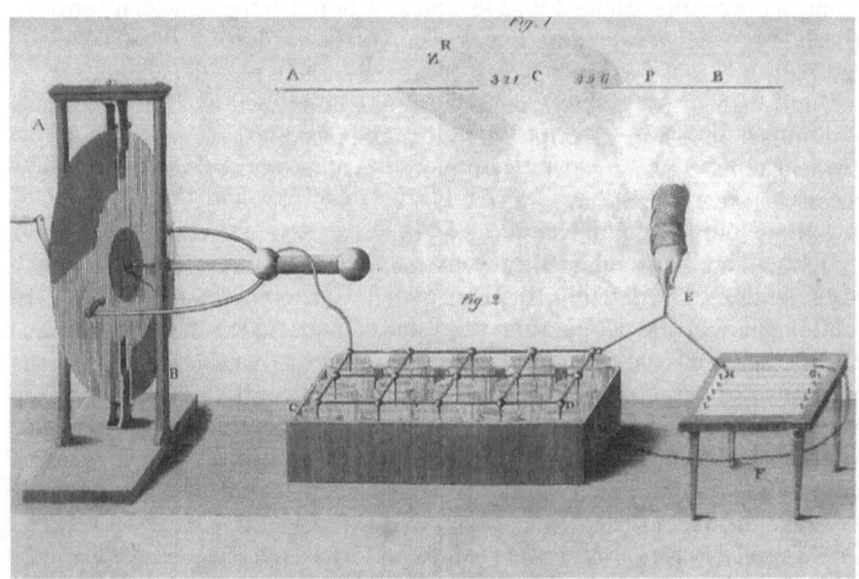

FIG. 7.3.   An early apparatus for exploding gold wires (Wilkinson, 1804, Vol. 2, Plate 9).

its very low internal resistance (Bottone, 1902). Thus, the simplicity of Faraday's first deflagrations may be only apparent. Perhaps he had also tried earlier and failed, until he could provide a suitable current source. Furthermore, he was not entirely happy with the results obtained with the Grove's cells. Even Wilkinson (1804) had indicated that wires could be exploded along their entire length. Perhaps the more elaborate setup shown in Fig. 7.3 was necessary?

How can a current of electricity be generated swiftly enough to bring the temperature of the metal to the point that a portion or the entirety of the wire explosively vaporizes? Because current is a function of both time and voltage, one must have a circuit that not only can be rapidly closed but also has very low resistance. When we set out to replicate this process, our initial attempts used various combinations of automotive storage batteries as a current source and a knife switch that could be rapidly closed. We were able to melt wires with this setup, but not deflagrate them, probably because the internal resistance of such batteries is relatively high (which meant, in turn, that the heating effects on the wire were too slow). We eventually had to use a bank of capacitors as a source of low-resistance current. Two heavy brass mounting brackets were therefore attached to a parallel bank of seven capacitors. A direct-current generator with a maximum output of 250 volts charged the capacitors, and a heavy utility knife switch was used to open and close the circuit. All the connections had to be made with heavy gauge copper wire to minimize the resistance of the circuit. After test trials with copper wire, pure gold wire (0.025 in. [0.64 cm] diameter, 99.99%) was mounted in the 2.5-cm gap between the brackets, and a glass microscope slide was placed beneath the wire. We calculated that when the circuit was closed, an amount of energy equivalent to that used by a 100-watt light bulb in 0.3 seconds would pass nearly instantly between the brackets containing the gold wire.

When the circuit was closed, the specimen wires exploded with a flash of light and a sharp cracking noise. Enough energy was generated by the deflagration to propel bits of unvaporized metal to distances of several feet. Most important, there was a small cloud of vaporized gold, manifested as a deposit on the glass slide. We repeated this procedure several times with gold wire, obtaining similar results each time. In each case, one example of which is shown in Fig. 7.4, the wire was vaporized only at certain points along the wire before the circuit was broken and current was no longer able to pass through the wire. This pattern was similar to that found on some of the first deflagrated slides produced by Faraday (note that Fig. 7.4 shows only one deflagrated spot, at greater magnification than the image of Faraday's slides in Fig. 7.2).

In Faraday's first deflagrations, as in ours, only portions of the wire were vaporized, probably because of the lower voltage potentials pro-

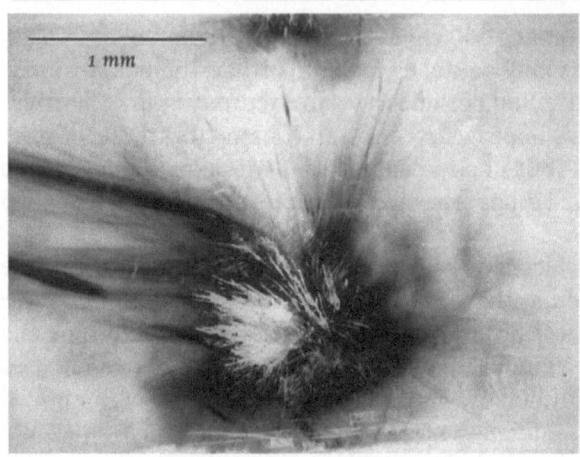

FIG. 7.4. Blue and red deposit of exploded gold (our replication).

duced by the Grove's batteries; one cell produces 1.94 volts, so a battery of 10 such cells could not have produced more than 20 volts. This accounts for the fact that Faraday later used a setup like that shown in Fig. 7.3, a battery of Leyden jars, charged by a frictional electricity machine. Such a battery could store thousands of volts of potential. Using such an apparatus, Faraday, like Wilkinson (1804), was able to vaporize wires along their entire length (Entry 14699, April 14, 1856); see the example slide on the right in Fig. 7.2. Also, the result was immediate and convincing: "Here we obtain the general tints obtained by other processes [i.e., the manipulations of chemically produced gold films]—supporting the conclusion that these other process[es] yield divided gold" (Entry 14699, April 14, 1856). The next day, a careful optical examination in daylight confirmed his belief: "Every part of the [deflagrated] gold film reflected yellow light .... The transmitted light was very various, from a rose on the outside by a green or green violet to ruby at the place of the wire and heat" (Entry 14708, April 15, 1856).

Color images of some of these slides, numbered by Faraday from 319 to 340, can be seen at the following Web site: http://personal.bgsu.edu/~tweney, along with the corresponding images of our own slides. We were, of course, pleased by the resemblance between his slides and ours, but the more important comparison is between those slides Faraday prepared by deflagration and those he prepared by chemical means. Both are metallic and shiny, yellow by reflected light and green or blue or ruby by transmitted light. Because Faraday was convinced the deflagrations are particulate, what evidence could he still invoke to argue that the chemical films are continuous? At this point, we can see that his quest for continuity is breaking apart. His "confusions" have passed into the realm of true anomalies!

DISCUSSION

Many cognitive accounts have emphasized search through a problem space as the fundamental feature of scientific thinking (Klahr & Simon, 1999). For example, Kulkarni and Simon (1988) successfully modeled the problem-solving activities of Hans Krebs using such an approach, basing their model on the historical analysis of Krebs's diaries provided by Holmes (1980), whereas Tweney and Hoffner (1987) examined the problem space of Faraday's 1831 diary account of his discovery of electromagnetic induction. Later work—for example, that by David Klahr and his students (e.g., Klahr, 2000)—suggested separating problem spaces into those specific to experimentation design and those centered on hypothesis search, and there have been proposals that have extended this idea beyond two spaces to as many as four (e.g., Schunn & Klahr, 1995).

Other analyses have concentrated on the heuristics used in scientific research. Thus, Tweney and Hoffner's (1987) analysis of Faraday's 1831 research demonstrated how Faraday's experimental strategy involved a two-step process. First, he used a narrow search to find evidence confirming newly developed ideas, without paying particular attention to potentially disconfirming evidence. Second, in later steps, the disconfirming evidence received more of Faraday's attention; at this stage, he explicitly attempted to disconfirm ideas generated and supported by his initial searches for confirmatory evidence. Earlier work in our laboratory had found similar patterns of heuristic use among some student participants and working scientists attempting to discover the rules that governed an "artificial universe" (Mynatt, Doherty, & Tweney, 1978; Tweney & Doherty, 1983). A similar "confirm early–disconfirm late" strategy was observed in studies of individuals attempting to discover the function of an obscure control key on a robot toy (Klahr, 2000) and by Dunbar (1995) in an *in vivo* study of laboratory molecular biologists.

Despite the success of the account of scientific thinking as problem space search, and of the investigation of the heuristics of scientific discovery, other aspects of scientific thinking seem resistant to such description (see, e.g., Kurz & Tweney, 1998). David Gooding (1990), for instance, explored in great detail the process of Faraday's 1821 work on the principles of electromagnetic rotations. For Gooding, the discovery of the electromagnetic rotations, and their consolidation within a simple theoretical scheme, was part of a dynamic "eye–hand–brain interaction." According to Gooding, Faraday had to construct the meaning underlying experimental results that would have otherwise appeared only chaotic. Similarly, Elizabeth Cavicchi (1997) replicated some of Faraday's experimental work on diamagnetism and found that Faraday's discovery process depended on attention to both anomalies and ambiguities, which then led to successively more refined exploration. In other words, Faraday's developing understandings were part of a critical pattern-finding stage of the discovery process. Cavicchi argued that Faraday's experimentation proceeded "not by progressively refining explanations, but by exposing previously unnoticed ambiguities in the phenomena, and uncertainties in interpretation. This exposing deepens the space of [his] confusions" (p. 876). For Cavicchi, these "confusions," like Gooding's "construals," are a crucial aspect of the pattern-finding involved in discovery; both Gooding and Cavicchi were attempting to describe the construction of a phenomenology, to sort out a mass of experience and sensation into more or less relevant domains, and to formulate what is strange or puzzling in the domain. Cavicchi was thus able to show that Faraday's "confusions" resembled those of a student exploring the relationships between bar magnets and iron needles. One implication of her account is that the cognitive analysis of anomalous results has so far been incomplete, because most such analyses focus on later stages of inquiry in which already-sophisticated expectations are violated by phenomena, a process that contrasts markedly with the wholly unexpected "confusions" that characterize the pattern-finding stages of inquiry.

The use of anomalous data has received much attention in the psychology of science.[7] For example, Trickett, Trafton, Schunn, and Harrison (2001) showed that astrophysicists presented with optical and radio data of ring galaxies paid attention to anomalies in the data and used such anomalies to search for analogies with other features of the data. Dunbar (1995) paid special attention to the use of analogies to explore anomalous findings, in the analysis of protocols gathered during his *in vivo* study of molecular biologists. He found that the relative "nearness" or "farness" of the analogies was related to the relative success of a given laboratory. Our account of Faraday's gold research is fully consistent with all of these other accounts. Thus, uses of analogy are present, imagery is extensive, and much of the record could be interpreted as search through a problem space (see, e.g., Dunbar, 2001; Gentner et al., 2001; Gorman, 1992; Langley & Jones, 1988; Nersessian, 1999; Tweney, 2001). Yet none of these processes taken singly can fully capture the way in which Faraday interacted with the materials and objects of his laboratory, and thus none of these, taken singly, can fully account for his creative discovery processes. His "confusions" must be part of a complete account.

An earlier examination of two of Faraday's papers—one on acoustic vibrations and one on optical illusions of motion—explored the development of a series of representations in Faraday's work and suggested

---

[7]See also the accounts of anomaly finding and anomaly resolution provided by Darden (1992). By contrast, Nersessian's (1999) account of the role of generic abstraction in anomaly resolution is closer to the level we are emphasizing here.

that his constructive perceptual processes imply a continuum of developing representative explicitness (Ippolito & Tweney, 1995; Tweney, 1992b). Beginning with what appeared to be little more than the perceptual rehearsal of remembered events, Faraday used these and his first experimental efforts to construct "inceptual" representations, that is, representations that abstracted away potentially irrelevant features, with an effort to "see" what the results would look like. Only toward the end could he be said to have developed a mental model of the phenomena. Faraday clearly appeared to be using an eye–hand–mind dynamic in constructing new spaces for both thought and action, much as he had done earlier in his discovery of electromagnetic rotations in 1821 (Gooding, 1990). Similarly, Nersessian (1999) argued that Maxwell used analogies and imagery in a process of generic abstraction, a process by which intangible and vague "hunches" became explicit mental models.

Accepting such a view of the nature of scientific discovery requires that one focus on the process by which meaning is *made* (Gooding, 1990). It is just here that the "situatedness" of cognition is most manifest, precisely because it is here that a representational system is under construction. In the case of experimental research in particular, as in the case of technological development, a full understanding of the process requires close attention to the cognitive and epistemic artifacts in the arena of interest. Thus, whether in the mind of one investigator or of a team of investigators, there is a need to understand what Nersessian, Newstetter, Kurz-Milcke, and Davies (2002) referred to as the "biography of the object." The artifacts of scientific cognition are themselves contingent historical objects with both agency and a developmental past and, just as Faraday's slides shape the representation of the properties of gold, so too do our replications of Faraday's slides shape the representation of his discoveries.

In seeking to understand Faraday's achievements at the remove of a century and a half, it should be clear from our presentation that his epistemic artifacts and the practices of his experimentation are required elements in a full account. The term *epistemic artifact* is deliberately chosen here; we mean to imply that the artifacts (slides, colloids, precipitates) were made by Faraday precisely because they can answer a question—or, for that matter, ask a question (Tweney, 2002). Like the related concept of a *cognitive* artifact, an artifact that must be regarded as conducting an externalized computation (Zhang & Norman, 1994), an *epistemic* artifact externalizes cognition, but it also serves as a source of new knowledge. In a real sense, although made by Faraday, the specimens become agents in his inquiry (see also Rheinberger, 1997).

Our replications of Faraday's work contribute to understanding the processes by which the vague becomes concrete—by which, to use

Gooding's (1990) terms, *construals* become *concepts*, which become, eventually, public demonstrations. "Phenomena are made visible and public through the *invention* [italics added] of observational practices." (Gooding, 1990, p. 214). The phenomena by themselves are without meaning; the experimenter sets the stage, first to enable his or her understanding and then so that others may participate. In all of Faraday's research one can see a determination to produce phenomena of such clarity that the explanations of the phenomena would be transparent to his audiences: "Seeing was believing" in a deep sense for him (see also Fisher, 2001) and was the ultimate criterion for the authority of a claim.

What Faraday "saw" in looking at his slides, precipitates, or colloids is private and varies in permanence and in the ease with which it can become public. The permanence of the slides, and the relative permanence of some of the colloids was, in fact, the initiating cause of our studies, whereas the impermanence of his precipitates in effect forced us to replicate them in order to see what he saw. Of course, "seeing" in the context of scientific discovery is not simple; without the context created by all three kinds of preparations—slides, colloids, and precipitates—and the experience of our own experimental practices, we could not have "seen" the artifacts in the proper fashion (even granting, of course, that we can never see them exactly as Faraday saw them). It thus makes sense to speak of Faraday as *negotiating* what he sees with the artifacts and of our seeing as being also a negotiation—with Faraday's text, his artifacts, and our own replicated specimens and activities. The term *negotiation* implies more than one active agent, of course, and that makes the metaphor even more apt. Faraday negotiated with "nature," in a sense, but his eventual audience is part of the negotiation as well, just as our potential audience is part of the negotiation we undertook in this project. The distinction between the private and the public is a blurred distinction indeed![8]

Just as Holmes (1998) suggested that the public side of science and the private side of science, as revealed in laboratory notebooks, were complementary, so also do we believe that the material side of science complements the private (mental) side. It is not right, however, to specify these three domains—the public, the private, and the material—as if they were separate and independent. Each depends on the others, and the classification is truly a nominal one. In practice, science is a system of activity, and the unit of

---

[8]Gooding (1990) noted an apt comparison between such an account of scientific discovery and a series of shadow box experiments conducted by Howard Gruber (1990). By placing participants into an epistemic context in which two individuals had to share descriptions of the same object producing two very different shadows, Gruber was able to study the negotiation process directly in a social context. There is a striking parallel between Gruber's account and our account of Faraday making sense out of his colloids as particulate in nature.

study ought to be not the individual scientist alone, not the social network alone, and not the laboratory (instruments and objects) alone but the individual scientist *in* the laboratory and *in* the social context.

ACKNOWLEDGMENTS

This research was partially supported by National Science Foundation Award 0100112. We thank Andy Wickiser for constructing the deflagration apparatus and Harry Brown for assistance with the chemistry and the photography. Yanlong Sun, Robert Gibby, and Neil Berg provided valuable help and discussion. Elke Kurz-Milcke and David Gooding provided careful readings of a draft of this chapter. Finally, we are deeply grateful to the staff at the Royal Institution of Great Britain for making this project possible, and we acknowledge with special thanks the efforts of Frank A. J. L. James, Reader in the History and Philosophy of Science at the Royal Institution, for his support and encouragement.

REFERENCES

Andersen, H. (2002). The development of scientific taxonomies. In L. Magnani & N. J. Nersessian (Eds.), *Model-based reasoning: Science, technology, values* (pp. 95–112). New York: Kluwer Academic/Plenum.

Bostock, J. (1818). *An account of the history and present state of galvanism.* London: Baldwin, Cradock, & Joy.

Bottone, S. R. (1902). *Galvanic batteries: Their theory, construction and use.* London: Whittaker.

Cavicchi, E. (1997). Experimenting with magnetism: Ways of learning of Joann and Faraday. *American Journal of Physics, 65,* 867–882.

Chen, X. (2000). *Instrumental traditions and theories of light: The uses of instruments in the optical revolution.* Dordrecht, The Netherlands: Kluwer Academic.

Darden, L. (1992). Strategies for anomaly resolution. In R. N. Giere (Ed.), *Minnesota Studies in the Philosophy of Science: Vol. 15. Cognitive models of science* (pp. 251–273). Minneapolis: University of Minnesota Press.

Dunbar, K. (1995). How scientists really reason: Scientific reasoning in real-world laboratories. In R. J. Sternberg & J. Davidson (Eds.), *Mechanisms of insight* (pp. 365–396). Cambridge, MA: MIT Press.

Dunbar, K. (2001). What scientific thinking reveals about the nature of cognition. In K. Crowley, C. D. Schunn, & T. Okada (Eds.), *Designing for science: Implications from everyday, classroom, and professional settings* (pp. 115–140). Mahwah, NJ: Lawrence Erlbaum Associates.

Faraday, M. (1857, February 5). Experimental relations of gold (and other metals) to light. *Philosophical Transactions,* 145–181.

Fisher, H. J. (2001). *Faraday's experimental researches in electricity: Guide to a first reading.* Santa Fe, NM: Green Lion Press.

Gentner, D., Brem, S., Ferguson, R. W., Markman, A. B., Levidow, B. B., Wolff, P., & Forbus, K. D. (1997). Analogical reasoning and conceptual change: A case study of Johannes Kepler. *Journal of the Learning Sciences, 6,* 3–40.

Golinski, J. (1992). *Science as public culture: Chemistry and the Enlightenment in Britain, 1760–1820.* Cambridge, England: Cambridge University Press.

Gooding, D. C. (1981). Final steps to the field theory: Faraday's study of magnetic phenomena. *Historical Studies in the Physical Sciences, 11,* 231–275.

Gooding, D. C. (1990). *Experiment and the making of meaning: Human agency in scientific observation and experiment.* Dordrecht, The Netherlands: Kluwer Academic.

Gooding, D. C. (1992). The procedural turn; or, why do thought experiments work? In R. N. Giere (Ed.), *Minnesota Studies in the Philosophy of Science: Vol. 15. Cognitive models of science* (pp. 45–76). Minneapolis: University of Minnesota Press.

Gorman, M. E. (1992). *Simulating science: Heuristics, mental models, and technoscientific thinking.* Bloomington: Indiana University Press.

Grove, W. (1839). On a small Voltaic battery of great energy. *Philosophical Magazine, 15,* 287–293.

Gruber, H. E. (1990). The cooperative synthesis of disparate points of view. In I. Rock (Ed.), *The legacy of Solomon Asch: Essays in cognition and social psychology* (pp. 143–157). Hillsdale, NJ: Lawrence Erlbaum Associates.

Heering, P. (1994). The replication of the torsion balance experiment: The inverse square law and its refutation by early 19th-century German physicists. In C. Blondel & M. Dörries (Eds.), *Restaging Coulomb: Usages, controverses et réplications autour de la balance de torsion* (pp. 47–66). Florence, Italy: Leo S. Olschki.

Holmes, F. L. (1980). Hans Krebs and the discovery of the ornithine cycle. *Federation Proceedings, 39,* 216–225.

Holmes, F. L. (1998). *Antoine Lavoisier—The next crucial year. Or, the sources of his quantitative method in chemistry.* Princeton, NJ: Princeton University Press.

Hutchins, E. (1995). How a cockpit remembers its speeds. *Cognitive Science, 19,* 265–288.

Ippolito, M. F., & Tweney, R. D. (1995). The inception of insight. In R. J. Sternberg & J. E. Davidson (Eds.), *The nature of insight* (pp. 433–462). Cambridge, MA: MIT Press.

James, F. A. J. L. (1985). "The optical mode of investigation": Light and matter in Faraday's natural philosophy. In D. Gooding & F. A. J. L. James (Eds.), *Faraday rediscovered: Essays on the life and work of Michael Faraday, 1791–1867* (pp. 137–162). Basingstoke, England: Macmillan.

Klahr, D. (2000). *Exploring science: The cognition and development of discovery processes.* Cambridge, MA: MIT Press.

Klahr, D., & Simon, H. A. (1999). Studies of scientific discovery: Complementary approaches and convergent findings. *Psychological Bulletin, 125,* 524–543.

Kulkarni, D., & Simon, H. A. (1988). The processes of scientific discovery: The strategy of experimentation. *Cognitive Science, 12,* 139–176.

Kurz, E. M., & Tweney, R. D. (1998). The practice of mathematics and science: From calculus to the clothesline problem. In M. Oaksford & N. Chater (Eds.), *Rational models of cognition* (pp. 415–438). Oxford, England: Oxford University Press.

Langley, P., & Jones, R. (1988). A computational model of scientific insight. In R. Sternberg (Ed.), *The nature of creativity* (pp. 177–201). Cambridge, England: Cambridge University Press.

Martin, T. (Ed.). (1936). *Faraday's diary: Being the various philosophical notes of experimental investigation made by Michael Faraday during the years 1820–1862* (Vol. 7). London: G. Bell.

Mynatt, C. R., Doherty, M. E., & Tweney, R. D. (1978). Consequences of confirmation and disconfirmation in a simulated research environment. *Quarterly Journal of Experimental Psychology, 30,* 395–406.

Nersessian, N. J. (1985). Faraday's field concept. In D. Gooding & F. A. J. L. James (Eds.), *Faraday rediscovered: Essays on the life and work of Michael Faraday, 1791–1867* (pp. 175–188). Basingstoke, England: Macmillan.

Nersessian, N. J. (1999). Model based reasoning in conceptual change. In L. Magnani, N. J. Nersessian, & P. Thagard (Eds.), *Model-based reasoning in scientific discovery* (pp. 5–22). New York: Kluwer/Plenum.

Nersessian, N. J., Newstetter, W., Kurz-Milcke, E., & Davies, J. (2002). A mixed-method approach to studying distributed cognition in evolving environments. In *Proceedings of the International Conference on Learning Sciences* (pp. 307–314). Mahwah, NJ: Lawrence Erlbaum Associates.

Parkes, S. (1822). *The chemical catechism, with tables, notes, illustrations, and experiments* (10th ed.). London: Baldwin, Cradock, & Joy.

Puddephatt, R. J. (1978). *The chemistry of gold.* Amsterdam: Elsevier Scientific.

Rheinberger, H.-J. (1997). *Toward a history of epistemic things: Synthesizing proteins in the test tube.* Stanford, CA: Stanford University Press.

Rudwick, M. J. S. (1985). *The great Devonian controversy: The shaping of scientific knowledge among gentlemanly specialists.* Chicago: University of Chicago Press.

Schunn, C. D., & Klahr, D. (1995). A 4-space model of scientific discovery. In J. D. Moore & J. F. Lehman (Eds.), *Proceedings of the 17th Annual Conference of the Cognitive Science Society.* Mahwah, NJ: Lawrence Erlbaum Associates.

Simon, H. A. (1996). *The sciences of the artificial* (3rd ed.). Cambridge, MA: MIT Press. (Original work published 1967)

Steinle, F. (1996). Work, finish, publish? The formation of the second series of Faraday's "Experimental Researches in Electricity," *Physis, 33,* 141–220.

Trickett, S. B., Trafton, J. G., Schunn, C. D., & Harrison, A. (2001). That's odd! How scientists respond to anomalous data. In *Proceedings of the 23rd annual conference of the Cognitive Science Society* (pp. 1054–1059). Mahwah, NJ: Lawrence Erlbaum Associates.

Tweney, R. D. (1991). Faraday's notebooks: The active organization of creative science. *Physics Education, 26,* 301–306.

Tweney, R. D. (1992a). Inventing the field: Michael Faraday and the creative "engineering" of electromagnetic field theory. In R. J. Weber & D. N. Perkins (Eds.), *Inventive minds: Creativity in technology* (pp. 31–47). Oxford, England: Oxford University Press.

Tweney, R. D. (1992b). Stopping time: Faraday and the scientific creation of perceptual order. *Physis, 29,* 149–164.

Tweney, R. D. (2001). Scientific thinking: A cognitive–historical approach. In K. Crowley, C. D. Schunn, & T. Okada (Eds.), *Designing for science: Implications from everyday, classroom, and professional settings* (pp. 141–173). Mahwah, NJ: Lawrence Erlbaum Associates.

Tweney, R. D. (2002). Epistemic artifacts: Michael Faraday's search for the optical effects of gold. In L. Magnani & N. J. Nersessian (Eds.), *Model-based reasoning: Science, technology, values* (pp. 287–304). New York: Kluwer Academic/Plenum.

Tweney, R. D. (in preparation). *Replicating scientific discovery: Faraday, phosphorus, and gold.*

Tweney, R. D., & Doherty, M. E. (1983). Rationality and the psychology of inference. *Synthese, 57,* 139–162.

Tweney, R. D., & Gooding, D. (Eds.). (1991). *Faraday's 1822 "Chemical Notes, Hints, Suggestions, and Objects of Pursuit."* London: Science Museum & Peter Peregrinus.

Tweney, R. D., & Hoffner, C. E. (1987). Understanding the microstructure of science: An example. In *Program of the ninth annual Conference of the Cognitive Science Society* (pp. 677–681). Hillsdale, NJ: Lawrence Erlbaum Associates.

Tweney, R. D., Mears, R. P., Gibby, R. E., Spitzmüller, C., & Sun, Y. (2002). Precipitate replications: The cognitive analysis of Michael Faraday's exploration of gold precipitates and colloids. In C. Schunn & W. Gray (Eds.), *Proceedings of the 24th annual conference of the Cognitive Science Society* (pp. 890–895). Mahwah, NJ: Lawrence Erlbaum Associates.

Wilkinson, C. (1804). *Elements of galvanism in theory and practice* (Vol. 2). London: John Murray.

Williams, L. P. (1965). *Michael Faraday: A biography*. New York: Basic Books.

Williams, L. P. (1966). *The origins of field theory*. New York: Random House.

Zhang, J., & Norman, D. A. (1994). Representations in distributed cognitive tasks. *Cognitive Science, 18*, 87–122.

Zsigmondy, R. (1909). *Colloids and the Ultramicroscope* (Jerome Alexander, Trans.). New York: John Wiley & Sons. (Original work published 1905)

8

How to Be
a Successful Scientist

Paul Thagard
University of Waterloo

Studies in the history, philosophy, sociology, and psychology of science and technology have gathered much information about important cases of scientific development. These cases usually concern the most successful scientists and inventors, such as Darwin, Einstein, and Edison. But case studies rarely address the question of what made these investigators more accomplished than the legions of scientific laborers whose names have been forgotten.

This chapter is an attempt to identify many of the psychological and other factors that make some scientists highly successful. I explore two sources of information about routes to scientific achievement. The first derives from a survey that Jeff Shrager conducted at the "Cognitive Studies of Science and Technology Workshop" at the University of Virginia in March 2001. He asked the participants to list "seven habits of highly creative people," and after the workshop he and I compiled a list of habits recommended by the distinguished group of historians, philosophers, and psychologists who had attended. My second source of information about the factors contributing to scientific success is advice given by three distinguished biologists who each won a Nobel prize: Santiago Ramón y Cajal, Peter Medawar, and James Watson. These biologists have provided advice that usefully supplements the suggestions from the workshop participants.

·HABITS OF HIGHLY CREATIVE PEOPLE

When Jeff Shrager asked the workshop participants to submit suggestions for a list of 7 habits of highly creative people, I was skeptical that they would come up with anything less trite than *"work hard"* and *"be smart,"* but the suggestions turned out to be quite interesting, and Jeff and I compiled and organized them into the list shown in Table 8.1. It is not surprising that we ended up with not 7 habits but rather with 27, organized into six classes.

The first class of habits concerns ways to make new connections, recognizing the fact that creativity in science and technology usually involves putting together ideas in novel combinations (Ward, Smith, & Vaid, 1997). Many successful scientists do not restrict their reading to the particular area of research on which they are currently focused, but instead read widely, including work outside their own field. This enables them to grasp analogies between current problems they face and established ones that may suggest new solutions (Dunbar, 2001; Holyoak & Thagard, 1995). Visual representations may facilitate analogical and other kinds of inference (Giere, 1999; Nersessian, 1992). Working on multiple projects with multiple methods may make possible new approaches to those projects. As Herbert Simon often recommended in conversation, researchers should not follow the crowd and work on what everyone else is doing, because it is difficult to do anything novel in such situations. Because scientific explanations and technological breakthroughs often involve the discovery and manipulation of mechanisms, seeking novel mechanisms is often a good strategy (Bechtel & Richardson, 1993; Machamer, Darden, & Craver, 2000).

The second class of habits recognizes the fact that work in science and technology often does not proceed as expected. When anomalous experimental results arise, it is important for investigators to take them seriously and not brush them aside. There is then the possibility of learning from failed expectations—not by giving up but by recovering from the local failure and moving on to research suggested by the anomalous findings. Much can be learned from both successful and unsuccessful experiments (Gooding, 1990; Gooding, Pinch, & Schaffer, 1989).

Given the frequent difficulties and setbacks in research in science and technology, it is important that investigators have a third class of habits involving persistence. They need to focus on key problems rather than being distracted by peripheral issues, and they need to pursue their research systematically, keeping detailed records about successes and failures. The injunction to "confirm early, disconfirm late" goes against the methodological advice of Karl Popper (1959) that scientists should set out to refute their own ideas, but it allows a research project to develop without being destroyed prematurely by apparent disconfirmations that may arise from difficulties in getting good experimental research underway.

TABLE 8.1
Habits of Highly Creative People

1. Make new connections.
   Broaden yourself to more than one field.
   Read widely.
   Use analogies to link things together.
   Work on different projects at the same time.
   Use visual as well as verbal representations.
   Don't work on what everyone else is doing.
   Use multiple methods. Seek novel mechanisms.

2. Expect the unexpected.
   Take anomalies seriously.
   Learn from failures.
   Recover from failures.

3. Be persistent.
   Focus on key problems.
   Be systematic and keep records.
   Confirm early, disconfirm late.

4. Get excited.
   Pursue projects that are fun.
   Play with ideas and things.
   Ask interesting questions.
   Take risks.

5. Be sociable.
   Find smart collaborators.
   Organize good teams.
   Study how others are successful.
   Listen to people with experience.
   Foster different cognitive styles.
   Communicate your work to others.

6. Use the world.
   Find rich environments.
   Build instruments.
   Test ideas.

The first three classes all involve cognitive habits, that is, ones tied to basic thinking processes of problem solving and learning. The fourth class suggests that successful scientists are ones who also possess a set of emotional habits that get them intensely involved with their research projects (Feist &

Gorman, 1998). It is rarely possible for scientists to do a cost–benefit analysis of what projects to pursue, but following their noses to work on projects that are fun and exciting can keep them motivated and focused. Scientific research is not just a matter of doing experiments and forming hypotheses but in its early stages requires formulating a project that will answer a question that is interesting for theoretical or practical reasons. Interest, excitement, and the avoidance of boredom provide motivation to work hard and do work that is creative rather than routine. Playing with ideas and instruments can be inherently enjoyable. Taking risks to do nonstandard research can provoke fear of failure, but this emotion must be managed if a scientist is to move in highly novel directions. For further discussion of the role of emotion in scientific thinking, see Thagard (2002a, 2002b) and Wolpert and Richards (1997).

Cognitive and emotional habits both concern individual psychology, but no scientist is an island. The fifth class of habits is social, concerning ways in which working with others can foster a scientific career. Most scientific research today is collaborative, so having smart collaborators organized into effective teams is crucial (Galison, 1997; Thagard, 1999). Teams should not be homogeneous but should have members who combine a variety of areas of knowledge and methodologies (Dunbar, 1995). Scientists can also benefit from observing how other researchers have managed to be successful and by listening to the advice of mentors about how to conduct their research. Finally, there is little point to doing research if one does not devote time to communicating it effectively to others by well-written articles and interesting presentations.

The sixth and final class of habits acknowledges that science is not just a psychological and sociological process but also involves interactions with the world (Thagard, 1999). Scientists can benefit from finding rich environments to study and building instruments to detect features of those environments. Testing ideas is not just a logical matter of working out the consequences of hypotheses but involves interacting with the world to determine whether it has the properties that the hypotheses ascribe to it (Hacking, 1983).

It would take an enormous amount of empirical research to establish that the habits I have just described really are ones that lead to scientific success (see the subsequent discussion of Feist & Gorman, 1998). One would have to produce a substantial database of scientists, ordinary as well as illustrious, with records of the extent to which they exemplify the different habits and degrees of professional success. Here I can attempt only a much more modest kind of validation of the list of habits of successful scientists, by comparing it with the advice given by three distinguished biologists.

RAMÓN Y CAJAL

Santiago Ramón y Cajal was a Spanish biologist who won a Nobel prize in medicine and physiology in 1906 for important discoveries about nerve cells. While still an active scientist, he wrote in 1897 *Reglas y Consejos Sobre Investigacion Cientifica,* which was translated into English with the title *Advice for a Young Investigator* (Ramón y Cajal, 1897/1999). The book is rich with many kinds of recommendations for pursuing a career in biology.

Ramón y Cajal's book (1897/1999) begins by rejecting advice from philosophers such as Descartes and Bacon, insisting that "the most brilliant discoveries have not relied on a formal knowledge of logic" (Ramón y Cajal, 1999, p. 5). Rather, they arose from an "acute inner logic that generates ideas." In chapter 2 he warns beginning scientists against traps that impede science, including excessive admiration for the work of great minds and a conviction that the most important problems are already solved. He also recommended cultivating science for its own sake, without considering its application (p. 9). Ramón y Cajal doubted that superior talent is required for good scientific work. Even individuals with mediocre talent can produce notable work if they concentrate on information pertinent to an important question.

Concentration is one of the intellectual qualities that Ramón y Cajal (1897/1999) described in chapter 3 as being indispensable for the researcher: "All great work is the fruit of patience and perseverance, combined with tenacious concentration on a subject over a period of months or even years" (p. 38). Other intellectual qualities include independent judgment, taste for scientific originality, passion for reputation, and patriotism. Concentration and taste for originality are partly cognitive qualities, but they are also emotional, because they involve desire and motivations. Ramón y Cajal strongly emphasized the emotional side of scientific thought: "Two emotions must be unusually strong in the great scientific scholar: a devotion to truth and a passion for reputation" (p. 40). The passion for reputation is important, because eagerness for approval and applause provides a strong motivational force; science requires people who can flatter themselves that they have trodden on completely virgin territory. Similarly, Ramón y Cajal sees patriotism as a useful motivating force, as researchers strive to make discoveries in part for the glory of their countries. Most motivating qualities, however, are more local: "Our novice runs the risk of failure without additional traits: a strong inclination toward originality, a taste for research, and a desire to experience the incomparable gratification associated with the act of discovery itself" (p. 48). Discovery is an "indescribable pleasure—which pales the rest of life's joys" (p. 50).

Chapter 4 is of less general interest, as it is directed primarily at newcomers to research in biology. Ramón y Cajal (1897/1999) pointed to the value of having a general education, with philosophy particularly useful because it "offers good preparation and excellent mental gymnastics for the laboratory worker" (p. 54). However, specialization is also necessary if a researcher is to master a particular scientific area. Ramón y Cajal also provided advice about the importance of learning foreign languages, reading monographs with special attention to research methods and unsolved problems, and mastering experimental techniques. He stressed the "absolute necessity of seeking inspiration in nature" (p. 62) and urged patient observations designed to produce original data. Researcher should choose problems whose methodology they understand and like.

Chapter 5 provides negative advice—qualities to avoid because they militate against success. The chapter is evocatively entitled "Diseases of the Will," and it divides unsuccessful scientists into a number of types: contemplators, bibliophiles and polyglots, megalomaniacs, instrument addicts, misfits, and theorists. For example, theorists are wonderfully cultivated minds who have an aversion to the laboratory, so that they can never contribute original data. According to Ramón y Cajal (1897/1999), "enthusiasm and perseverance work miracles" (p. 94).

Chapter 6 describes social factors beneficial to scientific work, including material support such as good laboratory facilities. Ramón y Cajal (1897/ 1999) also made recommendations about marriage that presuppose that researchers are male: "We advise the man inclined toward science to seek in the one whom his heart has chosen a compatible psychological profile rather than beauty and wealth" (p. 103).

Chapter 7 returns to more cognitive advice, concerning the operations in all scientific research: observation and experimentation, working hypotheses, and proof. Observation is not to be done casually: "It is not sufficient to examine; it is also necessary to observe and reflect: we should infuse the things we observe with the intensity of our emotions and with a deep sense of affinity" (Ramón y Cajal, 1897/1999, p. 112). Experiments should be carried out repeatedly using the best instruments. Once data have been gathered, it is natural to formulate hypotheses that try to explain them. Although hypotheses are indispensable, Ramón y Cajal (1897/1999) warned against "excessive attachment to our own ideas" (p. 122). Hypotheses must be tested by seeking data contrary to them as well as data that support them. Researchers must be willing to give up their own ideas, as excessive self-esteem and pride can prevent improvements.

Chapters 8 and 9 return to more practical advice concerning how to write scientific papers and how to combine research with teaching.

PETER MEDAWAR

To my knowledge, the only other book-length advice for scientists written by a distinguished scientist is Peter Medawar's (1979) *Advice to a Young Scientist*. The title is similar to the English translation of Ramón y Cajal's (1897/1999) much earlier book, but the translators must have been mimicking Medawar rather than Medawar mimicking Ramón y Cajal, whose original title lacked the word *young*. Medawar was a British biologist who shared the Nobel prize in 1960 for work conducted in 1949, when he was still in his 30s. There is no indication in Medawar's book that he had read Ramón y Cajal's book.

Medawar (1979) begins by asking the following question: How can I tell if I am cut out to be a scientific research worker? He says that most able scientists have a trait he called "exploratory impulsion" (p. 7), which is a strong desire to comprehend. They also need intellectual skills, including general intelligence and particular abilities required for specific sciences, such as manipulative skills necessary for many experimental sciences.

Medawar's (1979) next chapter provides advice concerning: On what shall I do research? His main recommendation is to study important problems, ones whose answers really matter to science generally or to humankind. Young scientists must beware of following fashion, for example, by picking up some popular gimmick rather than pursuing important ideas.

Chapter 4 concerns how scientists can equip themselves to be better scientists. Medawar (1979) recognizes that the beginner must read the literature, but he urges reading "intently and choosily and not too much" (p. 17). The danger is that novices will spend so much time mastering the literature that they never get any research done. Experimental researchers need to get results, even if they are not original at first. The art of research is the "art of the soluble" in that a researcher must find a way of getting at a problem, such as a new measuring technique, that provides a new way of solving the problem.

Whereas Ramón y Cajal (1897/1999) assumed that scientists were male, Medawar (1979) included a chapter on sexism and racism in science. Medawar sees no difference in intelligence, skill, or thinking style between men and women and has little specific advice for female scientists. Similarly, he sees no inborn constitutional differences in scientific prowess or capability between different races or nationalities.

Medawar (1979) stated that nearly all his scientific work was done in collaboration, and he emphasized the importance of *synergism*, when a research team comes up with a joint effort that is greater than the sum of the several contributions to it. Young scientists who have the generosity of spirit to collaborate can have much more enjoyable and successful careers than loners. Scientists should be willing to recognize and admit their mistakes: "I cannot give any scientist of any age better advice than this: the intensity of the conviction that a hypothesis is true has no bearing on whether it is true or not" (p. 39). Medawar suggested that creativity is helped by a quiet and untroubled life. Scientists concerned with priority may be inclined toward secretiveness, but Medawar advised telling close colleagues everything one knows. Ambition in young scientists is useful as a motive force, but excess of ambition can be a disfigurement.

Like Ramón y Cajal (1897/1999), Medawar (1979) encouraged scientists to make their results known through publications and presentations. He recommended a policy generally followed in the sciences (but not, unfortunately, in the humanities) that presentations should be spoken from notes rather than read from a script. Papers should be written concisely and appropriately for the intended audience.

Medawar (1979) also provided advice about the conducting and interpretation of experiments. He advocated "Galilean" experiments, ones that do not simply make observations but rather discriminate among possibilities in a way that tests hypotheses. He warned against falling in love with a pet hypothesis. Young scientists should aim to make use of experiments and theories to make the world more understandable, not just to compile information. A scientist is a "seeker after truth" (p. 87), devising hypotheses that can be tested by practicable experiments. Before scientists set out to convince others of their observations or opinions, they must first convince themselves, which should not be too easily achieved. Medawar prefers a Popperian philosophy of science based on critical evaluation to a Kuhnian philosophy based on paradigms.

JAMES WATSON

My third eminent biologist is James Watson, who shared a Nobel prize in 1962 for his role in the discovery of the structure of DNA. In 1993 he gave an after-dinner talk at a meeting to celebrate the 40th anniversary of the discovery of the double helix and later published the talk in a collection of occasional pieces (Watson, 2000). The published version, titled "Succeeding in Science: Some Rules of Thumb," is only 4 pages long, much more concise than the books by Ramón y Cajal (1897/1999) and Medawar (1979). Watson wrote that to succeed in science one needs a lot more than luck and intelligence, and he offered four rules for success.

The first rule is to learn from the winners, avoiding dumb people. To win at something really difficult, you should always turn to people who are brighter than you are. The second rule is to take risks, being prepared to get into deep trouble. Big success requires taking on a very big goal that you are not prepared to pursue and ignoring people, including your mentors, who tell you that you are not ready for it. Watson's (2000) third rule, however, is to have someone as a fallback when you get into trouble. He described how important it was to his career to have John Kendrew and Salvador Luria behind him at critical moments (Watson, 2000).

Watson's (2000) final rule is "Never do anything that bores you" (p. 125). It is much easier to do well things that you like. Watson (2000) also remarked on the importance of having people around you who care about you and to whom you can go for intellectual help. It is also valuable to have people with whom you can expose your ideas to informed criticism; Watson (2000) suggested that his main competitors in the search for the structure of DNA, Rosalind Franklin and Linus Pauling, both suffered from a lack of people who could usefully disagree with them. People should not go into science as a way of avoiding dealing with other people, because success in science requires spending time with other scientists, both colleagues and competitors. Watson's success was partly the result of knowing everyone he needed to know.

DISCUSSION

To what extent does the advice offered by these three distinguished biologists coincide with the habits of creative scientists summarized in Table 8.1? There is clearly some overlap—for example, with Watson's (2000) suggestion to take risks and Medawar's (1979) suggestion to have good collaborators. However, the three biologists also made many recommendations that were not reported by the participants in the 2001 "Cognitive Studies of Science and Technology Workshop." I have listed the additional recommendations in Table 8.2, which should be read as a supplement rather than as a replacement for Table 8.1.

The three biologists do not add a lot to the cognitive advice in Table 8.1, although there are valuable nuggets, such as Medawar's (1979) advice about new techniques making problems soluble, Ramón y Cajal's (1897/1999) and Medawar's concerns about giving up one's own ideas when necessary, and Ramón y Cajal's recommendation to concentrate tenaciously. The emotional additions are more interesting, particularly Ramón y Cajal's and Medawar's discussion of the kinds of passion that foster scientific success, such as strong desires for truth, reputation, discovery, and comprehension. The correlate to Table 8.1's advice to have fun is Watson's (2000) advice to avoid boredom, which was a major impetus behind his own work on the double helix (Thagard, 2002b; Watson, 1969).

The three biologists also have a wealth of social and environmental advice that goes well beyond that found in Table 8.1. Ramón y Cajal (1897/1999), Medawar (1979), and Watson (2000) all have useful social recommendations, summarized in Table 8.2, ranging from marrying appropriately to communicating well with colleagues and the general scientific community. Ramón y Cajal and Medawar were more emphatic than philosophers, psychologists, and historians of science usually are about the importance of using experiments and instruments to interact effectively with the world. Whereas the workshop participants did an excellent job of identifying cognitive factors in scientific success, the three biologists who have provided advice seem stronger on the relevant emotional, social, and environmental factors.

TABLE 8.2
More Habits of Successful Scientists

1. Make new connections.

   Find new ways of making problems soluble, for example, by new techniques (Medawar).

2. Expect the unexpected.

   Avoid excessive attachment to your own ideas (Ramón y Cajal).

   Be willing to recognize and admit mistakes (Medawar).

3. Be persistent.

   Concentrate tenaciously on a subject (Ramón y Cajal).

4. Get excited.

   Have a devotion for truth and a passion for reputation (Ramón y Cajal).

   Have an inclination toward originality and a taste for research (Ramón y Cajal).

   Have a desire for the gratification of discovery (Ramón y Cajal).

   Have a strong desire to comprehend (Medawar).

   Never do anything that bores you (Watson).

5. Be sociable.

   Marry for psychological compatibility (Ramón y Cajal).

   Tell close colleagues everything you know (Medawar).

   Communicate research results effectively (Ramón y Cajal, Medawar).

   Learn from winners. (Watson).

   Have people to fall back on when you get into trouble (Watson).

6. Use the world.

   Seek inspiration in nature (Ramón y Cajal).

   Have good laboratory facilities, and use them (Ramón y Cajal).

   Observe and reflect intensely (Ramón y Cajal).

   Perform experiments that rigorously test hypotheses (Medawar).

7. Other.

   Avoid excessive admiration for the work of great minds (Ramón y Cajal).

   Cultivate science for its own sake (Ramón y Cajal, Medawar).

   Study important problems (Medawar).

   Don't read too much (Medawar).

   Have a quiet and untroubled life (Medawar).

Note. Medawar = Medawar (1979); Ramón y Cajal = Ramón y Cajal (1897/1999); Watson = Watson (2000).

Table 8.2 also lists a set of other factors that do not seem to fit into any of the six classes in Table 8.1. Both Ramón y Cajal (1897/1999) and Medawar (1979) recommended studying science for its own sake without worrying too much about practical applications. Ramón y Cajal's counsel to avoid being too impressed with great minds fits with Watson's (2000) injunction to take risks by deviating from established opinion. Medawar's suggestion not to read too much seems to contradict the suggestion in Table 8.1 to read widely. Studying important problems and avoiding life disruptions seem like good general advice.

Another possible source of information about what makes scientists successful comes from psychological studies of personality. Feist and Gorman (1998) reviewed a large literature that compares personality characteristics of scientists to nonscientists. Their major conclusions are:

- Scientists are more conscientious.
- Scientists are more dominant, achievement oriented, and driven.
- Scientists are independent, introverted, and less sociable.
- Scientists are emotionally stable and impulse controlled.

They also reviewed literature that compares eminent and creative scientists with less eminent and creative ones, concluding that:

- Creative scientists are more dominant, arrogant, self-confident, or hostile.
- Creative scientists are more autonomous, independent, or introverted.
- Creative scientists are more driven, ambitious, or achievement oriented.
- Creative scientists are more open and flexible in thought and character.

From this literature, one could conclude that to become a successful scientist it helps to be dominant, independent, driven, and flexible.

Even without the personality findings, between Tables 8.1 and 8.2 we now have assembled close to 50 pieces of advice for scientific success. Surely some of these recommendations are much more important than others for fostering the creative breakthroughs that contribute most to scientific success, but we have no way of knowing which ones are most influential. Perhaps a large-scale psychological and/or historical survey might be able to provide some ideas about which factors are most important (cf. Feist, 1993).

There is also the possibility of providing practical advice about how to conduct a scientific career at a much lower level—for example, how to deal with career choices, work–family conflicts, and becoming an aging scientist (see, e.g., Sindermann, 1985; Zanna & Darley, 1987). In this chapter I have not attempted to provide a definitive list of what traits and activities it takes to become a creatively successful scientist, but I hope I have provided both a framework and a set of factors for understanding scientific success.

The broad range of the factors for scientific success discussed in this chapter demonstrates the great diversity of what needs to be taken into account in explanations of the growth of scientific knowledge. Science studies need to go beyond the traditional concerns of particular disciplines, such as philosophical attention to modes of reasoning, psychological attention to cognitive processes, and sociological attention to social interactions. All of these concerns are legitimate, but they need to be complemented by understanding how emotions, personality, and intelligent interactions with the world also contribute to the development of science.

ACKNOWLEDGMENTS

I am grateful to Jeff Shrager for permission to include here the habits of highly creative people, and to various workshop members who made contributions: Michael Gorman, Bob Hanamann, Vicky Dischler, Michael Hertz, David Gooding, David Klahr, Jim Davies, and an anonymous participant. I thank also David Gooding for extensive comments on a draft of this chapter. The Natural Sciences and Engineering Research Council of Canada provided financial support.

REFERENCES

Bechtel, W., & Richardson, R. C. (1993). *Discovering complexity*. Princeton, NJ: Princeton University Press.

Dunbar, K. (1995). How scientists really reason: Scientific reasoning in real-world laboratories. In R. J. Sternberg & J. Davidson (Eds.), *Mechanisms of insight* (pp. 365–395). Cambridge, MA: MIT Press.

Dunbar, K. (2001). The analogical paradox: Why analogy is so easy in naturalistic settings, yet so difficult in the laboratory. In D. Gentner, K. Holyoak, & B. K. Kokinov (Eds.), *The analogical mind* (pp. 313–334). Cambridge, MA: MIT Press.

Feist, G. J. (1993). A structural model of scientific eminence. *Psychological Science, 4,* 366–371.

Feist, G. J., & Gorman, M. E. (1998). The psychology of science: Review and integration of a nascent discipline. *Review of General Psychology, 2,* 3–47.

Galison, P. (1997). *Image & logic: A material culture of microphysics*. Chicago: University of Chicago Press.

Giere, R. N. (1999). *Science without laws*. Chicago: University of Chicago Press.

Gooding, D. (1990). *Experiment and the nature of meaning*. Dordrecht, The Netherlands: Kluwer.

Gooding, D., Pinch, T., & Schaffer, S. (Eds.). (1989). *The uses of experiments*. Cambridge, England: Cambridge University Press.

Hacking, I. (1983). *Representing and intervening*. Cambridge, England: Cambridge University Press.

Holyoak, K. J., & Thagard, P. (1995). *Mental leaps: Analogy in creative thought*. Cambridge, MA: MIT Press/Bradford Books.

Machamer, P., Darden, L., & Craver, C. F. (2000). Thinking about mechanisms. *Philosophy of Science, 67,* 1–25.

Medawar, P. B. (1979). *Advice to a young scientist.* New York: Harper & Row.

Nersessian, N. (1992). How do scientists think? Capturing the dynamics of conceptual change in science. In R. Giere (Ed.), *Cognitive models of science* (Vol. 15, pp. 3–44). Minneapolis: University of Minnesota Press.

Popper, K. (1959). *The logic of scientific discovery.* London: Hutchinson.

Ramón y Cajal, S. (1999). *Advice for a young investigator* (N. S. Swanson & L. W. Swanson, Trans.). Cambridge, MA: MIT Press. (Original work published 1897)

Sindermann, C. J. (1985). *The joy of science.* New York: Plenum.

Thagard, P. (1999). *How scientists explain disease.* Princeton, NJ: Princeton University Press.

Thagard, P. (2002a). Curing cancer? Patrick Lee's path to the reovirus treatment. *International Studies in the Philosophy of Science, 16,* 179–193.

Thagard, P. (2002b). The passionate scientist: Emotion in scientific cognition. In P. Carruthers, S. Stich, & M. Siegal (Eds.), *The cognitive basis of science* (pp. 235–250). Cambridge, England: Cambridge University Press.

Ward, T. B., Smith, S. M., & Vaid, J. (Eds.). (1997). *Creative thought: An investigation of conceptual structures and processes.* Washington, DC: American Psychological Association.

Watson, J. D. (1969). *The double helix.* New York: New American Library.

Watson, J. D. (2000). *A passion for DNA: Genes, genomes, and society.* Cold Spring Harbor, NY: Cold Spring Harbor Laboratory.

Wolpert, L., & Richards, A. (1997). *Passionate minds: The inner world of scientists.* Oxford, England: Oxford University Press.

Zanna, M. P., & Darley, J. M. (Eds.). (1987). *The compleat academic.* New York: Random House.

9

Seeing the Forest
for the Trees:
Visualization, Cognition,
and Scientific Inference

David C. Gooding
University of Bath, United Kingdom

This chapter addresses the ways that scientists construct and manipulate mental images, sometimes called *spatial cognition*. Case studies of innovation and discovery in science and technology confirm what we have known all along but have barely begun to describe and theorize in a systematic way: that visual modes of representation are essential to the generation and dissemination of new knowledge.[1] Can we develop a model of how scientists use images to devise solutions to problems? I will identify a schema that is widely used in a range of contexts and at different stages in the development of visual images, models, and instruments and in discourse about these cognitive artifacts. This schema elucidates the notion of cognition by showing

---

[1]Studies include Rudwick's (1976) study of the emergence of a visual language for geology; Gooding's (1982, 1985, 1992) work on the interaction of visual and verbal reasoning in Faraday's experiments; Henderson's (1991, 1999) studies of sketches and diagrams in engineering design; Miller's (1986, 1996) studies of the distinction between visualization and (nonsensory) visualizability in20th-century physics; Lynch and Woolgar's (1990) collection of sociological studies of representation. More recent historical studies include Beaulieu (2001) on brain imaging, de Chadarevian and Hopwood (in press) on modeling of complex molecules, Giere's (1999b) study of visual models in geology , Gooding's (1996, 1998) comparative studies, and Jones and Galison's (1998) collection on visualization in art and science.

how image-based thinking is a mental activity that is embodied and both technologically and socially situated (Clark, 1997, 2002).

Psychologists have of course been interested in image-based reasoning. Gregory (1981) treated the interpretability of visual stimuli as an analogy for an abductive model of hypothesis generation, whereas cognitive psychologists generally investigate visualization through experimental studies of problem-solving tasks. Historical studies provide suggestive, coherent narratives about detailed cases but are not meant to support general, cognitive theories about the processes at work.[2] Sociologists treat images only as public objects used to construct and negotiate the facticity of scientific results.[3] Philosophers generally assume the priority of propositional over visual modes of representation and have, until recently, ignored visual aspects of reasoning.[4]

In addition to Gregory's (1981) work on perceptual hypotheses and Kosslyn's (1994) analysis of mental manipulation governed by learned constraints, there have been studies of mental transformations of visualized objects (Cooper, 1976; Cooper & Shepard, 1984) and on the roles of mental images in creativity (Shepard 1978) and on the dialectical play of sketches and visual images (Van Leeuwen, Verstijnen, & Hekkert, 1999). Tversky (2002) surveyed graphical communication and visual depiction and explored some implications of the situated and embodied character of perception for a theory of spatial cognition (Tversky, 1998). Nevertheless, our understanding of visualization as depiction and of visual reasoning remain vague, with no decisive empirical support for any particular approach.

Scientific discovery is a complex process that involves many types of activity. These activities are mediated by personal experience; the availability of resources, such as investigative technologies; problem-solving techniques and expertise in their use; and interactions with other people. The complexity and long duration of many of these processes of discovery place them beyond the scope of experimental (*in vitro*) or observational (*in vivo*) studies by psychologists. Newell (1990) proposed four hierarchical bands of cognition: (a) biological, (b) cognitive, (c) rational, and (d) social. These are differentiated by the time each requires for cognitive processing and problem solving and, by implication, the quantity and explicitness of the

---

[2]Kemp's (2000) comparative study of visualization in art and science introduces the suggestive but rather vague notion of "structural intuitions."

[3]Like the essays in Jones and Galison (1998), Kemp emphasizes similarities between representational practice in the two domains. The seminal collection of sociological studies of representation is Lynch and Woolgar's (1990); recent studies are Henderson (1999) and Beaulieu (2001).

[4]Philosophers who regard visualization as important include Giaquinto (1992), Giere (1999), Hanson (1972), Magnani (2001), Miller (1996), Nersessian (1988), and Shelley (1996).

knowledge used or produced. This approach assumes that processes of many hours duration at the social end of the spectrum are wholly decomposable into mental processes lasting minutes, and these in turn into evolutionarily engendered (close-coupled) processes lasting only milliseconds. Anderson (2002) pointed out that the time intervals required to relate the biological and social aspects of cognition differ by over seven orders of magnitude. In practice, however, the cognitive bands are differentiated by the duration of the experimental tasks used to investigate each kind of process. Psychologists have tended to focus their experiments on biologically engendered processes of short duration, which involve little or no social interaction, so the identification of the type of process according to its supposed temporal duration remains hypothetical.

The real-time creativity that historians and biographers can document involves cognitive processes that could be located in three of Newell's (1990) four bands, taking minutes, hours, days, or weeks to complete. Other well-documented cases of problem solving have continued over several decades, far longer than Newell envisaged (Carlson & Gorman, 1990; Gooding, 1990a; Gruber, 1974/1981; Holmes, 1991; Nersessian, 1984; Tweney, 1992; Westfall, 1980). Moreover, some problem solving in discovery draws on several parallel investigations that have taken months or even decades to complete (Gruber, 1974/1981, 1994; Gruber & Davis, 1988; Tweney, 1989). These case studies also show that spatial cognition often integrates a range of experiences that both vary in duration and originate in different sensory modalities (Gooding, 1992; Tweney, 1992). An adequate cognitive theory should address both the integrative role of visual experience and the ways it permits analogical transfers between synchronous yet distinct research problems.

Given such complexity and temporal span, the neglect of real world examples by psychology is hardly surprising. Psychologists are like the physiologist and the geneticist in the horse-racing story about theorizing complex processes (see chap. 1, p. 4). Confronted by many variables and uncertainties, psychologists place much of what is involved in making visual inferences beyond the current reach of investigation. Yet we risk losing sight of the cognitive forest for the task-specific trees. Are there ways of spanning at least the larger temporal categories in Anderson's (2002) seven orders of magnitude? Every science has faced this sort of difficulty at some stage in its development. Because there are plenty of precedents, we can look to other sciences for examples of how temporally minute, embodied, and embedded cognitive capacities are brought into interaction with the more decoupled, abstract domain of rule- and representation-based reasoning and problem solving. These cannot yet be explained by reduction to cognitive processes grounded in experimental studies (Klahr, 2000), but this does not justify ignoring what they can show about creative human problem solving. Alternatively, these cases could be taken as counterexamples, anomalies that show that scientists' use of visualization cannot be theorized by decomposition via a cognitive hierarchy. If cognitive scientists cannot explain these anomalies they could, eventually, bring about its downfall. There are historical precedents for this outcome as well. To theorize discovery in the larger context of scientific practice we need another approach that includes the cognitive sciences but transgresses the disciplinary boundaries they have set (Gardner, 1985, pp. 43–45).

There are several nonexperimental ways of exposing cognitive factors relevant to a psychological theory of visualization. One way is to trace and compare the development of primary modes of representation, such as numerical (digital) and visual–verbal (analog).[5] Many episodes in the history of science suggest that both are essential and that neither has supremacy (see Galison, 1997; Gooding, 2003a; LeGrand, 1990). Another way is to examine transitions in science and art between perception-based depictions and visualizations of states that cannot be objects of sense experience, even indirectly (Miller, 1986, 1994, 1996). A third method, the subject of this chapter, is to identify and compare visualization strategies in different discovery domains.

PHENOMENA, THEORIES, AND MODELS

Researchers conducting cognitive studies need to find an appropriate level of abstraction at which to describe, model, and theorize the phenomena they seek to explain. The history of science shows that in the development of any scientific field there is a searching back and forth between abstract, simplifying models and real world phenomenological complexity to achieve an appropriate level of description—not overly simple, capturing enough to be representative or valid, yet not so complex as to defeat the existing problem-solving strategies of a domain. There is always a tension between complexity and realism on the one hand and simplicity and solvability on the other. This point applies as much to the design of experiments (which abstract certain controllable features or variables; Gooding & Addis, 1999) as it does to models, theories, and symbolic representations such as diagrams (Stenning & Oberlander, 1995). A key function of representations is to resolve the tension between the demands of complexity and abstraction through successive reformulations of a problem.

Many scholars have introduced special terms to describe the abstractive, representation-changing work of science, replacing everyday notions of things that purport to describe aspects of the world (such as sketches, drafts,

---

[5]The terms of the debate about the primacy of modes of representation are set in Kosslyn (1981) and Pylyshyn (1981).

essays, explorations, models) by more specialized notions. Examples include Cartwright's prepared descriptions and her distinction between models of phenomena and theoretical models (Cartwright, 1983, pp. 133–134; Cartwright, Shomar, & Suarez, 1995); models as mediating instruments (Morrison & Morgan, 1999); Pickering's (1989) distinction among phenomenological models, models of instrumentation, and theoretical models; and Gooding's construals, which are phenomenological models that may develop into theoretical models or be derived from such models (Gooding, 1982, 1986, 1990c, 1992). Tweney and his colleagues also have distinguished emergent "inceptions" from insights and concepts (Ippolito & Tweney, 1995; Kurz & Tweney, 1998; chap. 7, this volume). Ethnographic and *in vivo* studies identify representations such as "collaborative utterances" (Goodwin, 1990, 1995), "inscriptions" (Latour & Woolgar, 1986) and "translations" (Latour, 1990), elastic "boundary objects" used in trading zones (Fujimura, 1992; Galison, 1996; Star & Griesemer, 1989), and "noticings" (Trickett, Fu, Schunn, & Trafton, 2000, chap. 5, this volume).

These different concepts denote distinctions that are not purely descriptive; on the contrary, they often address different concerns. Cartwright (1999a) and Giere (1988, 1999b) are concerned with issues of truth and realism in relation to the semantic view of theories. Johnson-Laird (1983), Gentner and Stevens (1983), and Nersessian (1999) consider mental models as an alternative to purely syntactic accounts of reasoning, whereas work by Gooding (1986), Henderson (1999), Tweney (1992), and Trickett et al. (2000) aims to show how plasticity and implicit meaning enable representations to support creative, interpretative thinking. Sociological studies are primarily concerned with showing how representations support the closure of controversies through the negotiation of factual status especially across conceptual, methodological and other boundaries (e.g., Galison, 1997; Henderson, 1991; Latour, 1990; Pickering, 1989; Star & Griesemer, 1989). Nevertheless, the functions of the representations identified in these studies are often complementary and interdependent. Henderson (1999, p. 198 ff) argued that much of the power of visual representations lies in the fact that they can carry information both explicitly and implicitly, giving scope for negotiation in the fixing of meaning so that, as with Gooding's (1986) construals, creative construction, interpretation, communication, and negotiation of meaning go together. Fujimura (1992, p. 169) introduced the notion of *standardized packages*, inclusive representations for collective work leading to the stabilization of facts between groups using divergent investigative practices. Cartwright (1999b) replaced her phenomenal–theoretical models distinction with another that is between phenomenological models and those that represent very specific, local, real world implications of theories (Suarez, 1999). This reflects a more constructivist, postsemantic conception of theorizing as a process in which the world is engineered to fit the models at specific times and places (similar views were developed by Gooding, 1990c; Goodman, 1978; Hacking, 1983, 1992; Pickering, 1989, 1990).[6]

Giere prefers to use the term *model* for all types of representation, from the initial, tentative construals of data to the well-articulated, unambiguous hypothesis or theory (1999a). He is therefore obliged to conclude that theory construction in science is models "almost all the way up" from the phenomenology and that theory evaluation is also "models almost all the way down" (Giere, 1999a, pp. 54–56). This last formulation suggests that there is a continuum of representations, from initial phenomenology and classification of particulars through to the most general and abstract representations of properties of generic systems and processes (Giere, 1994). The term *model* is used "more promiscuously" than any other, according to Nelson Goodman (1978, p. 171). For our purposes, Giere's inclusive and promiscuous use of the term *model* is adequate, provided it is remembered that *model* denotes a broad class of representations that scientists need (visual, material, analogical, verbal, and symbolic; phenomenological, interpretative, representational, idealized, etc.) and that cognitive functions differentiated on Newell's (1990) hierarchy as biological, rational, or social are often conflated. In practice, as our examples show, models sometimes enable several functions at once.

SITUATING VISUAL INFERENCE

Visualization is widely used to create the models that scientists use to negotiate the tension between simplicity and solvability on the one hand, and phenomenological complexity and real world application on the other. Giere (1999b) suggested that visual models serve "as an organizing template for whatever other potentially relevant information" is possessed by theorists (p. 132) but is reluctant to go further. He remarked that such suggestions are "in line with current thinking in the cognitive sciences" and show "at least that images *could* play a significant role in scientific reasoning" (Giere, 1999, p. 132).

Given the wide range of image-based strategies for constructing interpretations, arguments, and theories in science, Giere's (1999) conclusion is too cautious. It is often possible to describe the process of image-based reasoning in sufficient detail to show that it is dialectical and progressive, involving many versions and iterations of a whole panoply of cognitive resources: interpretations, models, instruments, theories of instruments, and theories

---

[6]This pragmatic view abandons the semantic distinction between theories (as representations) and the world (as what is represented) in favor of models as one type of mediating representation (Cartwright 1999b, pp. 241–242; 1999a, p. 34).

(for an example, see Gooding, 1990b, 1990c; Nersessian [2002] also emphasized the iterative process whereby models are enhanced). One can then characterize the process in other terms, such as the sequences of decisions, actions, and outcomes making up the procedures used (Gooding, 1992) and, at a higher level of abstraction, characterize the process very generally, in terms of the range of cognitive capacities required to execute these procedures at each stage (Gooding, 2003; Van Leeuwen et al., 1999).

Such general schemes enable one to grapple with the glorious complexities of scientific thinking rather than compartmentalize it as *either* problem solving *or* hypothesis testing *or* concept-based discovery (Dunbar, 1999), or in purely sociobehavioral terms, as the social construction of facts. A scheme of inference is needed that shows how visual information relates to models as socially shared interpretative constructs and how the latter are, in turn, elaborated and related to other knowledge to form general, explanatory theories. The scheme proposed here shows how visual inference and modeling strategies are recast as derivations whereby the originating "source" phenomenology can be a consequence of the constructs proposed along the way. Thus, it integrates the personal, cognitive processes with the social processes by which new methods, concepts, and facts are established.

This scheme deals with visualization and how it relates to verbal representation, so it is just one of several kinds of inferential strategy needed in a complete account of scientific inference. Its importance is that it helps us to explain the cognitive impact of an observation, experiment, or novel theoretical insight by situating it in terms of processes of inference and argumentation, without reducing it to an idealized logical or algorithmic procedure. As I show, logical features remain, but they are situated within a larger set of processes that includes both informal reasoning, such as visualization, and proposition-based inference. Whereas semantic and network theories treat scientific theories as systems of propositions in which any proposition must bear some well-defined logical relationship to other propositions, this scheme enables us to place any particular image in the context of associated images in a way that clarifies its contribution to a larger empirical–theoretical enterprise. This clarification is provided by identifying the informal transformation rules that guide inferential moves between images that abstract features from particulars, constructs that generalize and explain these features, and further consequences derived from the constructs.

Finally, this scheme has two methodological advantages: (a) It requires us to unpack cognitive case studies systematically, so as to support comparison between cases, and (b) it also requires us to differentiate clearly between the areas of creative thinking to which logical models of inference apply and those for which cognitive psychology has yet to provide an account. It therefore provides empirical constraints on *in vitro* and *in silico* methods, which have tended to impose logical or statistical models of inference on their subject matter.

VISUAL INFERENCE

The constructive role of graphic representation is a particularly important feature of the cognitive dimension of scientific discovery. Complex phenomena can be decomposed into simpler visualizations; conversely, simple visual images can be manipulated and combined to construct complex representations having considerable explanatory power. In many instances, this process can be represented as the following iterative schema:

$$\text{pattern} \rightarrow \text{structure} \rightarrow \text{process} \rightarrow \text{pattern} \rightarrow \dots$$

where each arrow indicates an as-yet-unspecified type of inference. This will be abbreviated as the *PSP schema*. The process begins with new or anomalous phenomena imaged as patterns. There is first a *reduction* of complex, real-time phenomena to an image. This is usually an abstract pattern or set of patterns, such as a drawing of a magnetically induced distribution of iron filings, or a visualization of numerical data. Such two-dimensional (2D) images are themselves partial abstractions. However, as we shall see, the image may be a veridical representation of pattern as observed in nature. This 2D image is then *enhanced* by "adding" dimensions. This is the visual–perceptual analogue of including more information in a representation. The first enhancement creates a three-dimensional structure that can be imagined, drawn, or built. Because most explanations relate to real-time processes, a further enhancement produces a temporal, four-dimensional model. This scheme is found wherever there is a need to resolve, order, and communicate experience that is anomalous, ambiguous and particular—in other words—experience that requires interpretation through the construction of a phenomenology.[7] In the language of Kant's philosophy, such representations are visualizable representations that stand for more complex experience (Miller, 1986, pp. 127–128; 1996, pp. 44–47).

Sequences of transformations like these enable scientists to draw on and extend an existing repertoire of descriptors and concepts.[8] The scheme can be developed to show how a particular visual representation is related to structural and process models. In practice, it usually displays two further features. The first feature is that each enhancement has an as-

---

[7]*Dimensional reduction* is always necessary when recording real world processes as, say, sketches in a notebook. Dimensional enhancement therefore depends on a prior abstraction or *reduction*.

[8]See Gooding (1989, 1996, 1998) for examples and further references.

sociated reductive move: Three-dimensional (3D) and four-dimensional (4D) constructions are used as sources from which to derive images of lower dimensionality. This extends the empirical base for the models by constructing a larger phenomenology. These are usually 2D images but may include alternative structural representations (or even physical models) of the process model. The second feature is that some of the derived images are subsequently presented as validating the structural and process models by reasoning that has the form "If the unobservable structure/process has the form described, then these are the patterns/sections we would expect to observe." In this way, a generative visual schema provides a way of integrating images into a verbal, possibly deductive rationale. These cognitive processes clearly coexist, so if it is possible to relate these types of visualization to levels of cognition in Newell's (1990) classification, then that can no longer be considered to be a hierarchy in which processes of greater duration are decomposable into shorter ones. The examples I provide next show that the scheme applies both to work by individuals and work by collaborators or groups.

Figure 9.1 diagrams the different combinations of moves typically displayed during iterations of the scheme. The arcs depicted in Fig. 9.1 allow for different sequences of moves. Typical variations on the PSP schema include reductions from structure to pattern:

$$\text{pattern} \rightarrow \text{structure} \rightarrow \text{pattern} \rightarrow \ \dots$$

and from process to structure:

$$\text{pattern} \rightarrow \text{structure} \rightarrow \text{process} \rightarrow \text{structure} \rightarrow \text{pattern} \rightarrow \ \dots$$

Different patterns of transformation will occur, depending on the nature of the immediate problem or objective, changing as work moves from an exploratory, interpretative phase into model-building, empirical verification, formal validation, predictive testing, and so on.

Dimensional reduction is involved when one is recording real world processes as, say, sketches in a notebook. Dimensional enhancement therefore depends on a prior abstraction or *reduction*. An *enhancement* (a move to three or four dimensions) resembles an abductive inference because it is generative: The resulting structure incorporates and integrates new information not present in the source pattern.[9] A *reduction* (a move to fewer dimensions) always produces a particular instance of a pattern or structure from a more general representation. Reductions include prediction of new

---

[9]This scheme therefore shows how abductive inference (as defined by Gregory, 1981, Hanson, 1972, and others) works in concert with other kinds of inference—see Gooding (1996).

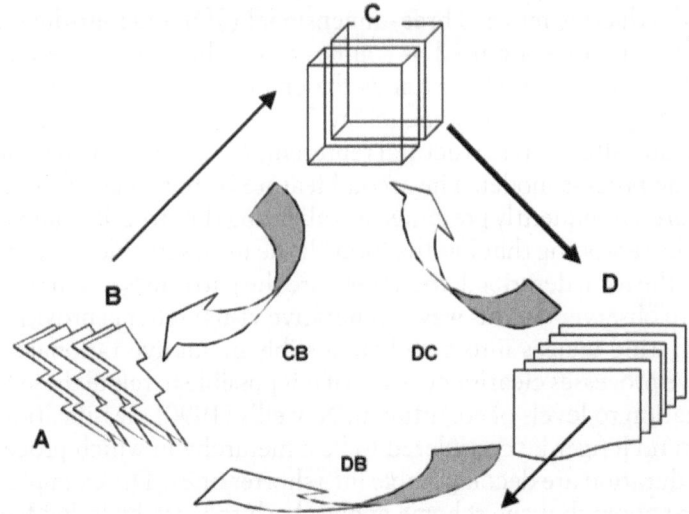

FIG. 9.1.   The pattern–structure–process schema. Visual inference schema showing
the range of transformations typically found in a series of iterations. Moves from pattern to structure (Arc BC) and from structure to process (CD) are generative and ex-
pansive (enhancements). Moves from unresolved phenomena to a pattern, from
process to structure or from structure to pattern (AB, CB, DC) are also generative but
usually reduce the raw information content of an image (reductions).

phenomena and retrodiction (where one of the derived images is equivalent
to the source image). Thus, a reduction is functionally equivalent to a deductive inference (even though the inference rules for images are not explicit). As indicated in Figure 9.1, the process can proceed in a progressive,
cyclical manner (and is often written up for publication in this form). However, it is often necessary to test an enhancement—say, from A to B, by
making reductive moves to 2D patterns, before moving from B to C.

This scheme summarizes key features of reasoning with and about phenomena and data identified in a number of cognitive case studies (Gooding,
in press).[10] In the following section I summarize the phenomenology of some
cases, showing how the visual inference scheme applies to public, commu-
nication-oriented transformations. I then show how the study of laboratory
notebooks allows us to investigate the mental processes behind the production of such images.

---

[10]These include crystallography (inferring the structure of complex molecules from interference patterns), paleobiology (reconstruction of fauna of the Burgess shales), hepatology
(explanation of liver function in terms of modular structure), electromagnetism (invention of
the electric motor and the origins of electromagnetic field theory), geophysics (mechanisms
to explain sea-floor features that also explain continental drift), and diagnostic imaging (validation of the early use of x-rays).

Paleobiology: Reanimation of the Burgess Shales

Gould (1989) introduced his chapter on Whittington's reconstruction of the life forms fossilized in the Burgess shales with the a remark that "I can't imagine an activity further from simple description than the reanimation of a Burgess organism. You start with a squashed and horribly distorted mess and finish with a composite figure of a plausible living organism" (p. 100). This involves making careful *camera lucida* drawings of both positive and negative impressions of the flattened, fossilized animals, as, for example, in Figure 9.2.[11] These 2D images are then interpreted in terms of what they suggest about possible organisms (see Fig. 9.3). In some cases, depending on the nature of the fossil material, these interpretations can be further tested by "dissecting" the rock matrix. Although the images of some imprints might be interpreted by analogy to modern counterparts, most Burgess organisms are extinct and have no modern counterparts. Moreover, the organisms were "fixed" in mud at many different angles, so that the cleavage planes in the fossil-bearing shale may cut the already distorted organisms at any orientation whatever. This creates the problem of determining which impressions image the same organism. However, neither the number of organisms in a cache nor the identity of any organisms is known beforehand (Whittington, 1985).

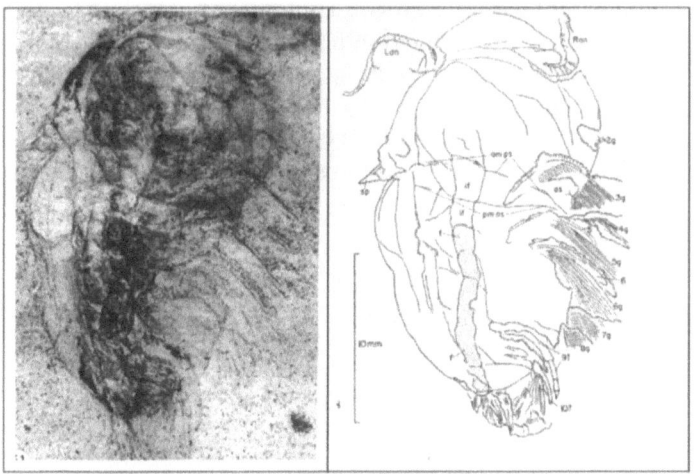

FIG. 9.2.   An imprint of the arthropod *Naraoia*, photographed under alcohol, and its explanatory diagram (right). From H. B. Whittington, (1977), "The Middle Cambrian Trilobite *Naraoia*, Burgess Shale, British Columbia, *Philosophical Transactions of the Royal Society*, B 280, 409–443, Plate XI fig. 61 and Text-fig. 58. Reprinted by permission of the Royal Society.

[11]The *camera lucida* was invented by W. H. Wollaston in 1806 as a drawing aid. It is a prism-based device that used refraction and internal reflection to allow the subject and the drawing to be seen simultaneously, as if the drawn image is hovering over the subject. Here, technology superimposes the representation on what it depicts.

To be identified as a section of a particular organism, an impression must be associated with other impressions. Thus, whereas my second example shows a method that creates images of parallel sections, here every fossil impression must be mentally reimaged as if from several points of view. Each of the resulting images is evaluated as a possible match for other known impressions. There is a painstaking, time-consuming process of elimination as possible matches are identified. One investigator (Morris, cited in Gould, 1989) reported having drawn

> specimens that had been found in various orientations, and then passing count-less hours "rotating the damned thing in my mind" from the position of one drawing to the different angle of another, until every specimen could be moving without contradiction from one stance to the next. (Gould, 1989, p. 92)

Morris's phrase "without contradiction" points to the existence of informal, largely implicit rules for transforming and comparing mental images. However, this is only a part of the process. Possible structures for the animals are envisaged as 3D structures generated from 2D images of a set of matched or correlated impressions, as in Fig. 9.3. These images integrate several types of information, because they must be consistent both with the set of images and with working assumptions about the physiology of such organisms. These are derived in part from what is known about the animals ecology (food sources, predators, etc.). Finally, it must be possible to derive the original impression(s) from the 3D model. There is a dialectical process of moving back and forth between images of the flattened layers and solid objects. These 3D objects may be imaged or physical models.

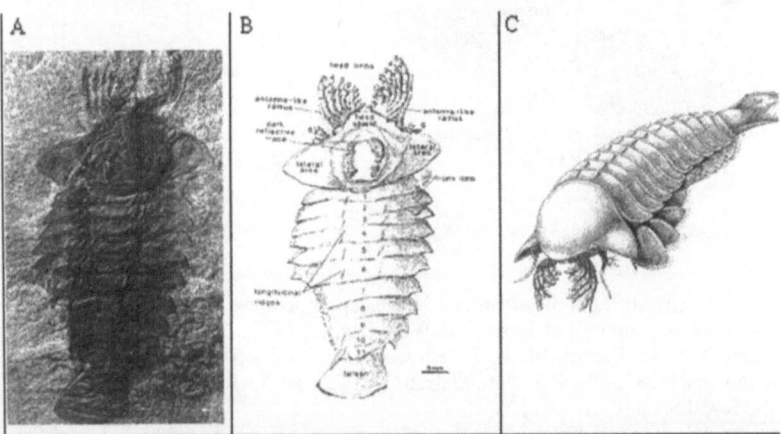

FIG. 9.3.   A selection of images from the reconstruction of the arthropod *Sanctacaris uncata*. 9.3A: Photograph of a counterpart (dry specimen, dorsal view). 9.3B: Labeled drawing based on the photograph. 9.3C: Reconstruction drawing based on several fossil 'sections.' From D. Briggs & D. Collins (1988), A middle Cambrian Chelicerate from Mount Stephen, British Columbia, *Palaeontology, 31*: 779–798, Text-fig. 1B; Plate 71 fig. 3; Text-fig. 6. Copyright The Palaeontological Association.

Figure 9.4 illustrates a sectional model of a different species (*Sidneyia*) used in the reconstruction process.

This example illustrates the use of a technology (the *camera lucida*) to produce images with which scientists then work. It also illustrates considerable skill in the manipulation of images and models representing impressions in inert pieces of stone. Gould (1989) noted that Whittington's ability to visualize in three dimensions was exceptional and unusual:

> to reconstruct three-dimensional form from flattened squashes, to integrate a score of specimens in different orientations into a single entity, to marry disparate pieces on parts and counterparts into a functional whole—these are rare and precious skills. (p. 92)[12]

Complex skills such as making transformations between 2D and 3D representations, matching via the features generated, and integrating the results into a 4D model, may be beyond the current reach of cognitive psychology. Nevertheless, one can discern features that this process has in common with the interpretation and explanation of data in many other areas of scientific research. The first important feature was noted by Gould (1989):

> The consistent theme is unmistakable: a knack for making three-dimensional structures from two-dimensional components and, inversely, for depicting

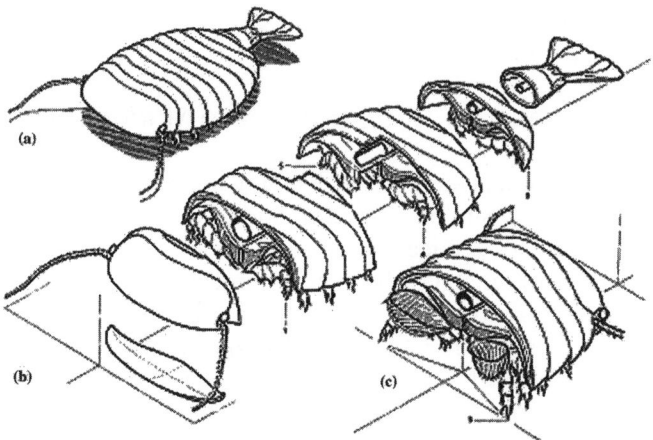

FIG. 9.4.   Reconstruction of the arthropod *Sidneyia inexpectans* as a 3-D model built in sections derived from camera lucida drawings such as those in figs. 9.2 and 9.3, from D. L. Bruton (1981), "The arthropod *Sidneyia inexpectans*, Middle Cambrian, Burgess Shale, British Columbia, *Philosophical Transactions of the Royal Society*, B 295: 619–656, fig. 107. Reprinted by permission of the Royal Society.

---

[12]When shales have split along cleavage planes to reveal a fossil imprint, *part* and *counterpart* denote the matching positive and negative imprints. The matching imprints are often separated by geological process or into different geological collections by paleontologists who find only one member of the pair.

solid objects in plane view. This ability to move from two to three dimensions and back again, provided the key for reconstructing the fauna of the Burgess shales. (p. 101)

These movements "from two to three dimensions and back again" are inferences between Nodes B and C in Fig. 9.1. The structure implicit in 2D images is interpreted or explicated by constructing a 3D structure. The 2D image can then be reconstrued as evidence for the 3D structure of which it is shown to be a section or imprint. Evaluating the capability of this structure to support life processes involves moves among Nodes B, C, and D.

This example of the manipulation of visual representations enables one to discriminate between inferences based on enhancing or reducing images and those involving other kinds of cognitive process, such as pattern matching or rule-based inference. A possible sequence of manipulations is set out in Table 9.1. Sequences like this would be repeated many times over in order to achieve a model of anatomical structure (Fig. 9.4) that can support physiological process and the interpretation of the 2D images as representations of flattened anatomical features in various orientations.

The ability to move between two and three dimensions is a key to interpretation and explanation of phenomena and is evident in many other scientific domains. In some cases, as the next example illustrates, what begins as a cognitive process of organizing and interpreting thousands of data points by constructing 3D models from 2D arrays is itself reconstituted and presented as a definitive method of discovery.

Hepatology: Vascular Structures From Modular Sections

To develop an empirical method for deciding between different models of the relationship between liver function and its structure, a large numerical data set about biochemical processes in the liver is organized first into 2D and then 3D arrays. These become a framework for theorizing the invisible physiological structures that support the biochemistry. In both these cases, the final result of the reconstructive method is a 4D or process model, which explains the original source images of visualized data. Ten years' work by Harald Teutsch and his collaborators (Teutsch, in press) produced a fully functional model of primary and secondary structures. These cannot be observed directly. Nevertheless, as implied by the title of their final article, reconstructive visualization is presented as a method for demonstrating the superiority of one theory over others (Teutsch, Shuerfeld, & Groezinger, 1999).

One function of the liver is to regulate the supply of glucose to the blood, given major fluctuations in the nutritional supply. Regulation is achieved by opposing systems of enzymes that act to release glucose or to reduce the up-

TABLE 9.1
Paleobiology: A Possible Sequence of Visual Manipulations by Dimensional Enhancement, Reduction, and Matching

Outcome/Dimensions	2D	3D	4D
Operation			
Reduction ⇓ From material objects and images of objects (Fig. 9.2) to line drawings	§1a. sections: Drawings of impressions in shale (Fig. 9.3A).	§1b. Dissection: Separation of layers of fossil specimen (not shown).	
Enhancement ⇒ Drawings to 3D visualization		§2. Initial construction of possible anatomical structure relating features in §1 to knowledge of modern counterparts.	
Reduction ⇓ 3D visualization to 2D section images	§3. New sections derived by mental rotation of possible structure in §2.		
Matching: Comparison of features in 2D images. (biological function of features not yet determined)	§4. Initial matching of actual sections in §1 with and possible sections in §3 (derived in §2).		
Enhancement ⇒ Selected 2D sections to 3D model		§5. Sectional anatomical model constructed (Fig. 9.4)	

(continued on next page)

TABLE 9.1 (continued)

Outcome/Dimensions	2D	3D	4D
Operation			
Enhancement ⇒ Interpret 3D structure in terms of process.			§6. Initial evaluation of anatomy of model in §5 in terms of biological process.
Reduction ⇓ Infer identity of 2D features via function identified in 4D process	§7. Interpretation of anatomical features in §1 and §3 via constructs §5 and §6.		
Enhancement ⇒ Evaluate 2D images in 7 as sections of revised 3-D model		§8. Revised possible structure based on plausible matches in 5 and analogy to modern counter- parts and biology (from §6) (Figs. 9.3B and 9.4).	§9. Validated physiological model based on §§6–7–8.

Note. Cell positions locate an image or construct in the larger sequence of inferences. Numbers in each cell cross-reference each visualization to its source operation. 2D = two dimensional; 3D = three dimensional; 4D = four dimensional.

take of glucose into the blood. The enzyme chemistry is understood, but in some mammalian livers there is a long-standing puzzle as to the structure (physiology) that supports this function, because the structure could not be observed by the usual methods of observation based on close visual inspection of sections. The appearance of regular features in sections of liver that suggest modularity had been known since the 17th century (see Fig. 9.5 [left panel]; Teutsch, in press). This is a typical example of a hypothesized property that cannot be observed in whole livers and which in some mammals dissection does not actually support. The problem is therefore to construct a physiologically plausible structure using clues provided by nonvisual information.

To achieve a spatial separation of the opposing catalytic actions of the enzymes, investigators made a close study of rates of glucose uptake and release in cryosections of many rat livers, with a view to envisioning the process from a spatially distributed set of values. Further visualizations represent gradients of activity rather than discrete values (see Fig. 9.6). As the quantity of data was increased, so too did the "resolution" of these images increase, in at attempt to infer submodular structures consisting of populations of cells (see Fig. 9.7).

Working over a period of 10 years, Teutsch and his colleagues showed that the modularity and the vascular structures that support it can be demonstrated by the construction of 3D images from virtual "stacks" of images of very thin cryosections (Teutsch, Altemus, Gerlach-Arbeiter, & Kyander-

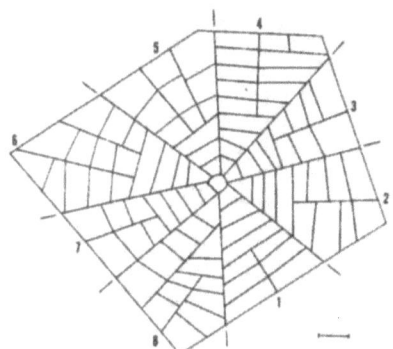
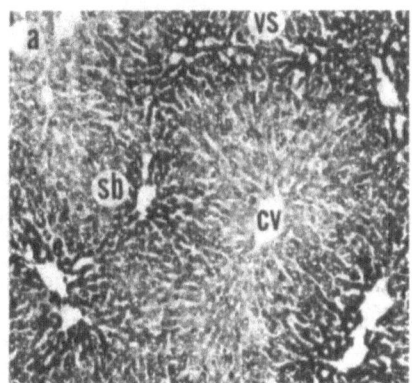

FIG. 9.5.   Microdissection map (left) showing subdivision of a liver lobule cross-section into eight adjacent sectors. Histochemically stained sections (right) are used to guide this subdivision. The abbreviated labels indicate sections of entities whose structure is not yet known but will emerge (approximately) perpendicular to the plane of the imaged section. From "Regionality of Glucose-6-Phosphate Hydrolysis in the Liver Lobule of the Rat," by H. Teutsch, 1988, *Hepatology, 8,* pp. 313 and 314. Copyright 1988 by American Association for the Study of Liver Diseases. Reprinted by permission of Wiley-Liss, Inc., a subsidiary of John Wiley & Sons, Inc.

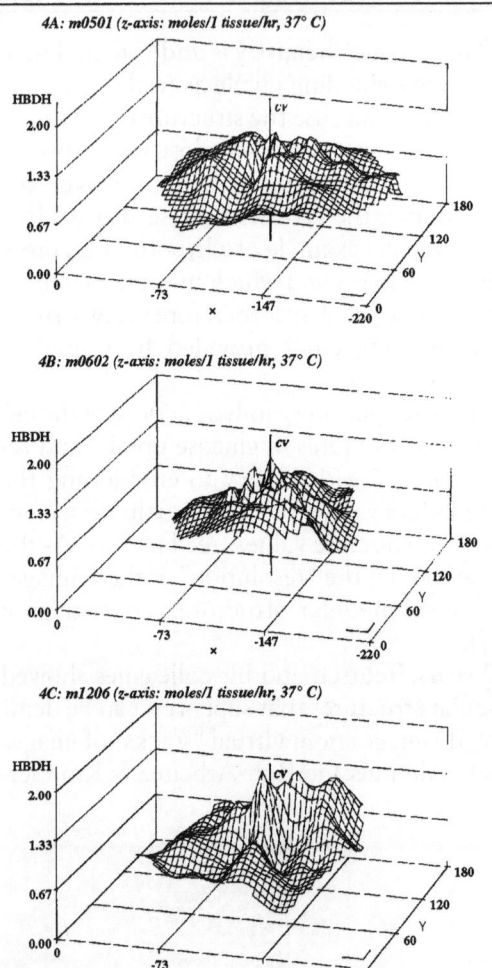

4A: m0501 (z-axis: moles/l tissue/hr, 37° C)

4B: m0602 (z-axis: moles/l tissue/hr, 37° C)

4C: m1206 (z-axis: moles/l tissue/hr, 37° C)

FIG. 9.6. Plots of the distribution of enzyme activity over cross-sectional areas of a primary modular structure of rat liver, reconstructed as 3-D surfaces. The central vertical line shows the location of the central venule (labelled 'CV' in Fig. 9.5). Reproduced by permission of the Histochemical Society, from Teutsch, H., Altemus, S., Gerlach-Arbeiter, S., & Kyander-Teutsch, T. "Distribution of 3-Hydroxybutyrate Dehydrogenase in Primary Lobules of Rat Liver," *Journal of Histochemistry and Cytochemistry*, 40: 213–219, 1992, figure 4.

Teutsch, 1992; Teutsch, 1988). Sample drawings of these are shown in Fig. 9.7 (left panel). A particular structure of the modules and of the ducts that supply and drain them was produced and validated by creating a 3D image (using 3D image printing technology). From this image, models of paste and plastic were created to work out how the modular structure incorporates portal and venal veins (see Figs. 9.7 and 9.8). Just as drawing is used to enhance attention to features, physical modeling is frequently used to check the efficacy of a structure with regard either to function (as here and in the case of the Burgess shale fauna) or to phenomenology, as in the case of x-ray crystallography (de Chadarevian & Hopwood, in press). The complex vascular structures that service the modules are now considered to be facts established by reconstruction. Figure 9.8 shows a 3D model that is also an artistic reconstruction comparable to the drawing of *Sanctacaris* in Fig. 9.3.

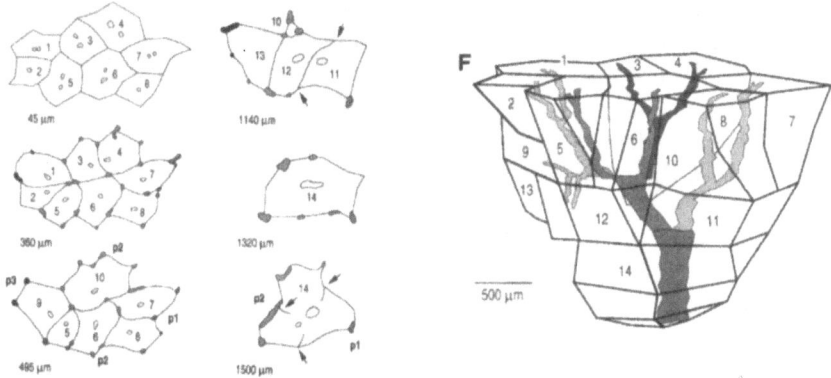

FIG. 9.7.   Schematic drawings of some of the 146 cross-sections or planes of a secondary unit showing sections of the portal tracts and septa, from which a three-dimensional structure (shown at right and in Fig. 9.8) was constructed. From "Three-Dimensional Reconstruction of Parenchymal Units in the Liver of the Rat," by H. Teutsch, D. Shuerfeld, and E. Groezinger, 1999, *Hepatology, 29*, pp. 488, 501. Copyright 1992 by American Association for the Study of Liver Diseases. Reprinted by permission of Wiley-Liss, Inc., a subsidiary of John Wiley & Sons, Inc.

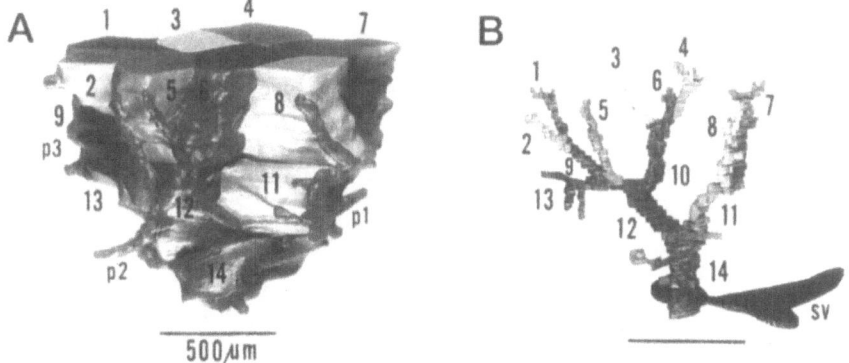

FIG. 9.8.   These images show a model that is also an artistic reconstruction comparable to the drawing of *Sanctacaris* in Fig. 9.3. A and B show the front view of a three-dimensional reconstruction of a secondary parenchymal unit (left) and of the central venular tree that drains it (right). In the original, color coding was used to differentiate the primary units of which the whole is composed. This coding is lost in monochrome reproductions. From "Three-Dimensional Reconstruction of Parenchymal Units in the Liver of the Rat," by H. Teutsch, D. Shuerfeld, and E. Groezinger, 1999, *Hepatology, 29*, p. 499. Copyright 1999 by American Association for the Study of Liver Diseases. Reprinted by permission of Wiley-Liss, Inc., a subsidiary of John Wiley & Sons, Inc.

191

These methods of reconstruction combine veridical images, schematic images, 2D and 3D plots, 3D stacks of 2D plots, and physical models. The 3D images and physical models are the physical counterpart of the virtual or mental procedures of visualization or depiction based on the accumulation and structuring of data. The use of images to represent mental visualization is illustrated by the next two examples.

Geophysics: Anomalies, Maps, and Continental Drift

When thinking in verbal or numerical mode people conceptualize observational regularities as temporal conjunctions of events or as correlations between events or properties, yet there are many examples of scientists observing regularities that take a visual form. Examples known since the 18th century are the structuring effect of magnetism on iron filings and of water on sand. A widely discussed example is the pattern of shading produced by greyscale plots of numerical data produced by magnetometer scans of the seabed of the northwest Pacific ocean (Giere, 1999b; Gooding, 1996; LeGrand, 1990). When a threshold is imposed to eliminate values outside a specified range, this reduces the resolution. The resulting binary image shows clear striping, which corresponds to the alternating North–South sense (polarity) of the residual magnetization of the basalt (compare the images in Fig. 9.9). These maps represent numerical data in a form that

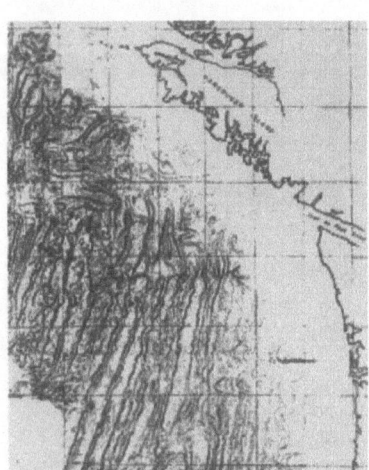 

FIG. 9.9.   Greyscale image of magnetometer data (left); binary image of magnetometer data (right). Although the binary image contains less information (data) than the greyscale, its simplicity displays a pattern and is more suggestive of a phenomenon needing explanation. From "Magnetic Survey Off the West Coast of North America 40-520N," by A. Raff and R. Mason, 1961, *Geological Society of America Bulletin, 72*, p. 1268. Copyright 1961 by the Geological Society of America.

displays pattern. Although they do not convey enough information to suggest an explanation, they display a form of regularity that is unexpected and therefore invites explanation. The striping is an anomaly like the "noticings" of astronomers interpreting optical and radio data of a ring galaxy (Trickett, et al., chap. 5, this volume).

This kind of mapping is another widely used representational technique. It displays patterns in data accessed through instruments and can be used to incorporate other, possibly relevant phenomena and features, such as centers of volcanic activity or earthquakes. LeGrand (1990) called representations that integrate these different kinds of information *symbolic maps* (see Fig. 9.10).[13]

Figure 9.9 is a flat 2D image that visualizes information obtained by the process of recording the position, intensity, and polarity of terrestrial magnetism. Scientists typically regard flat images such as these as the results of more complex processes or configurations of causes existing in space and time. Vine and Matthews (1963) argued that the alternating stripes indicated physical differences in the strips of basalt delineated by the plots. They constructed a 3D structural model showing how the strips could be related to an underlying geological structure (Vine, 1966; Vine & Wilson, 1965). From this, together with the knowledge that the terrestrial field is subject to periodic reversals, they could infer that each strip was extruded as molten basalt from along a fissure in the sea floor. Reversals of the terrestrial

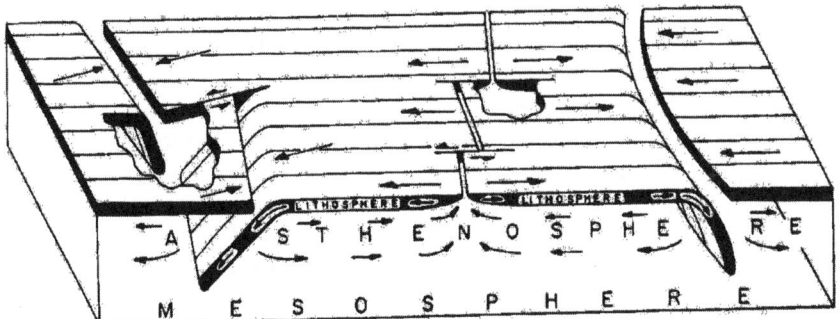

FIG. 9.10.   Block diagram in which arrows are used to indicate the underlying process that extrudes molten basalt to form the seafloor of which Fig. 9.9 is a magnetic snapshot. From Isacks, B., Oliver, J., & Sykes, L. R. (1968), "Seismology and the new global tectonics." *Journal of Geophysical Research, 73*: 5855–5899, figure 1 on p. 5857.

---

[13]They are analogous to *sections* or 3D representations of geological strata, constructed to display, by analogy to coastal cliffs, what would be revealed by cutting through strata. With the construction of canals in the 18th century and railways in the 19th century, many new sections were revealed; see Rudwick (1985, p. 46).

field between the time each strip of basalt had formed and cooled would explain the observed alternations in the polarity of the residual magnetism of each strip. The block diagram in Fig. 9.10 images only a temporal instant of a much longer process. This 4D model enhanced the physical meaning of the 2D pattern because it now showed that basalt had been extruded at intervals over a long period of time, linking the visualized phenomenon to larger problems, such as explaining the movement of continental plates.

As the previous examples show, the process of constructing visual models goes through several iterations and includes moves that reduce the complexity of the model (see Fig. 9.1 and Table 9.1). This is because although a more complex model has greater explanatory power, it is validated with respect to data visualized in two dimensions, so derivation of the original "source" phenomenon (the sea-floor stripes) was crucial to the validation of the emerging explanation. As indicated in the schematic representation of this process in Fig. 9.1, the 2D patterns are derived from the 4D process model via a 3D sectional or block diagram in Fig. 9.10. This corresponds to a reductive move from Node D to Node C in Fig. 9.1, and thence to Node B. Vine and Matthews (1963) also derived a new, previously unobserved property of the visualized basalt strips. Measured intensities of residual magnetism should display symmetry along an axis perpendicular to the axis of extrusion. A plot of these intensities is a 2D representation of a section across the 3D (structural) model. This corresponds to another move from Node C to a new 2D representation at Node B. The expected symmetrical profile was generated first by simulation methods. Real-data profiles were soon found to match it. LeGrand (1990) showed that images of these profiles were crucial to the acceptance both of the fact of sea floor spreading and the new explanation of it.

These three examples share the following features: having observed or produced a 2D image or having constructed a 3D image, our practitioners (a) enhance the information content and the explanatory potential of the image by adding a dimension and (b) simplify a 3D or 4D model by removing a dimension, in order to generate an originating source pattern or a previously unknown feature.

Table 9.2 shows how these examples instantiate the PSP scheme introduced in Fig. 9.1.

VISUAL INFERENCE AND MENTAL PROCESS

These examples identify as visual inferences two kinds of transformation I call *enhancement* and *reduction*. The PSP schema in Fig. 9.1 and the sequence of inferences set out in Tables 9.1 and 9.2 show how these inferences are used together in the course of interpreting data, constructing a

TABLE 9.2

Summary Array of Images Used for Discoveries in Paleobiology, Hepatology, and Geophysics as Instantiations of the Pattern–Structure–Process Scheme

Case	Source phenomenon	2D initial	3D	4D	2D derived
Paleobiology	Visual: Fig. 9.2	Fig. 9.3A	Fig. 9.4	Fig. 9.3B	Fig. 9.3A
Hepatology	Numerical: visualized as in Fig. 9.5B	Fig. 9.5A	Figs. 9.6 and 9.7	Fig. 9.8	Figs. 9.5A and 9.5B
Geophysics	Numerical: visualized as in Fig. 9.9A	Fig. 9.9B	Fig. 9.10	Fig. 9.10	Figs. 9.9A and 9.9B

Note. 2D = two dimensional; 3D = three dimensional; 4D = four dimensional.

theory, and validating that theory as an explanation of the source data and of implications (new predicted features). All of these images are used in the interpersonal or social domain for discussion, evaluation, criticism, and argument. Can this general scheme be developed so as to include the personal domain of mental processes? A representational scheme for human cognition should show how it can be generative (produce novelty) and systematic (enable consistency) and that it enables movement of representations between personal cognitive processes and socially situated ones.[14]

Generativity and systematicity may be enabled both syntactically (by features of a shared representational scheme) and socially (by the ways in which its users regulate acceptable transformations and outcomes). It is clear that the sort of image manipulation described here is generative. It is also clear that general and simple rules of manipulation are at work, although few of these are articulated. Such rules could be exposed by a closer, *in vivo* study of, for example, paleobiologists' manipulations of 2D abstractions from *camera obscura* drawings of fossil imprints. Where transformations are rejected by colleagues and peers one would expect to find discussion invoking criteria for evaluating the transformations themselves. Some of these criteria would be domain specific; those that are not would point to domain-independent processes.

By allowing us to identify such regular features of image-using behavior, these examples hint at cognitive processes that cannot be observed. Never-

---

[14]Productivity and systematicity are much debated in connection with the implications of connexionist models for a representational theory of mind—see Fodor and Pylyshyn (1988).

theless, one can go beyond the generalizations summarized in Tables 9.1 and 9.2 to open a window on this process in the mind of a creative scientist. Detailed records of problem-solving behavior such as those found in diaries and laboratory notebooks allow us to take the analysis further, from transformations on images and concepts in the public domain to mental processes on objects that have not yet "gone public."

Electromagnetism: Abstracting Structure— Theorizing Process

As chapter 7 of this volume also shows, two aspects of Faraday's work make it particularly informative for a cognitive approach. First, the detail of Faraday's records allows one to investigate some of the mental processes at work. The long duration of these records, covering many attempts to solve problems and improve on previous solutions, shows that that visualization works in conjunction both with biologically endowed capacities such as proprioception or kinesthetic awareness and with deliberative experimental manipulation. The importance of these material manipulations as a window into related mental processes cannot be underestimated. It enables one to identify some of the cognitive strategies that Faraday used. Like the work of other scientists described earlier in this chapter, Faraday developed procedures both to maximize the capacities of ordinary modes of perception and representation *and* to transcend their limitations. Tweney (1992) made the point succinctly, summarizing the significance of Faraday's 1831 studies of acoustical vibrations and of optical illusions in relation to his discovery of the elusive phenomenon of electromagnetic induction that same year:

> The task, which he succeeded in mastering, was to place the relatively slow acting perceptual system, the "eye," in a position to see what might be (and turned out in fact to be) fast acting events. Faraday succeeded (where others had failed) in this much-sought-after discovery precisely because, for him, the creation of a new order of natural law depended upon both cognitive and physical knowledge. (p. 164)

The goal of experimental design and manipulation is, therefore, complex. It should not be surprising that this produces representations whose cognitive (generative) and social (communicative and critical) functions are closely linked.

Faraday also used the method illustrated in the previous examples. The existence of detailed records over such a long period shows that the same pattern of inference is at work over the long term as in his short-term problem solving. In other words, the visual inference strategies characterized in Fig. 9.1 and in Table 9.2 are used both for making particular discoveries and for integrating several distinct domains of phenomena over the long term.

Faraday's work is recorded in sufficient detail to show how different elements of the inference scheme are used, both together (as in the previous examples) and separately.[15] Visualization works in conjunction with investigative actions to produce a phenomenology of interpretative images, objects, and utterances.[16] The associated linguistic framework provides the basis for inferences about processes behind the phenomena. These proto-representations merge images and words in tentative interpretations of novel experience. This experience is created through the interaction of visual, tactile, sensorimotor, and auditory modes of perception together with existing interpretative concepts, including mental images. These word–image hybrids integrate several different types of knowledge and experience. Mental models developed "in the mind's eye" (i.e., without physically engaging the world), or by rule-based transformation of existing images, would lack such integrative power.

As I remarked earlier, the PSP scheme begins with a *reduction*, an image that simplifies and abstracts from some real-world, real-time process or phenomenon (see Fig. 9.1 and Table 9.2). Sometimes regularities are apparent —as with sand patterns on a beach—but it is usually necessary to construct 2D images, such as the *camera lucida* drawing in Fig. 9.2 or the plots in Fig. 9.9. Visualizations of large quantities of data, as in Figs. 9.6 and 9.9, are reductive abstractions. Visualized phenomena such as these initiate a visual inference process. In Faraday's case, such visualized phenomena played a particularly important role in shaping and developing his theory of electromagnetism. The earliest of these are shown in chronological order as Figs. 9.11A–C. Between 1820 and 1852, he moved from phenomena described in terms of lines of force to a general explanatory theory of space-filling systems of force that obey both empirical laws and principles of conservation and economy. To do so, he made many moves between fairly concrete visualizable representations and more abstract, theoretically informed visualizations such as those shown

---

[15]In addition to Faraday's laboratory notebooks (Martin, 1932–1936), a large number of extant instruments, and many samples of materials used by Faraday, we also have information obtained from the reconstruction of certain of his experiments (Gooding, 1989, chap. 7, this volume). I have discussed elsewhere methods for achieving consistency in identifying and interpreting the procedures that make up Faraday's investigations (Gooding, 1990b, 1992).

[16]I call them *construals* (Gooding, 1990, chaps. 1–3); Magnani (2001, pp. 53–59) described them as *manipulative abductions*.

in Figs. 9.11D–F. Our account begins with images derived from his earliest experiments on electromagnetism (Gooding, 1990a, chap. 2).

Experience and Technique

It is possible to identify the process by which images such as these are first constructed. One finds the same progression "up" the dimensions, followed by reduction or simplification to derive the original source phenomena and, through further iterations, new variations. In what follows I examine the process that produced the image of electromagnetic rotation (see Fig. 9.11B).

Electrical and magnetic effects are mixed in a way that the eye simply cannot see. So, Davy and Faraday practiced a method of *accumulation*, that is, they combined discrete images obtained over time into a single geometrical structure and, conversely, spatially distributed effects into a single image (Gooding, 1990a, chap. 2). By September of 1821 they had developed experimental methods of integrating discrete experimental events (or rather, integrating the images depicting them; Davy, 1821). A structure of needles arranged in a spiral around the wire and examined after discharging a current through it made a 3D magnetic "snapshot" of the magnetizing effect of the current. They also created a physical structure of sensors with which to record the effects of a single event at different points of space.[17] A typical procedure involved carefully positioning one or more needles in the region of a wire, connecting the circuit to a battery, and observing the effect on the needle(s). Finally, continuous manual exploration of the space around the wire would produce many discrete observations of needle positions. Davy and Faraday combined these results into a 3D model: a horizontal disc with needles arranged around its perimeter, which displayed the magnetization effects of a current. This setup integrates temporally distinct observations into a single spatial array, which is a possible model of certain features of the phenomenon depicted in Fig. 9.11A.

Sketches of Faraday's apparatus are just as informative as his images of phenomena. They show that Faraday construed many of his experiments as capturing a temporal slice—a "snapshot"—of the effect of some more complex, but hidden physical process. His experimental setups and procedures organize phenomena through construction and manipulation into phenomenological models. This involves more than visualization. At this stage, these models are complexes of material things, active manipulations, effects, and visualized interpretations of the outcomes (compare the examples in chap. 7, this volume). These visualized phenomena are inspected for features that give clues about process (cf. chap. 5, this volume). The conjectures may be expressed as images, which guide further exploration of struc-

---

[17]For a detailed, illustrated account see Gooding (1990a, chaps. 1–3).

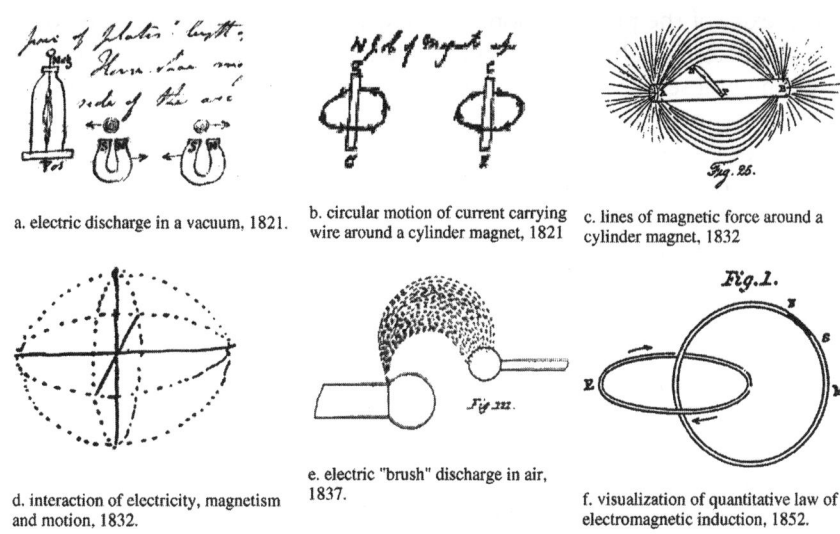

a. electric discharge in a vacuum, 1821.

b. circular motion of current carrying wire around a cylinder magnet, 1821

c. lines of magnetic force around a cylinder magnet, 1832

d. interaction of electricity, magnetism and motion, 1832.

e. electric "brush" discharge in air, 1837.

f. visualization of quantitative law of electromagnetic induction, 1852.

FIG. 9.11. Top row (a–c): three crucial images from Faraday's laboratory notes between 1821 and 1832. All attempt to depict in two dimensions a more complex three-dimensional phenomenon produced by manipulation. Bottom row (d–f): the key structural models of 1832, 1837, and 1852. Panels d and f are visual models that integrate most of the learned phenomenal properties of electricity and magnetism shown by the behaviors of the objects and structures in Panels a–c and e.

tures hidden from ordinary human vision. An important example is Davy's explanation of phenomena observed in an experiment carried out in May of 1821. Assisted by Faraday, he passed a current through a vacuum to produce a luminous glow discharge. Davy reported that when "a powerful magnet [was] presented to this [luminous] arc or column, having its pole at a very acute angle to it, the arc, or column, was attracted or repelled with a rotatory motion, or made to revolve by placing the poles in different positions, … as described in my last paper" (cited in Gooding, 1990a, chap. 2). Davy and Faraday construed the filamented structure of the arc as indicating the structure of the conduction process.

In September 1821, Faraday returned to the search for structure that could elucidate the nature of this process. During the summer, he had repeated many electromagnetic experiments by Oersted, Davy, Biot, and others. In these experiments he developed the image of a circle, created by accumulating small motions of a magnetized needle near a current. At the end of this period he made notes for experiments on September 3 and 4, 1821. These begin with a reexamination of the original magnet–wire interactions because, as he explained in his *Historical Sketch* of electromagnetic experiments, Faraday was convinced that Davy and others had missed something important in the complexity of the phenomenon (Faraday, 1821–1822). This work used the circular image as a heuristic for subsequent exploration with more complex experimental setups. To show how, I describe the steps in his exploration of the interaction of magnets and currents.

Figures 9.12A–G are a sequence of sketches extracted from 4 pages of Faraday's notes of his work on electromagnetic rotations. These sketches display the same progression from phenomena to pattern and from pattern to structure to animated model found in the previous examples.

1. Initial reductions: In Figs. 9.12A–B the arrows relate apparent needle *motions* to needle *positions* relative to the wire. Here Faraday was mapping attractions and repulsions as properties of the space around the wire. He used similar methods in 1831–1832 to map magnetism and in 1837 to map the electrostatic potential around charged conductors.

2. Enhancement: The images in Fig. 9.12C are produced by a move into three dimensions, by a mental rotation of the images in Fig. 9.12A through 90°. The images in Figs. 9.12C and 9.12D "accumulate" the preceding sets of images. Because these are already distillations of many manipulative actions, they incorporate learning based on proprioception as well as visualization.

3. Enhancement: In Fig. 9.12D the two circles indicate mental construction of possible motion in three dimensions from the motions as represented in Figs. 9.12A–C, so this (static) image is really a trajectory of a (possible) dynamical process. To be realized in the world, an appropriate apparatus would have to be built.

4. Reduction: Figures 9.12E–F show the various configurations of wires and magnets that were partially successful in producing unambiguously circular motion, but this was hand-assisted, not independent and continuous (see Gooding, 1990b).

5. Enhancement: Figure 9.12G is a sketch of the configuration of the first device in which the tip of a current carrying wire could move continuously in a circle around a vertical magnet. The circles drawn here depict the effect to be produced by this design.

6. Enhancement: The prototype motor constructed at the end of the day (Fig. 9.12G) realizes an arrangement of elements of these images. In particular, it reproduces the circular images directly and materially as the circular path traced out by the tip of the moving wire suspended in the dish of mercury. The circles Faraday drew in Fig. 9.12G now depict an *actual* effect (in contrast to those in Figs. 9.12D–F). The working apparatus constitutes the process behind the images.

7. Reduction: The motions sketched in Fig. 9.12D could now be reconstrued as tendencies to continuous motion, suitably constrained by a physical setup.

An explanation of images indicating the manipulation of patterns, structures and process models in Faraday's discovery of the electromagnetic rotation motor.

12a. In these sketches the arrows indicate the position of a magnetized needle viewed from the side; the wire is indicated by the vertical lines. Faraday's annotations label the resulting motion relative to the current-carrying wire.

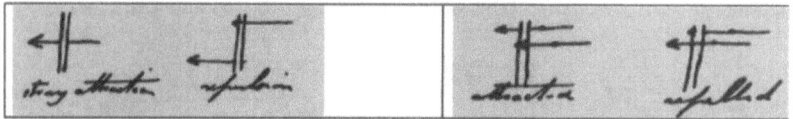

12b. Here the arrows again indicate positioning of a magnetized needle relative to a wire, now drawn at right angles to the page so the wire is seen as if from above, in section. The tendency of the resulting motion is written below each case.

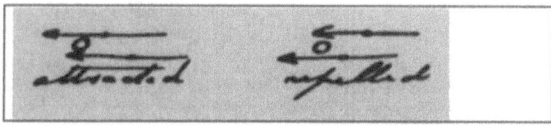

12c. In these sketches the eight different observed tendencies to motion of either a magnet or a wire are 'accumulated' or summarized in just two images. The arrows now indicate the position of the magnet while the tendency of motion of the wire (seen from above) is indicated by an 'A' or an 'R'.

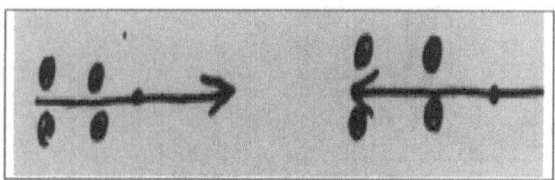

12d. The following sketch shows further compression of his observations. Here Faraday superimposes the two images above to make a single image of the eight detected positions of attraction and repulsion. He construes this as indicating the possibility of circular motion of wire or magnet.

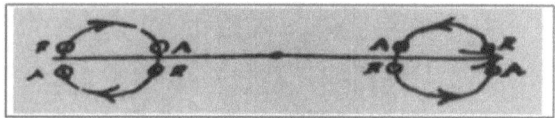

FIG. 9.12. (continued on next page).

12e. The next sketches describe configurations of wires (shown as curved vertical objects), the trajectories followed by the outermost point on a curved wire (where 'N' and 'S' denote the polarity of the current). As a mnemonic to heolp Faraday read his own sketches consistently, he notes that a 'rod' drawn as two vertical lines does not denote a wire but is "merely put there to shew the front and back part" (paragraphs 8-9).

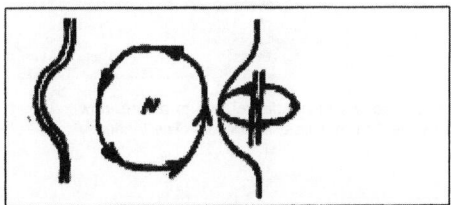

12f. The sketches below record an 'assisted' motion of the wire, that is, a circular trajectory produced by positioning the pole of a bar magnet perpendicular to the vertical wire. Reversing the polarity of the current or of the magnet produces opposing motions. This suggested to Faraday that "the effort of the wire is always to pass off at a right angle from the pole, indeed to go in a circle around it ..." (paragraph 11). The inscription in the left hand circle reads "the N pole being perpendicular to the wire".

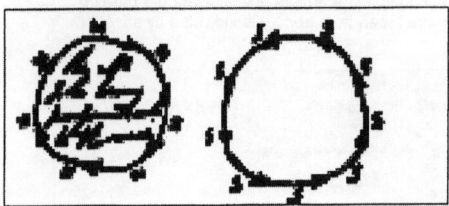

12g. The previous manipulation redefined the problem. Faraday reconfigured the device so as to arrange magnet and wire so that the latter could move while carrying an unbroken current, all without human intervention. This is shown in the final sketch in the sequence for 3 September 1821. The apparatus consists of a magnet fixed vertically in a basin of mercury into which a suspended mobile wire has been immersed. The labels 'c' and 'z' denote connections to the copper and zinc poles of a battery consisting of a single pair of plates having a very large surface area.

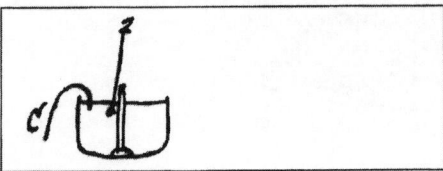

FIG. 9.12.   An explanation of images indicating the manipulation of patterns, structures, and process models in Faraday's discovery of the electromagnetic rotation motor.

Images and Inference

The images of circles in Figs. 9.12D–F represent possibilities elicited from a closely observed and manipulated world. To make this connection between image, apparatus, and phenomenon required pictorial and practical thinking of a high order. Except for the last sketch (Fig. 9.12G), these images do not depict objects already differentiated in experience from mental images. They are still tools for thinking, not images that depict its outcomes. In other words, the depictive role of the final set of images is something accomplished during the day's work. Faraday's notes show that he was trying to visualize a set of changing relationships. His sketches purport to represent something that is complex, dynamic, and emergent. They are early manifestations of a process of establishing a basis for shared experience and for communication about that experience. Moreover, each image itself stands for an accumulation of practical and theoretical knowledge. The meaning and function of such images therefore varies, depending on how it is used in relation to others that represent earlier and later work, just as the meaning of the sea-floor magnetization patterns (Fig. 9.9) was enhanced after the basalt extrusion model had been accepted.

Using the two dimensions of a matrix we can show how these images relate to each other. Table 9.3 represents the sequence as a series of dimensional enhancements and reductions. Faraday first added dimensionality (moving from left to right in Table 9.3). As explained earlier, the 2D representation of 3D structure in the top row of column 3 was constructed by rotating what is represented in Column 2 through 90°. These images are then "accumulated" by mental superposition to produce the hypothesis (in Row 3) that more than four discrete indicating positions should be detected. An image of continuous motion is implied in Column 3. This is visualized in Row 4, Column 2, as a pattern of points at which attractions and repulsions should occur. Further accumulation or integration into the image in Column 4 gives the new hypothesis that the structure of the motion should be circular. This is generalized in Columns 4 and 5. Faraday thus envisaged a physical device by making various physical manipulations as in Row 5. As a real working device, this is a kinesthetic representation of a phenomenon. The final rows illustrate how Faraday repeated the process in order to derive the continuous circular motion sought and to produce variations on the new phenomenon.

In discussing earlier examples, I pointed out that the status and meaning of an image changes—for example, the patterns in Fig. 9.9—are at first a suggestive (source) phenomenon and later become evidence for the visual theory constructed from them. This suggests that what may appear to be the same image or object can function in different areas of Newell's (1990) bands of cognition—both to generate new concepts (cognitive band) and

TABLE 9.3
Dimensions of Creativity

Dimensions:	2	3	4	Derivation
I. Representation	Four positions of magnetic attraction & repulsion	Structural counterpart of these positions	Complex process that cannot be observed until row VI	Pattern?
II. Initial reduction and enhancement to new feature	"strong attraction" "repulsion"	Viewed from above, in 3D		Symmetry in relationship of position to attractive or repulsive effect
III. Inference				More than four indicating positions should be found
IV. Re-presentation: second reduction and enhancement	Eight positions of attraction and repulsion	"Or"—this word indicates the positions being accumulated visually, as:	"Hence the wire moves in opposite circles"	Circulation of wire around magnet is possible

204

	First experimental setup	Inhibited, incomplete motions obtained	Reconfigure the setup
V. Representation depicts real world feature			
VI. New device produces new phenomenon implied by 2D and 3D interpretations of previous phenomena		 "the N pole being perpendicular to the ring"	 The first electric motor
VII. Use general features of original phenomena to derive other forms of rotation	 "attracted" "repelled"		Further new phenomena produced on September 4–6, 1821

"attracted"   "repelled"

Note. This table illustrates the changing representational power and meaning of the images that Faraday sketched in his laboratory book as he explored the spatial distribution of electromagnetic action and discovered a working configuration of wires, contacts, and magnets for the rotation motor. 2D = two dimensional; 3D = three dimensional.

to enlist support for their existence and interpretations of them (rational and social band). It is therefore significant that Faraday selected certain images and constructs that he had used to develop the prototype rotation motor (cognitive and rational band) and reproduced them as means of communicating aspects of the phenomena that indicate the invisible structure of electromagnetism (rational and social bands). Similarly, the motor was first sketched as a reconfiguration of wires, magnet, and other components that should produce continuous motion of the wire (cognitive and rational bands), but it was soon produced and disseminated as a working kinesthetic demonstration device (rational and social bands). Closer examination of cases where visual and kinesthetic representations serve this dual role by moving from one cognitive function to another may clarify whether the relationship between these different cognitive processes is, as Newell maintained, reductive (every learning process being decomposable into smaller and smaller units of knowledge and cognition).

Displaying the sketches in a 2D array allows one to view each image in context rather than as a self-contained depiction. Each image is part of a continuous process involving different kinds of manipulation and inference. We have seen that Faraday's constructive method involved moving from 2D patterns to 3D structures that could then be animated either as thought-experiments in time or as material, benchtop simulations of the invisible processes. To summarize: Faraday took an image to express patterns discernible in a process; he construed such patterns as indicative of hidden process. To investigate the latter involved imagining a process as transformations of a structure. If "frozen" in time, this process would have a structure that could generate the observed patterns as sections. The rotation motor produced a process (in four dimensions) that linked the patterns of needle–wire interaction (two dimensions) via a structural model (three dimensions).

Situating Other Images

Visual inference tables such as Table 9.3 can be constructed for many of Faraday's other discoveries, including his powerful visualization of the interconnectedness of electricity, magnetism, and motion (see Fig. 9.11D). However, such work extends over many months and years rather than the day or so recorded in Fig. 9.12. This suggests that the PSP scheme captures features of visual inference both at the micro level of discovery and at the macro level. The former involves coordinating visualization of highly situated physical and mental manipulations, such as the rotation motor (Fig. 9.12G), whereas the latter involves organizing and coordinating distinct problem-solving enterprises. Using this scheme, one can identify particular points at which visualization assists the analogical transfer of ideas between experimental domains.

Consider the structuring, or *striation,* of luminous electrical discharge, which Faraday first observed with Davy in 1821 (see Fig. 9.11A). In 1836–1837, Faraday mapped the structure of electrostatic potential around conductors, building up the picture shown in Fig. 9.13 by recording and plotting observations with an electroscope. Faraday took the 2D patterns produced by this exploration of 3D space to be sections through a 3D structure. Not satisfied with the static image of a possible structure, Faraday analyzed luminous electrical discharge (seen in 1821, Fig. 9.11A) using stroboscopic techniques developed with Charles Wheatstone in the 1820s. The images in Fig. 9.14 show what Faraday took to be the dynamical or process equivalent of the statical structure implied by the bottom image in Fig. 9.13. These images of the filamented "electric brush" display structure and changes in configuration produced by manipulation (varying the distance between the terminals, the conductivity or density of the medium, etc.).

The phenomenon imaged in Figs. 9.13 and 9.14 extends to electrostatics the same sort of dynamical relationship Faraday had already worked out for

Mapping the electrostatic field, 1835, Faraday's Diary volume II, pp. 412-14.

Each dot represents a point of equal electrostatic intensity, as indicated by the deflection of the leaves of a Bennett's electrometer.

Lines of equal electrostatic intensity (or induced charge) constructed by many scans such as the one illustrated above.

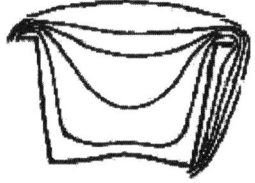

FIG. 9.13. Images generated by mapping lines and surfaces of potential as sketched in Faraday's notes of December 1835. Faraday aimed to produce a three-dimensional structural representation. He believed he could observe the transformations of this structure as imaged in Fig. 9.15. From *Faraday's Diary* (Vol. 2, pp. 412–414), T. Martin (Ed.), 1932–1936, London: Bell.

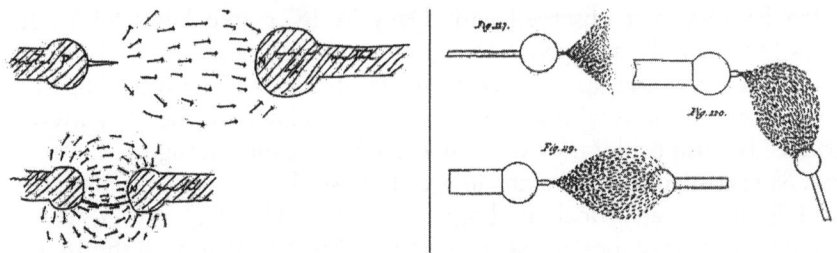

FIG. 9.14.   Left: Faraday's sketch of luminous electric discharge—the "electric brush." From *Faraday's Diary* (Vol. 2, paras. 3435–3436), T. Martin, (Ed.), 1932–36, London: Bell. Right: an engraving of the variable form of the brush. From *Experimental Researches in Electricity* (Vol. 2, plate X, figures 119—121), by M. Faraday, 1839–1855, London: Taylor and Francis. Copyright 1838–1835 by Taylor and Francis.

electromagnetic induction by 1832. This is shown in Fig. 9.15 with the accompanying instructions for "mental animation." This single image integrates what he had learned from many experiments about the interactions of electricity, magnetism, and motion. This image, which is arguably the most important visual model of his early scientific work, integrates information obtained through the different modalities of sight, touch, and proprioception.

When animated according to the instructions, this image is a visual theory. Faraday combined text and image to generalize certain learned properties of electricity and magnetism. This is a higher order visualization of a set of relationships that is implied by what happens to the distribution of lines of filings when magnets are manipulated (Fig. 9.11C) and to lines of discharge when the terminating points of an electric spark are manipulated (Fig. 9.14). It is also more abstract and more general than, for example, Faraday's visualization of rotating lines of force, an image he used to derive (from the

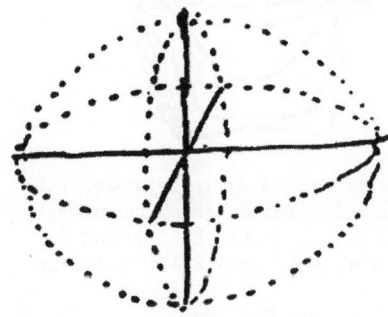

FIG. 9.15.   Faraday's sketch of March 26, 1832, of a central unifying image of the mutual relations of magnetism, electricity, and motion. It is accompanied by verbal instructions that describe how to animate the elements of the drawing to show the mutual dependence of electricity, magnetism, and motive force. Taken together, image and text comprise a visual hypothesis about electromagnetism. From *Faraday's Diary* (Vol. 1, p. 425), T. Martin (Ed.), 1932–1936, London: Bell.

physical independence of the system of lines from the originating magnet) the phenomenon of unipolar induction. This thought experiment was soon realized on the bench by making a cylinder magnet rotate within a conducting circuit of which it formed a part. This visualizes the assumption of the physical independence of a system of magnetic lines and deduces from it a physical consequence for a particular situation.

Faraday's 1832 "animation" of the interactions of electricity, magnetism, and motion—learned by active manipulation of magnets and currents both in wires and in luminous form (Fig. 9.11D)—could explain (or so he thought) both the static patterns and the dynamical effects produced by electricity and magnetism. Many years later, and after a substantial amount of further work, Faraday produced a simpler and still more general image that shows how the relationship between electric and magnetic action can be expressed quantitatively, by construing them as fluxes (quantities of action measured across a unit area or section; see Fig. 9.11F). This is the highest level of abstraction at which these images work, allowing the integration of different domains (static and current electricity, electrochemistry, magnetism, light, etc.). Images of lines making up structures involved in processes enabled Faraday to compare and make analogies between features of the phenomena he observed at the highest level of complexity that his skills of material and mental manipulation could support (Wise, 1979).

SITUATING IMAGES IN INFERENCE

Images have a context that they themselves help to generate: They are produced, interpreted, manipulated, and applied. Faraday's well-known magnetic lines (Fig. 9.11C) can now be seen in the context of the experiential knowledge that produced them, knowledge that they also represent. However, images such as this are also generative, as shown by his use of the two rings on an axis of attraction or repulsion (Fig. 9.12D) to derive the possibility of continuous circular motion (Fig. 9.12G), or his visualization of iron filings as a system of lines to deduce the phenomenon of unipolar induction. Faraday's theoretical aspirations are expressed as much through these images as through the many words that he wrote.

Far from functioning in isolation from other modes of perception or from other persons as sources of experience, visual perception integrates different types of knowledge and experience. There is a complex interaction of different kinds of doing: hands-on manipulation, looking, listening, imagining, thinking, and argumentation. This type of visualization depends on and interacts closely with other forms of perception; it is a sort of *perceptual inference* that is not reducible to visual perception. Much of the cognitive power of images resides in this integrative capability, which is central to inference in many sciences. The integrative role of images may also explain their value in facilitating analogical transfer between different phenomenological domains and, more abstractly, between different problems.

Visual Inference and Newell's Bands of Cognition

An important feature of creative processes, identified in many case studies, is the ability to maintain several lines of inquiry and transfer fruitful ideas and methods between them. In many cases this involves (or at least resembles) analogical transfer. This can be achieved in several ways at different levels of abstraction. The most concrete and intuitive would be "direct" recognition of a similarity at the perceptual level, such as Robert Young's recognition of similarities between the behavior of water waves and ray-diagrams of optical interference (Miller, 1996). This would also rely on fast cognitive processing that is closely coupled to evolved neural structures. At a more abstract level are similarities whose significance derives from being features generated by structural or process models in different domains. Examples are Faraday's recognition of the potential significance of relationships between the behavior of electric filaments (Fig. 9.14A), electrostatic lines (Figs. 9.11E and 9.13) and magnetic lines (Figs. 9.11C and 9.14). At a still higher level of abstraction is his recognition of the significance of changes in the appearance of high-frequency processes (e.g., acoustical patterns) with changes such as frequency and the density of the ambient medium (e.g., pitch of the sound made by the electric brush). Here, as we saw earlier in the case of paleobiology and geophysics, the ability to see the significance of similarities that are perceived directly depends on deliberative kinds of reasoning with and about visual models of inferred or hidden process. Communicating the meaning of these representations involves both deliberation and negotiation about them.

The PSP scheme relates the recognition and use of similarities at these different levels. It therefore suggests an alternative, nonhierarchical way of relating task-based studies of cognitive processes of varying duration. If one takes direct perception of similarities as pattern matching to be biologically engendered and operating in Newell's (1990) biological band, then the cases described earlier suggest that processing in the biological, cognitive, and rational bands can be at work simultaneously and not sequential, as the decomposition thesis implies. The integrative role of images and the relative ease of visually supported analogical transfers between distinct but simultaneous lines of inquiry also suggests simultaneity of process or, in Newell's terms, considerable overlap and interaction of processes in the biological, cognitive, rational, and social bands. To return to the schematic representation of visual inference set out in Fig. 9.1: Recognition of patterns, and of similarities between patternlike features at Node B, would involve biological-band processes, but some of those 2D representations are generated by transformations on representations of structures or processes, for example, derivation of the 2D sea-floor striping pattern (Fig. 9.9) as a consequence of the 4D basalt extrusion process (Node D) as modeled in the 3D block diagram in Fig. 9.10 (Node C). These transformations are motivated (hence rational-band processes) and require image manipulation and interpretation that I take to be cognitive-band processes.

It should not be supposed that the PSP schema can be mapped onto Newell's (1990) cognitive hierarchy (or vice versa); rather, the attempt to locate the types of cognition involved in producing and using images of the sort one finds at Nodes B, C, and D of Fig. 9.1 indicates how cognitive processes of differing duration and representational complexity work together to enable complex problem solving in the cases described in the VISUAL INFERENCE and VISUAL INFERENCE AND MENTAL PROCESS sections.

CONCLUSIONS

In their static, printed form, images inevitably appear much less process-like than scientists such as Morris, Whittington, Vine, Matthews, Teusch, Davy, or Faraday intended. Yet the examples I have presented show that the dimensionally reduced image is construed as imaging a momentarily arrested continuous process. In all of the examples, the goal is to explain certain regularities in phenomena by discovering and modeling a process. We have seen that discovery goes hand in hand with the visualization of structures that can relate the observations (patterns) to processes. This method allows scientists to vary the level of abstraction to suit what they perceive to be constant about the physical relationships between the variables involved. The scheme is at work both in the personal realm of the mind's eye and embodied interaction with the world as well as in the interpersonal domain of collaboration and negotiation.

The PSP scheme describes a method whereby scientists move between images of varying complexity and other kinds of representation, such as propositions. The interplay of verbal and visual representations is shown in Table 9.3. This highlights four different roles for images. Each role corresponds to a different stage of the process of constructing a new representation and integrating it into an argument. Images:

- may be instrumental in decomposing complex problems into simpler ones, by *generating new representations* or in extending the use of existing ones;

- enable the transfer of knowledge between domains by *promoting analogical transfers* between different lines of inquiry that make up a network of enterprise;
- enable construction of a complex model from simpler elements to produce an *integrated model* of a process that involves many more variables than the eye or the mind could otherwise readily comprehend. In these two cases, visualization is essential to the construction and use of interpretative and analytical concepts;[18]
- *enable empirical support* for the theory embodied by the model, usually through the dissemination of images in 2D form. Here the visualization of observations or data provides part of the content of a verbal argument developed to generalize and communicate the models produced by dimensional enhancements.

In this chapter we have seen that the PSP schema describes how scientists in different domains use images generatively to develop and establish the depictive capability of other images. Establishing the depictive status of an image goes hand in hand with establishing its relevance to an explanation of the phenomena it depicts. The PSP schema is an empirical generalization, which we can develop into a qualitative theory about how visualization works to integrate information from different sensory sources and of differing degrees of abstractness.

ACKNOWLEDGMENTS

I am grateful to Mike Gorman, Chris Philippidis, Ciara Muldoon, Frank James, Barbara Tversky, and Ryan Tweney for their comments and suggestions. Completion of this work was supported by a Leverhulme Research Fellowship.

REFERENCES

Anderson, J. R. (2002). Spanning seven orders of magnitude: A challenge for cognitive modelling. *Cognitive Science, 26,* 85–112.
Beaulieu, A. (2000). *The space inside the skull: Digital representations, brain mapping and cognitive neuroscience in the decade of the brain.* Unpublished doctoral dissertation, University of Amsterdam, Amsterdam, Holland.
Beaulieu, A. (2001). Voxels in the brain. *Social Studies of Science, 31,* 635–680.

---

[18]There is a strong analogy here to the visualized mental models of Edison and Bell—see Carlson and Gorman (1990) and Gorman (1997).

Carlson, B., & Gorman, M. E. (1990). Understanding invention as a cognitive process. *Social Studies of Science, 20*, 387–430.

Cartwright, N. (1983). *How the laws of physics lie.* Oxford, England: Oxford University Press.

Cartwright, N. (1999a). *The dappled world: A study of the boundaries of science.* Cambridge, England: Cambridge University Press.

Cartwright, N. (1999b). Models and the limits of theory: Quantum Hamiltonians and the BCS model of superconductivity. In Morgan and Morrison (Eds.), *Models as mediators: Perspectives on natural and social science* (pp. 241–281). Cambridge, England: Cambridge University Press.

Cartwright, N., Shomar, T., & Suarez, M. (1995). The tool box of science. In L. Nowack & W. Herfel (Eds.), *Theories and models in scientific processes* (pp. 137–149). Atlanta, GA: Rodopi.

Clark, A. (1997). *Being there: Putting brain body and world together again.* Cambridge, MA: MIT Press.

Clark, A. (2002). Towards a science of the bio-technical mind. *Cognition and Technology, 1*, 21–33.

Cooper, L. A. (1976). Demonstration of a mental analog of an external rotation. *Perception & Psychophysics 19*, 296–302.

Cooper, L. A., & Shepard, R. N. (1984). Turning something over in the mind. *Scientific American 251*, 106–114.

Davy, H. (1821). On the magnetic phenomena produced by electricity. *Philosophical Transactions of the Royal Society, 11*, 7–19.

de Chadarevian, S., & Hopwood, N. (Eds.). (in press). *Models: The third dimension of science.* Stanford, CA: Stanford University Press.

Dunbar, K. (1999). Scientific thinking and its development. In R. Wilson & F. Keil (Eds), *The MIT encyclopedia of cognitive science* (pp. 730–733). Cambridge, MA: MIT Press.

Faraday, M. (1821–1822). Historical sketch of electromagnetism. *Annals of Philosophy, 18*, 195–222, 274–290, and 19, 107–121.

Faraday, M. (1839–1855). *Experimental researches in electricity* (3 vols.). London: Taylor and Francis.

Faraday, M. (1859). *Experimental researches in chemistry and physics.* London: Taylor and Francis.

Faraday, M. (1859b). On a peculiar class of optical deceptions. In M. Faraday (Ed.), *Experimental researches in chemistry and physics* (pp. 291–309). London: Taylor and Francis. (Original work published 1831)

Fodor, J., & Pylyshyn, Z. (1988). Connexionism and cognitive architecture: A critical analysis. *Cognition, 28*, 3–71.

Fujimura, J. (1992). Crafting science: Standardized packages, boundary objects and "translation." In A. Pickering (Ed.), *Science as practice and culture* (pp. 168–214). Chicago: Chicago University Press.

Galison, P. (1996). Computer simulations and the trading zone. In P. Galison & D. Stump (Eds.), *The disunity of science: Boundaries, contexts, and power* (pp. 118–157). Stanford, CA: Stanford University Press.

Galison, P. (1997). *Image and logic. A material culture of microphysics.* Chicago: Chicago University of Chicago Press.

Galison, P., & Stump, D. (1996). *The disunity of science: Boundaries, contexts, and power.* Stanford, CA: Stanford University Press.

Gardner, M. (1985). *The mind's new science*. New York: Basic Books.

Gentner, D., & Stevens, A. (Eds.). (1983). *Mental models*. Mahwah, NJ: Lawrence Erlbaum Associates.

Giaquinto, M. (1992). Visualizing as a means of geometrical discovery. *Mind and Language, 7*, 382–401.

Giere, R. (1988). *Explaining science: A cognitive approach*. Chicago: University of Chicago Press.

Giere, R. (1994). The cognitive structure of scientific theories. *Philosophy of Science, 64*, 276–296.

Giere, R. (1999a). Using models to represent reality. In L. Magnani, N. Nersessian, & P. Thagard (Eds.), *Model-based reasoning in scientific discovery* (pp. 41–57). New York: Kluwer Academic.

Giere, R. (1999b). *Science without laws*. Chicago: University of Chicago Press.

Gooding, D. C. (1982). Empiricism in practice: Teleology, economy and observation in Faraday's Physics. *Isis, 73*, 46–67.

Gooding, D. C. (1985). In nature's school: Faraday as a natural philosopher. In D. Gooding & F. James (Eds.), *Faraday rediscovered* (pp. 105–135). New York: Macmillan/ American Institute of Physics.

Gooding, D. C. (1986). How do scientists reach agreement about novel phenomena? *Studies in History & Philosophy of Science, 17*, 205–230.

Gooding, D. C. (1989). History in the laboratory: Can we tell what really went on? In F. James (Ed.), *The development of the laboratory* (pp. 63–82). London: Macmillan.

Gooding, D. C. (1990a). *Experiment and the making of meaning*. Dordrecht, The Netherlands: Kluwer.

Gooding, D. C. (1990b). Mapping experiment as a learning process. *Science, Technology and Human Values, 15*, 165–201.

Gooding, D. C. (1990c). Theory and observation: The experimental nexus. *International Studies in Philosophy of Science, 4*, 131–148.

Gooding, D. C. (1992). Mathematics and method in Faraday's experiments. *Physis, 29*, 121–147.

Gooding, D. C. (1996). Creative rationality: Towards an abductive model of scientific change. *Philosophica: Creativity, Rationality and Scientific Change, 58*, 73–101.

Gooding, D. C. (1998). Picturing experimental practice. In M. Heidelberger, & F. Steinle (Eds.), *Experimental essays—Versuch zum experiment* (pp. 298–323). Baden-Baden, Germany: Nomos.

Gooding, D. C. (2003a). Varying the cognitive span: Experimentation, visualisation and digitalization. In H. Radder (Ed.), *The philosophy of scientific experiment* (pp. 255–283). Pittsburgh, PA: University of Pittsburgh Press.

Gooding, D. C. (2003b). Visualization, inference and explanation in the sciences. In R. Paton & D. Leishman (Eds.), *Visual representations and interpretation*. Dordrecht, The Netherlands: Kluwer Academic.

Gooding, D. C., & Addis, T. (1999). A simulation of model-based reasoning about disparate phenomena. In L. Magnani, N. Nersessian, & P. Thagard (Eds.), *Model-based reasoning in scientific discovery* (pp. 103–124). New York: Kluwer Academic.

Goodman, N. (1978). *The languages of art*. New York: Hackett.

Goodwin, C. (1990). Embodied skills in a geochemistry lab. In D. Gooding (Ed.), *Bath 3: Rediscovering skills* (pp. 69–70). Bath, England: University of Bath.

Goodwin, C. (1995). Seeing in depth. *Social Studies of Science, 25*, 237–274.

Gorman, M. (1997). Mind in the world: Cognition and practice in the invention of the telephone. *Social Studies of Science, 27*, 583–624.

Gould, S. J. (1989). *Wonderful life: The Burgess shale and the nature of history.* New York: Penguin.

Gregory, R. L. (1981). *Mind in science.* New York: Penguin.

Gruber, H. E. (1974). *Darwin on man. A psychological study of scientific creativity.* Chicago: University of Chicago Press. (Original work published 1974)

Gruber, H. E. (1994). Insight and affect in the history of science. In R. Sternberg, & J. Davidson (Eds.), *The nature of insight* (pp. 397–431). Cambridge, MA: MIT Press.

Gruber, H. E., & Davis, S. N. (1988). Inching our way up Mount Olympus: The evolving-systems approach to creative thinking. In R. Sternberg (Ed.), *The nature of creativity: contemporary psychological perspectives* (pp. 243–270). Cambridge, England: Cambridge University Press.

Hacking, I. (1983). *Representing and intervening.* Cambridge, England: Cambridge University Press.

Hacking, I. (1992). The self-vindication of the laboratory sciences. In A. Pickering (Ed.), *Science as practice and culture* (pp. 29–64). Chicago: University of Chicago Press.

Hanson, N. R. (1972). *Patterns of discovery.* Cambridge, England: Cambridge University Press.

Henderson, K. (1991). Flexible sketches and inflexible databases: Visual communication conscription devices and boundary objects in design engineering. *Science, Technology and Human Values, 16,* 445–1472.

Henderson, K. (1999). *On line and on paper: Visual representations, visual culture, and computer graphics in design engineering.* Cambridge, MA: MIT Press.

Holmes, L. (1991). *Hans Krebs. The formation of a scientific life, volume 1.* Oxford: Oxford University Press.

Ippolito, M. F., & Tweney, R. D. (1995). The inception of insight. In R. J. Sternberg & J. E. Davidson (Eds.), *The nature of insight* (pp. 433–462). Cambridge, MA: MIT Press.

Isacks, B., Oliver, J., & Sykes, L. R. (1968). Seismology and the new global tectonics. *Journal of Geophysical Research, 73,* 5855–5899.

Johnson-Laird, P. (1983). *Mental models.* Cambridge, MA: Harvard University Press.

Jones, C., & Galison, P. (1998). *Picturing science: Producing art.* New York and London: Routledge.

Kemp, M. (2000). *Visualizations: The nature book of art and science.* Oxford, England: Oxford University Press.

Klahr, D. (2000). *Exploring science: Cognition and the development of discovery processes.* Cambridge, MA: MIT Press.

Kosslyn, S. (1981). The medium and the message in mental imagery: A theory. In N. Block (Ed.), *Imagery* (pp. 207–244). Cambridge, MA: MIT Press.

Kosslyn, S. (1994). *Image and brain.* Cambridge, MA: MIT Press.

Kurz, E. M., & Tweney, R. D. (1998). The practice of mathematics and science: From calculus to the clothesline problem. In M. Oaksford & N. Chater (Eds.), *Rational models of cognition* (pp. 415–438). Oxford, England: Oxford University Press.

LeGrand, H. E. (1990). Is a picture worth a thousand experiments? In H. E. LeGrand (Ed.), *Experimental inquiries* (pp. 241–270). Dordrecht, The Netherlands: Kluwer Academic.

Latour, B. (1990). Drawing things together. In M. Lynch & S. Woolgar (Eds.), *Representation in scientific practice* (pp. 19–68). Cambridge, MA: MIT Press.

Latour, B., & Woolgar, S. (1986). *Laboratory life: The construction of scientific facts.* Princeton, NJ: Princeton University Press.

Lynch, M., & Woolgar, S. (Eds.). (1990). *Representation in scientific practice*. Cambridge, MA: MIT Press.

Magnani, L. (2001). *Abduction, reason and science*. Dordrecht, The Netherlands: Kluwer Academic Press.

Magnani, L., Nersessian, N., & Thagard, P. (Eds.). (1999). *Model-based reasoning in scientific discovery*. New York: Kluwer Academic Press.

Martin, T. (1932–1936). *Faraday's diary* (7 vols.). London: Bell.

Miller, A. I. (1986). *Imagery in scientific thought*. Cambridge, MA: MIT Press.

Miller, A. I. (1994). Aesthetics and representation in art and science. *Languages of Design, 2*, 13–37.

Miller, A. I. (1996). *Insights of genius: Imagery and creativity in science and art*. New York and London: Copernicus.

Morrison, M. S., & Morgan, M. (1999). Models as mediating instruments. In M. S. Morgan & M. Morrison (Eds.), *Models as mediators: Perspectives on natural and social science* (pp. 10–37). Cambridge, England: Cambridge University Press.

Nersessian, N. (1988). Reasoning from imagery and analogy in scientific concept formation. In A. Fine & J. Leplin (Eds.), *Philosophy of Science Association* (Vol. 1, pp. 41–48). East Lansing, MI:

Nersessian, N. (1999). Model-based reasoning in conceptual change. In L. Magnani, N. Nersessian, & P. Thagard (Eds.), *Model-based reasoning in scientific discovery* (pp. 5–22). New York: Kluwer Academic/Plenum.

Nersessian, N. (2002). The cognitive basis of model-based reasoning in science. In P. Carruthers, S. Stich, & M. Siegal (Eds.), *The cognitive basis of science* (pp. 133–153). Cambridge, England: Cambridge University Press.

Newell, A. (1990). *Unified theories of cognition*. Cambridge, England: Cambridge University Press.

Pickering, A. (1989). Living in the material world: On realism and experimental practice. In D. Gooding, T. Pinch, & S. Schaffer (Eds.), *The uses of experiment* (pp. 275–297). Cambridge, England: Cambridge University Press.

Pickering, A. (1990). Knowledge, practice, and mere construction. *Social Studies of Science, 20*, 652–729.

Pickering, A. (Ed.). (1992). *Science as practice and culture*. Chicago: University of Chicago Press.

Pylyshyn, Z. W. (1981). The imagery debate: Analogue media versus tacit knowledge. *Psychological Review, 88*, 16–45.

Raff, A., & Mason, R. (1961). Magnetic survey off the west coast of North America 40-520N. *Geological Society of America Bulletin, 72*, 1267–1270.

Rudwick, M. J. S. (1976). The emergence of a visual language for geology, 1760–1840. *History of Science, 14*, 149–195.

Rudwick, M. J. S. (1985). *The great devonian controversy*. Chicago: University of Chicago Press.

Shelley, C. (1996). Visual abductive reasoning in archaeology. *Philosophy of Science, 63*, 278–301.

Shepard, R. N. (1978). Externalization of mental images and the act of creation. In B. S. Ranhawa & W. E. Coffmann (Eds.), *Visual learning, thinking and communication* (pp. 133–189). New York/London: Academic Press.

Star, S. L., & Griesemer, J. (1989). Institutional ecology, translations and boundary objects: Amateurs and professionals in Berkeley's Museum of Vertebrate Zoology, 1907–39. *Social Studies of Science, 19*, 97–140.

Stenning, K., & Oberlander, J. (1995). A cognitive theory of graphical and linguistic reasoning: Logic and implementation. *Cognitive Science, 19,* 97–140.

Suarez, M. (1999). The role of models in the application of scientific theories: Epistemological implications. In M. S. Morgan & M. Morrison (Eds.), *Models as mediators: Perspectives on natural and social science* (pp. 168–196). Cambridge, England: Cambridge University Press.

Teutsch, H. (1988). Regionality of glucose-6-phosphate hydrolysis in the liver lobule of the rat. *Hepatology, 8,* 311–317.

Teutsch, H., Altemus, S., Gerlach-Arbeiter, S., & Kyander-Teutsch, T. (1992). Distribution of 3-hydroxybutyrate dehydrogenase in primary lobules of rat liver. *Journal of Histochemistry and Cytochemistry, 40,* 213–219.

Teutsch, H. E. (in press). Modular design of the liver of the rat. In R. Paton & D. Leishman (Eds.), *Visual representations and interpretations.* Amsterdam, The Netherlands: Elsevier.

Teutsch, H. E., Shuerfeld, D., & Groezinger, E. (1999). Three-dimensional reconstruction of parenchymal units in the liver of the rat. *Hepatology, 29,* 494–505.

Trickett, S. B., Fu, W.-T., Schunn, C. D., & Trafton, J. G. (2000). From dipsy-doodles to streaming motions: Changes in the representation of visual scientific data. In *Proceedings of the 22nd annual conference of the Cognitive Science Society* (pp. 959–964). Mahwah, NJ: Lawrence Erlbaum Associates.

Tversky, B. (1998). Three dimensions of spatial cognition. In M. Conway, S. Gathercole, & C. Coroldi (Eds.), *Theories of memory* (Vol. 2, pp. 259–275). London: Psychology Press.

Tversky, B. (2002). Spatial schemas in depictions. In M. Gattis (Eds.), *Spatial schemas and abstract thought* (pp. 79–112). Cambridge, MA: MIT Press.

Tweney, R. D. (1989). A framework for the cognitive psychology of science. In B. Gholson, W. R. Shadish, Jr., R. A. Neimeyer, & A. C. Houts (Eds.), *Psychology of science: Contributions to metascience* (pp. 342–366). Cambridge, England: Cambridge University Press.

Tweney, R. D. (1992). Stopping time: Faraday and the scientific creation of perceptual order. *Physis, 29,* 149–164.

Van Leeuwen, C. I., Verstijnen, X., & Hekkert, P. (1999). Common unconscious dynamics underlie common conscious effects: A case study in the iterative nature of perception and creation. In J. S. Jordan (Ed.), *Modeling consciousness across the disciplines* (pp. 179–219). Lanham, MD: University Press of America.

Vine, F. J. (1966). Spreading of the ocean floor: New evidence. *Science, 154,* 1405–1415.

Vine, F. J., & Matthews, D. (1963). Magnetic anomalies over oceanic ridges. *Nature, 199,* 949–950.

Vine, F. J., & Wilson, J. T. (1965). Magnetic anomalies over a young oceanic ridge off Vancouver Island. *Science, 150,* 485–489.

Westfall, R. S. (1980). *Never at Rest: A biography of Isaac Newton.* Cambridge, England: Cambridge University Press.

Whittington, H. B. (1985). *The Burgess shale.* New Haven, CT: Yale University Press.

Wise, M. N. (1979, March). The mutual embrace of electricity and magnetism. *Science, 203,* 1310–1318.

10

Problem Representation in Virginia Woolf's Invention of a Novelistic Form

Maria F. Ippolito

University of Alaska, Anchorage

> That [problem] representation makes a difference is a long-familiar point.... Solving a problem simply means representing it so as to make the solution transparent.... [Yet,] we have only a sketchy and incomplete knowledge of the different ways in which problems can be represented and much less knowledge of the significance of the differences. (Simon, 1981, pp. 153–154)

Getzels and Csikszentmihalyi (1976) studied artistic problem solving and designated the process of problem representation as *problem finding*. They, too, found that problem representation made a significant difference in the quality of problem solutions. In their oft-cited longitudinal study, art students were provided with various materials and videotaped as they created and drew a still life. Problem finding by the students with concurrent artistic ability (as determined by a panel of judges) and artistic success (i.e., continued employment as an artist 7 years later) was manifested in the initial selection of unusual sets of items; handling and repositioning of those items prior to the commencement of drawing; and the periodic reexamination and/or rearrangement of the still-life materials throughout the drawing process—giving consideration to qualities of the items other than their identities or utilities (e.g., their textures and the interplay of light with the objects). Student artists who devoted extended, thoughtful effort to problem representation (i.e., to the choice, arrangement, and characterizing of the items) produced artistic products that garnered high evaluations while they were in art school, and they continued to be employed in the highly competitive field of art for several years after graduation.

My aim in this chapter is, first, to characterize several heuristics that contribute to problem representation (or finding) and, second, to delineate the process of problem representation as to English novelist Virginia Woolf's (1882–1941) invention of a novelistic form. Before I proceed to a discussion of problem finding, it is important to note that the selection or construction (when necessary) of appropriate problem representations is one of the hallmarks of expertise—the end-product of 10 or more years of dedicated practice in a particular field (Ericsson, 1996; Richman, Gobet, Staszewski, & Simon, 1996; Simon, 1981; Walberg, 1988; Weisberg, 1991). Woolf, in fact, began writing her first experimental novel, *Jacob's Room*, in 1919—approximately 10 years after she wrote in her 1908 journal that in her writing she would like to "attain a different kind of beauty, achieve a symmetry by means of infinite discords, showing all traces of the mind's passage through the world; and achieve in the end, some kind of whole made of shivering fragments" (Woolf, 1990b, p. 393).

THE PSYCHOLOGICAL STUDY OF EXPERTISE AND WRITING PROBLEM SOLVING

Over a century ago, William James (1890/1950b) characterized experts as able to " 'see into the situation' … full of delicately differenced ingredients which their education has little by little brought to their consciousness but of which the novice gains no clear idea" (p. 344). Most studies of the psychology of writing in particular and the development of expertise in general have consisted of comparisons of the problem solutions of groups of amateurs and experts engaged in the solving of identical problems (e.g., Bereiter & Scardamalia, 1987; Chase & Simon, 1973; Chi, Glaser, & Farr, 1988; deGroot, 1946/1974; Ericsson, 1996; Ericsson & Smith, 1994; Flower & Hayes, 1980; Scardamalia & Bereiter, 1982). The findings of these studies suggest that, in essence, an expert becomes the analogue of a library scientist responsible for devising a means of cataloguing, maintaining, and arranging for the accessibility of a large quantity of related information. Experts necessarily organize or represent their relevant knowledge in ways that make knowledge germane to the solving of particular problems readily available and ease the acquisition of additional, pertinent information. This is achieved by means of an organization that emphasizes the interconnections of various aspects of domain-relevant knowledge, thus taking advantage of the critical role of meaningfulness in optimal human memory performance (e.g., chunking within one's domain of expertise). Expertise in various domains is thus typified by an enhanced capacity to perceive complex patterns (Bereiter & Scardamalia, 1987; Chi et al., 1988; Ericsson, 1996; Ericsson & Smith, 1994; Scardamalia & Bereiter, 1982). According to Woolf (1912/1986a), "a good novelist ... goes about the world seeing squares and circles where the ordinary person sees storm-drift" (p. 361).

Studies that compared the problem solving of experts and novices have found that experts typically have a history of dedicated practice and an enhanced ability to perceive relevant complex patterns as noted earlier. In addition, experts tend to exhibit superior metacognitive skills and apportion substantial time to problem analysis, including, in the case of expert writers, substantial persistent attention to the representation of even the most mundane problems across the duration of the writing task (Chi et al., 1988; Ericsson, 1996; Ericsson & Smith, 1994; Flower & Hayes, 1980; Scardamalia & Bereiter, 1991). That is, unlike experts in many fields where expertise brings with it greater ease in representing mundane tasks in their domain, "expert writers generally are found to work harder at the same assigned tasks than non-experts, engaging in more planning and problem solving, more revision of goals and methods, and in general more agonizing over the task" (Scardamalia & Bereiter, 1991, p. 172). Thus, writing expertise is defined by a greater emphasis on problem finding, making the study of writing an ideal source of information on possible strategies deployed in problem representation.

In summary, there are typically at least two distinctions between the problem solving of experts and novices: (a) greater ease by experts in representing and solving mundane problems in their domains and (b) higher quality of problem solutions within the domains or fields in which they are experts. As to expert writers, there are apparently no mundane or well-defined writing problems, although the higher quality of problem solutions by expert writers compared with novice writers is evident (Scardamalia & Bereiter, 1991). The generally ill-defined nature of writing problems, which research on the psychology of writing has confirmed, may have been what triggered American novelist, poet, and Pulitzer and Nobel prize winner Ernest Hemingway to remark of writers that "we are all apprentices in a craft where no one ever becomes a master" (quoted in Bryson, Bereiter, Scardamalia, & Joram, 1991, p. 81; see also Scardamalia & Bereiter, 1991).

Although the studies of expertise I have discussed achieved their goal of highlighting the distinctions between expert and novice problem solving—among these distinctions the differing problem representations typically employed by novices and experts—the design and intent of such nomothetic studies necessarily obscure the specifics of the strategies relied on in problem finding. Echoing the extended problem representation efforts that are characteristic of writing problem solving, it is my contention that what generally distinguishes expert from creative problem solving across domains is whether the problems targeted can be solved via what amounts to the *selection* of an appropriate problem representation (i.e., well-defined problems) or whether problem solving requires the construction or significant revision of an available problem representation (i.e., ill-defined problems). I further argue that the extensive construction and/or expansion of known problem representations that I have postulated as characteristic of the solving of ill-defined problems is due to the fact that the solving of these types of problems occurs in domains whose topographies have not been clearly defined, for example, the domain of stream-of-consciousness novels at the time Woolf set about trying to capture the flight of the mind in her writing. This is not dissimilar from the contrast Kuhn (1970) made between (intra-domain) puzzle solving and (domain-altering) paradigm shifts in science. He contended that scientists who generate creative products target problems that are not solvable within established, externally defined domains but instead require the restructuring or expansion of an established domain (in Woolf's case, the domain of the English novel; in Einstein's, the domain of Newtonian physics).

I should note before proceeding that the distinction Kuhn (1970) made between puzzle solving and paradigm shifts is more likely a continuum than a dichotomy, rather like the distinction made earlier between the solving of well- and ill-defined problems; dichotomies are used here for convenience in discussing the issues under consideration. The major point is that focusing on the solving of ill-defined problems should ensure the availability for study of an extended period of problem representation. Because writing problems are typically ill defined and expert writers particularly fluent, focusing on writing problem solving is likely to offer a fruitful opportunity to obtain information on the strategies deployed in problem representation.

THE NATURE AND ROLE OF PROBLEM FINDING IN PROBLEM SOLVING

As a result of many protocol-based and computer modeling studies of human cognition, Newell and Simon (1972) argued that "problem solving takes place via search in problem spaces. This principal is a major invariant to problem-solving behaviour that holds across tasks and across subjects" (p. 44; see also Simon, 1989a, 1989b). For them, a *problem space* is a "closed space of knowledge determined by the nature and goals of the task at hand .... [However,] problem spaces can be changed and modified during [problem] solving" (Newell & Simon, 1972, p. 44). Furthermore, human problem solving rarely consists of the kind of brute-force searches at which computers are facile; rather, human problem solvers use heuristics to represent or find the problem spaces to be searched (Simon, 1989b; see also Newell & Simon, 1972; Simon, 1989a). For example, Simon (1989b) pointed out that the relatively bounded domain of chess "admits some 10^{120} possible games .... [, yet] even chess-playing computer programs do not search spaces much larger than 10^6 and human chess players are only capable of searching problem spaces far less extensive" (p. 188). Like the development of expertise, the successful search of problem spaces is purposeful and knowledge-based and it relies on the problem solver's ability to recognize and utilize significant information to make progress. In essence, each domain of expertise encompasses a repertoire of problem representations or spaces wholly or partially suitable for the resolution of the types of problems targeted by the domain. An example of such problem representations in the field of psychology would be the research designs conveyed to students in psychology methods courses.

Perkins (1995; see also Perkins, 1981, 1991; Weber & Perkins, 1989) built on Newell and Simon's (1972) characterization of problem spaces and their potential malleability, suggesting particular features of the problem spaces searched and modified by individuals intent on solving puzzles versus those engaged in solving novel problems. Perkins (1995) designated problem spaces as "possibility spaces" and envisioned them as "a branching tree of possibilities," the exploration of which can lead to the resolution of well-defined problems as well as less structured problems, such as devising inventions or a new form of novel (p. 508). "In cases such as chess play, the end states are set in advance by the strict formal rules. In contrast, in many creative endeavours, the goals and, hence, what count as end states evolve along with the problem" (Perkins, 1995, p. 511).

Once Woolf had the germinal scene for a novel, she had to find a form to contain it. This form provided helpful signposts to assist her writing problem solving and was capable of evolving in response to the overarching scene as well as being adaptive to the terrain of the English language and the literary convention known as the novel. For example, the germinal scene of her fifth novel, *To the Lighthouse* (Woolf, 1927/1989b), centered on "father's character, sitting in a boat, reciting 'We perished, each alone,' while he crushes a dying mackerel" (A. O. Bell & McNeillie, 1980, pp. 18–19). The form Woolf found for this novel was a three-part structure, which she depicted schematically in her diary as two rectangles joined by a narrow passageway (A. O. Bell & McNeillie, 1980). The three parts Woolf envisioned were: "1. At the drawing room window; 2. seven years passed; 3. The voyage" (A. O. Bell & McNeillie, 1980, p. 36). "At the drawing room window" centers around a canceled trip to the lighthouse. Seven years (the narrow passage) later, the summer home is revisited and the long anticipated trip to the lighthouse finally takes place. For Woolf, the germinal scene and companion form of her novels served something of the role that hypotheses do in psychological research studies: They limited the scope of the investigation of human psychology that followed. Given the initial scene and form of *To the Lighthouse*, it needed to be enriched, thickened, and—in language that mirrors Perkins's (1995)—given the "branches and roots" that Woolf had yet to conceive (A. O. Bell & McNeillie, 1980, p. 36).

Perkins (1995), like Newell and Simon (1972), maintained that the targeted problem is instrumental in determining the topology of the problem space. Perkins (1995), like Kuhn (1970), specified critical differences between the possibility spaces that are generated by well-defined versus ill-defined problems, the solving of which require creative effort. Specifically, the possibility spaces of well-defined problems, like the mundane problems easily solvable by experts, are primarily "homing spaces" that are rich in clues "that point the way toward ... the goal .... By tracking these clues, the search mechanism can home in fairly easily on a resolution to the task at hand" (Perkins, 1995, p. 512). Successful solution of the targeted problem, while not guaranteed, is likely for the expert navigating a homing space; in many cases, algorithms can be relied on to rapidly advance the search of the homing space. The possibility space of an ill-defined problem, in contrast, was designated by Perkins (1995) as a *Klondike space*:

> A *Klondike space*, by definition [provides] no clear path of clues ... to goal states .... Imagine, for instance, searching for gold in the Klondike. Pockets of gold occur here and there in the gravel deposits and riverbed, but just where is hard to say. One needs to cast about, try downstream and over the hill, look for surface signs. Only when one gets close and detects the glitter among the gray pebbles can one home in .... That is, when one gets close enough to an end state in a Klondike space, one begins to see clues that point the way. (pp. 514–515)

> An effective search process needs to cast about exploring the space widely in order to hit on clues that point the way to an end state .... The human problem solver at work in a Klondike space needs, in addition to conventional clue-oriented expertise, another kind of expertise. An effective search process in a Klondike space involves casting a wide net, avoiding redundant coverage of some regions, searching for new regions altogether. (Perkins, 1995, pp. 514–515)

As Woolf (1916/1978a) pointed out: "Where everything may be written about, the difficulty is to know what to leave out" (p. 53). *Night and Day*, Woolf's (1919/1992) realistic novel that preceded her first stream-of-consciousness novel, *Jacob's Room*, taught Woolf "what to leave out by putting it all in" (Nicolson & Trautmann, 1980, p. 216). In *Jacob's Room*, Woolf (1922/1960) wrote: "In short, the observer is choked with observations.... The difficulty remains—one has to choose" (pp. 68–69).

"Because a Klondike space does not provide information about where to search next, search processes run long, with modest progress" (Perkins, 1995, p. 516). Thus, individuals at work on ill-defined problems, which have the potential to generate innovations, are at work in problem spaces that are unlikely to be well mapped, contain a significantly greater possibility for misdirected searches, and require comparatively lengthy searches. Initial searches of such problem spaces are less likely to benefit from reliance on algorithms (formulaic approaches) to shortcut the problem resolution process.

On the one hand, problem solving in homing spaces relies more heavily on responding to clues than on problem finding, hence the enhanced facility of experts versus novices in arriving at solutions to mundane problems. Experts are, by virtue of their extensive domain-specific knowledge, familiar with and likely to quickly select an appropriate problem representation and are better at spotting and recognizing the significance of the clues embedded therein. Conversely, the solving of ill-defined problems begins in Klondike spaces where there is greater reliance on problem finding (mapping the possibility space to identify promising areas to search). Certainly, expertise in a related field (e.g., Newtonian physics for Einstein) may be useful, but it is not sufficient to ensure success for the individual engaged in extra-domain problem solving. Einstein and Infeld (quoted in Hayes, 1989; see also Getzels & Csikszentmihalyi, 1976) contrasted problem representation to the clue-driven solution construction primarily utilized in solving well-defined scientific problems:

> The formulation of a problem is often more essential than its solution, which may be merely a matter of mathematical or experimental skills. To raise new questions, new possibilities, to regard old problems from a new angle, requires creative imagination and marks real advance. (p. 140)

Again, homing spaces and Klondike spaces are dichotomized here for the purpose of discussion; the structuredness of problems is probably most realistically conceived as a continuum. At one end of the continuum the representation of textbook-type problems typically requires only momentary consideration by an expert and quickly gives way to clue-driven solution construction. However, as I pointed out earlier, the relative unstructuredness of writing problems in general and, particularly, problems requiring creative solutions such as the re-formation of the English novel targeted by Woolf, engender a shift away from the prominence of solution construction toward an extended, initial emphasis on problem representation.

Once an ill-structured problem is appropriately represented, the search process is successively constrained and the problem space reconfigured until the problem solver finds him- or herself in a homing space(s). Thus, once

Woolf found a preliminary scene and viable form for a given novel, the subsequent search of the possibility space was significantly limited. The problem space was re-formed via successive constraints instituted until Woolf found herself amidst the tedium of rewriting, the meticulous matching of scene and form and language that Perkins (1995) designated as relying on clue-oriented expertise. The implication is not that solution construction is an unimportant feature of problem solving but that—given the extensive area of search in Klondike-type problem spaces (e.g., the numerous possible germinal scenes, the various forms for a given germinal scene, and the multitude of linguistic choices given a scene and associated form)—problem representation is of the utmost importance in limiting the search to potentially fruitful regions. The mapping of such possibility spaces is likely a heuristic-driven process.

Before I proceed with a discussion of how best to, in Perkins's words, "cast a wide net," that is, to a description of the several problem-finding heuristics designated as inceptual processing that Woolf and others utilized in representing ill-defined problems—I should point out that Woolf characterized her problem finding in a manner similar to Perkins's (1991, 1995; see also Ippolito & Tweney, 1995) conception, even using the same searching-for-gold analogy. While composing her second experimental novel, Woolf wrote:

> I'm working at [*Mrs. Dalloway*], and think it is a very interesting attempt; I may have found my mine this time I think. I may get all my gold out .... And my vein of gold lies so deep, in such bent channels. To get it I must forge ahead, stoop and grope. But it is a gold of a kind I think. (Nicolson & Trautmann, 1978, p. 292)

PROBLEM REPRESENTATION HEURISTICS (INCEPTUAL PROCESSING)

In accordance with the previous discussion, the solving of ill-defined problems involves constructing, periodically reformulating, and extensively searching the representation of a previously ill-defined problem to cordon off homing spaces or possibility spaces to arrange for "illuminations, matches struck unexpectedly in the dark, ... [that make] of the moment something permanent ... [and strike] this eternal passing and flowing ... into stability" (Woolf, 1927/1989b, p. 161) or, in accordance with Tweney's (1992b) characterization of the problem solving of physicist Michael Faraday, to "enhance the multiplicity of small events during a simultaneous moment of time" (p. 164).

Inceptual processing (see Ippolito & Tweney, 1995) refers to a set of interacting heuristics that potentially cordon off promising areas of the extended possibility spaces of ill-defined problems, thus differentiating potentially fruitful areas of an extensive Klondike space for further exploration. The heuristics that make up inceptual processing are viewed as cooperative contributors to problem representation. It is within the constraints that emerge from inceptual processing that a concentrated search that ultimately gives way to clue-driven solution construction takes place.

Clearly, the work of problem solving has only begun when problem representation yields to solution construction, and the exploration of homing spaces may necessitate periodic problem re-representation (see Karmiloff-Smith, 1992) to extend or revise the boundaries of the possibility space being searched. Specifically, on the basis of a number of psychological studies of human development, which she considers an ill-defined problem that requires creative effort by the developing individual, Karmiloff-Smith (1992) concluded that human development is characterized by the construction of increasingly flexible and manipulable representations of information about the way the world works. These representations arise from the back-and-forth of gathering knowledge and constructing successive re-representations based on the knowledge gleaned (see also Gooding, 1990; Shekerjian, 1990; Tweney, 1990).

In accordance with the distinction made earlier between problems that are easily solved by experts versus ill-defined problems, it is later, "as a result of the sharing of discoveries ... that the domains in which insightful discovering originated acquire the patina of conventionality" (Ippolito & Tweney, 1995, p. 434). In essence, a new domain, such as that of the stream-of-consciousness novel, emerges simultaneously with the development of its own repertoire of potentially useful problem representations. That is, the solving of the problem of writing a stream-of-consciousness novel by Woolf, James Joyce, and Dorothy Richardson suggested problem representations that might be utilized in subsequent stream-of-consciousness novels. Readers now take for granted that the novel they plan to read will provide access to the inner psychological lives of the characters, and writers have numerous exemplars of how to convey the distributed, nonlinear workings of the mind using the linear symbol system of the English language. In the sections that follow, I characterize the three heuristics that I contend contribute to problem representation: (a) perceptual rehearsal, (b) distillation of inceptions, and (c) construction of emulations.

Perceptual Rehearsal

Perception is generally viewed by researchers as a "reduction process whereby noisy, variable, and impoverished patterns of environmental energy become resolved into stable and consistent internal representations optimal for human performance" (Flowers & Garbin, 1989, p. 148). Thus, when microscopist Robert Hooke undertook the daunting task of constructing reproductions of some of the smallest of living creatures (e.g., lice, fleas) using the significantly flawed lenses of early versions of the compound microscope, Hooke (1987, third unnumbered page of Preface) contended that

> the first thing to be undertaken in this weighty work, is a watchfulness over the failings and an enlargement of the dominion, of the senses. To which end it is requisite, first, that there should be a scrupulous choice, and a strict examination, of the reality, constancy, and certainty of the particulars we admit. This is the first rise whereon truth is to begin, and here the most severe, and most impartial diligence must be employed; the storing of all, without any regard to evidence or use, will only tend to darkness and confusion.

On the one hand, perception in set ways can streamline one's performance and obviate the necessity of storing all collected perceptions; on the other hand, perceiving in the habitual way (and failing to enlarge the dominion of the senses) can interfere with the detection of information vital to the resolution of ill-defined problems. Schooler, Fallshore, and Fiore (1995) cited Bruner and Potter (1964) and their own unpublished nomothetic findings to support their claim that "recognition of out-of-focus pictures can be hampered by mental set" (p. 577); that is, Bruner and Potter (1964) asked research participants to identify pictures of common objects that slowly came into focus; the "greater or more prolonged the initial blur, the slower the eventual recognition, [suggesting that] interference may be accounted for partly by the difficulty of rejecting incorrect hypotheses based on substandard cues" (p. 424). The suggestion, then, is that habits or preconceived hypotheses of perception may sometimes place out of reach information that is vital to problem resolution.

A character in Woolf's (1925/1990a, p. 52) novel *Mrs. Dalloway* speaks of breaking the bonds of mental habit: "He had escaped! was utterly free— as happens in the downfall of habit when the mind, like an unguarded flame, bows and bends and seems about to blow from its holding." This sense that the absence of mental habits engenders the free flow of thought might lead one to "predict that some measures of perceptual performance that tap perceptual organization processes would be negatively related to measures of creative ability" (Flowers & Garbin, 1989, p. 148). However, if, as Gregory claimed, "perception must ... be a matter of seeing the present with stored objects from the past" (i.e., *must* be shaped by habit), and if attempts to store all perceived objects necessarily lead to confusion rather than understanding, as Hooke claimed, it is optimal not to abandon the effort to perceive in an organized way (i.e., leave the flame unguarded; quoted in John-Steiner, 1997, p. 22). Rather, one should strive to instantiate selected perceptual habits via purposeful practice at (or rehearsal of) the perceiving of particular events—to trim the wick and reposition the chimney, if you will. The argument here is, therefore, not that all perceptual processing can or should be deautomatized but that both consciously controlled and involuntary perceptual processes can be reshaped via purposeful practice as Stratton (1896; see also Csikszentmihalyi, 1991; Flowers & Garbin, 1989; Ippolito, 1994; Ippolito & Tweney, 1995) did when he wore lenses that turned his world upside down. Specifically, Stratton's experiment involved "wearing special eye glasses that turn the visual field through an angle of 180 degrees so that retinal images ... [were] inverted" (Goldenson, 1984, p. 715). At first, objects appeared to be upside down to Stratton, and his eye–hand coordination was significantly affected (Stratton, 1896). However, "by the third day things had thus been interconnected into a whole by piecing together parts of the ever-changing visual fields," and Stratton learned to perceive a world inverted by the lenses as upright (Stratton, 1896, p. 616). It is, therefore, not surprising when an accomplished musician detects features of a musical selection that cannot be heard by listeners who are less musically accomplished.

In fact, Gibson (1969) documented significant improvement in both the speed and accuracy of research participants' responses as a result of practice at various perceptual tasks, including unidimensional detection tasks (e.g., visual acuity, the two-point limen), complex detection tasks (searching for targets, finding embedded figures), and estimation of dimensions (of single or multidimensional stimuli). Ericsson and Faivre (1988) and Gibson (1969), furthermore, reported improvement in the discrimination of simple stimuli (e.g., differential limens, unlabeled hues, pitch) due to practice. That is, basic and complex perceptual skills across sensory modalities can be measurably altered with practice. There is even evidence that absolute threshold (the minimal energy necessary for an observer to detect a particular stimulus) can be improved with practice across perceptual modalities (Ericsson & Faivre, 1988; Gibson, 1969; Goldstein, 1989). In summary, perceptual set can be overcome by the willful instatement of a new set of perceptual habits.

The representation of all problems necessarily begins with perception—the collection of data from the world—but the finding of ill-structured problems may well require developing new perceptual habits. William James (1890/1950b) wrote more than a century ago that "genius, in truth, means little more than the faculty of perceiving in an unhabitual way" (p. 110). I would take issue with James and, instead, contend that genius *begins with* but is more than the faculty of perceiving in an unhabitual way. As a result of concerted effort, an individual can consciously circumvent habits of perception and thereby gain access to information that might typically be unavailable. This ability clearly becomes paramount when an ill-defined problem is targeted; in such cases—as Perkins (1995) pointed out—expectations may be violated frequently, and previously unnoticed details may be critical to successful problem solution. It was thus incumbent on Woolf to select different perceptions from her environment than those that had generally served as the fodder for the realistic novels from which she intended her novels to be a departure.

Ippolito and Tweney (1995) designated this retraining of the senses as *perceptual rehearsal*: "the saturation of one or more of the senses with all aspects of the phenomenon of interest to the discoverer ... [which is] a means of defeating the inherent biases of our perceptual apparatus and increasing the impact of unexpected sights" (p. 435; see also Eysenck & Keane, 1990; Ippolito, 1994). Perceptual rehearsal or practiced familiarity with particular classes of perceptions permits individuals engaged in the solving of ill-structured problems to move beyond initial, automatized perception of the phenomena of interest such that patterns invisible to the unrehearsed observer will be detected and remembered (Csikszentmihalyi, 1991; Ippolito & Tweney, 1995; Karmiloff-Smith, 1992). For example, the scientific notebooks of physicist Michael Faraday are exceedingly rich in examples of his efforts to choreograph particular perceptual events. When Faraday studied acoustical vibrations, he utilized various weights of particles and even liquids, placed obstructions on the vibrating metal plates, and gaps between the plates, and so on, to determine the origin and nature of the resulting patterns of movement (Tweney, 2003). English physicist Isaac Newton (quoted in Briggs, 1990, p. 60) "said when asked how he was able to make his discoveries that it was because 'I keep the subject constantly before me, and wait until the first dawnings open slowly little and little into the full and clear light.'" French novelist Guy De Maupassant (quoted in Allott, 1967) spoke of this process as "looking at everything one wants to describe long enough, and attentively enough .... In order to describe a fire burning or a tree in the field, let us stand in front of that fire and that tree until they no longer look to us like any other fire or any other tree" (p. 130). Perceptual rehearsal permits the observer to sense more, to be alert, as Einstein was, "to small signals in the large 'noise' of ... [the] experimental situation" (Holton, 1971–1972, p. 96).

As the result of perceptual rehearsal, "the observer's ability to 'see' the phenomena directly can [ultimately] be dispensed with *even in principle*" (Tweney, 2003). That is, perceptual rehearsal eventually allows the decoupling of perceptual mechanisms from sensory input. This extraction or disconnection of viable perceptions from their contexts permits the subsequent reconstruction of the workings of selected aspects of the world independent of sensory receptor input.

Whether the observer was Einstein, Faraday, Newton, or de Maupassant, and whether the observed was the nature of the physical world or human nature, the effort was to extract vital information from the world via perceptual rehearsal. Requisite to success in solving ill-defined problems—all of which begin with a vast, previously unexplored possibility space—is a perspective on reality not bound by the initial conclusions of the senses. However, the gathering of information by the senses is only the initial activity in detecting and utilizing the immutable patterns that are foundational to problem representations that yield creative products; that is, successful problem representation occurs only when perceptual rehearsal cooperates with the other heuristics described in the following sections (the distillation of inceptions and construction of emulations), which utilize the data gathered during perceptual rehearsal.

Distillation of Inceptions

The distillation of inceptions, the second heuristic I discuss in this chapter, provides potential advantages for individuals attempting to solve ill-defined problems. Subsequent to perceptual rehearsal, the mind's eye turns inward, and perceptions give way to inceptions. Thomas Hardy (quoted in Allott, 1967, p. 74) referred to the novel as "an artificiality distilled from the fruits of closest observation," and this phrase aptly defines an *inception*. I use the term *inception* to contrast it with *perception*—on the sensory–imagistic side—and *conception* or *categorization*—on the fully developed symbolic side. An *inception* is defined in the dictionary as "the beginning of something"; the word derives from the Latin "to take in" (Morris, 1969, p. 664). Inceptions of the phenomena of interest are the beginning of the subsequent construction of emulations, the next problem representation heuristic I discuss.

Inceptions are made from the raw material of perceptions taken into the mind. They embody the constraints of the real world perceptions from which they have been distilled, where *distillation* consists of being "separated from … unrelated or attenuating factors" (Morris, 1969, p. 382). Thus Faulkner constructed an imaginary community from his inceptions, Yoknapatawpha County, in his mind or "attic" (Millgate, 1989). "All his characters … had always been there, waiting … [and] ready to make another public appearance" (Millgate, 1989, p. 38). Similarly,

> Nersessian's (1992, p. 40) account of physicist James Maxwell's creative process implies exactly such entities: "Maxwell generated a representation of electromagnetism … at an intermediate level of abstraction: concrete enough to give substance to the relationships … and yet abstract enough to generate a novel representation. (Ippolito & Tweney, 1995, p. 443)

Physicist Richard Feynman (Feynman, Leighton, & Sands, 1964) similarly wrote of his conception of electromagnetic fields:

When I start describing the magnetic field moving through space .... I see some kind of vague shadowy, wiggling lines ... When I talk about the electro-magnetic field in space, I see some kind of superposition of all the diagrams which I've ever seen drawn about them. (pp. 20-9–20-10)

The viability of these reconstructed perceptions (sans their real world context) depends on emergence from the distillation process of inceptions that operate in ways veridical to reality. Rather than made representations of the real world, inceptions are the made beginnings of an imagined world. Inceptions are, as Woolf wrote, those "little creatures in my head which won't exist if I don't let them out" (Nicolson & Trautmann, 1978, p. 161); from "those invaluable seeds ... one can grow something that represents ... people's experiences" (Woolf, 1985, p. 135). So, to this point, perceiving in an unhabitual way (perceptual rehearsal) permits access to previously un-detected aspects of the environment, and the collected perceptions, if via-ble, can be separated from the complex external environment and manipulated sans their real world context.

It is not difficult to see how the ability to view real world processes inde-pendent of the real world provides the problem solver with an advantage (e.g., see Gooding, 1990; Gorman & Carlson, 1990; Shepard, 1978). In the thought experiment that served as the germ for his theory of relativity, Ein-stein chased "a point on a light wave"; by transcending the physical world, the properties of light became clear to him (Miller, 1989, p. 174). Such experi-ences can be so vivid they are perceived as physically occurring. Flaubert (quoted in Allott, 1967, p. 155) reported that when he "wrote about Emma Bovary's poisoning ... [he] had the taste of arsenic ... strongly in ... [his] mouth." Similarly, "the perfection and power of Beethoven's musical imagi-nation" outlived his ability to perceive sounds (Rolland, 1929, p. 268).

While perceptual rehearsal and the subsequent distillation of inceptions may seem an inefficient or wasteful approach in terms of a single problem, the inceptions that are the products of perceptual rehearsal may contribute, not just to extant, but also to future problem-finding efforts. For Faraday, as for Faulkner, inceptual scraps that did not meet the selection criteria for the problem at hand were not discarded but set aside. Faraday's carefully in-dexed diaries then provided him with a readily accessible verbal and picto-rial record of his perceptual *and* inceptual experiences (Tweney, 1991). As to Woolf's diaries, she intended them

to resemble some deep old desk or capacious hold-all, in which one flings a mass of odds and ends .... I should like to come back ... and find that the col-lection had sorted itself and refined itself and coalesced ... into a mold trans-parent enough to reflect the light of life and yet steady, tranquil, composed, with the aloofness of a work of art. (A. O. Bell, 1977, p. 266)

Like perceptual rehearsal, the distillation of inceptions is a heuristic that may be returned to throughout problem finding (and solving)—as Faraday did, as Pablo Picasso did during the construction of *Guernica,* and as expert problem solvers in general and expert writers do (Arnheim, 1962; Bereiter, Burtis, & Scardamalia, 1988; Bereiter & Scardamalia, 1987; Duncker, 1945; Gooding, 1990; Ippolito, 1999; Ippolito & Tweney, 1995; Scardamalia & Bereiter, 1982, 1991; Tweney, 1992a, 1992b, 1994). As Hooke (1987, 7th unnumbered page of the Preface; see also Karmiloff-Smith, 1992) contended, understanding must begin

> with the hands and eyes, and ... proceed on through the memory, to be continued by the reason; nor is it to stop there, but to come about to the hands and eyes again, and so, by a continual passage round from one faculty to another, it is to be maintained in life and strength, as much as the body of man is by the circulation of the blood through the several parts of the body.

With reference to writing problem solving, Scardamalia and Bereiter (1982, 1991; Bereiter & Scardamalia, 1987) found that writers engage in a dialectical process in which they repeatedly shift attention from reshaping the writing plan devised (problem re-representation) to the execution of the writing task and back again: "The writer represents general or superordinate goals at the beginning of the writing session ... [and] subgoals are constructed on-line during composing, which may modify, in significant ways, the nature of the initial goal structure" (Bryson et al., 1991, p. 66).

In the words of Anaïs Nin (quoted in John-Steiner, 1997), "if a stage is too cluttered, a description too heavy, too opaque, then certain elusive elements will be obscured. The process of distillation, of reduction to the barest essential ... is necessary to filter, to eliminate the upholstery" (p. 28). The larger generalization in which the distilled inceptions ultimately participate is designated here as a mental simulation or *emulation,* and it is to the characterization of the heuristic of constructing emulations that I turn next. The distillation of inceptions is an important precursor to the construction of emulations for, as Simon (1981) pointed out, "the more we are willing to abstract from the detail of a set of phenomena, the easier it becomes to simulate the phenomena" (p. 20).

Construction of Emulations

Emulations serve to cordon off potentially fruitful areas of vast, unfamiliar Klondike spaces to be meticulously surveyed. An emulation serves as a kind of divining rod to guide the exploration of the vast possibility space of an ill-defined problem. Specifically, I use the term *emulations* to designate the dynamic patterns that are the end products of the process that begins with perceptual rehearsal and later yields to the distillation of inceptions. In a real sense, emulations are not dissimilar from mental models or simulations in the physical world.

Mental models have been characterized in a number of ways (Eysenck & Keane, 1990). Johnson-Laird (1983) designated mental models as representations that serve as a kind of high-level programming language in the human brain. Mental models differ from images in that "images correspond to views of models, as a result of either perception or imagination ... [Mental models] represent the perceptible features of corresponding real-world objects" (Johnson-Laird, 1983, p. 157). Further, mental models can be "wholly analogical, or partly analogical and partly propositional" (Eysenck & Keane, 1990, p. 235). Like emulations, "these models can be run, in the sense that they can simulate the behaviour of a physical system" (Eysenck & Keane, 1990, p. 406). Thus, in the sense that mental models are more than just images or perceptions, are runnable for an extended period, and are distillations of particular patterns, Johnson-Laird's characterization is compatible with the notion of emulations presented here. However, the mental emulations discussed here are necessarily narrower in scope. They *must be* runnable and analogous to the real world phenomena they represent in that they abstract the essential structure and functioning of the phenomena of interest. In addition, emulations may simulate not just objects *but also processes* or combinations of objects and processes; that is, emulations must capture the essential elements and/or dynamic patterns of the phenomena/ process of interest. The emulation, which serves as a kind of dynamic template, suggests which areas should be fully explored within the vast, unstructured possibility space of, for instance, the new kind of novel Woolf intended to invent. Emulations are, by definition, more than images (as mental models are). However, emulations are presymbolic in form and are, therefore, not equivalent to high-level programming language in the brain or easily translated into symbolic form. The emphasis on the dynamic character of emulations is intended to convey a conception of mental models as more akin to organisms than to devices that can be reduced to the enumeration of structures or translated into machine language. Henry James (quoted in Allott, 1967) placed this restriction on the novel as well: "A novel is a living thing, all one and continuous, like every other organism, and in proportion as it lives will it be found, I think that in each of the parts there is something of each of the other parts" (p. 234).

The construction of emulations requires the arrangement and rearrangement of inceptions into a workable model or simulation. Einstein referred to the mental rearrangement of selected inceptions as *combinatory play,* which he identified as "an essential feature in productive thought—before there is any connection with logical construction in words or other kinds of signs which can be communicated to others" (Holton, 1971–1972, p. 103).

Poincaré (1946, quoted in Epstein, 1991) spoke of "ideas [that] rose in crowds" and collided "until pairs interlocked ... making a stable combination" (p. 366).

The construction of emulations may, in part, satisfy the drive toward simplicity and unity thought to underlie creative discoveries (Briggs, 1990; John-Steiner, 1997). In Einstein's words: "Man seeks for himself ... a simplified and lucid image of the world ... That is what the painter does, and the poet, the speculative philosopher, the natural scientist, each in his own way" (quoted in Holton, 1978, pp. 231–232). The major requirement of emulations is dynamic workability. Montgomery (1988) wrote of insight as having occurred when, through manipulation of the constructed mental model, one has achieved a "structural balance" (p. 86). In the words of novelist Henri Flaubert (quoted in Allott, 1967):

> This novel makes me break out in a cold sweat! ... Each paragraph is good in itself and there are some pages that are perfect, I feel certain. But ... *it isn't getting on*. It's a series of well-turned, ordered paragraphs which do not flow on from each other. I shall have to unscrew them, loosen the joints, as one does with the masts of a ship when one wants the sails to take more wind. (p. 231)

At this point, Flaubert's novel was not runnable. The paragraphs and pages of *Madame Bovary* had to be unscrewed, loosened, and reconstructed to achieve a viable emulation of human psychology. Emulations serve as dynamic templates that isolate particular venues of extensive possibility spaces. Emulations thus facilitate the selection and arrangement of features to be incorporated in the solutions of ill-structured problems.

In his discussion of perception and creativity, Cupchik (1999) contended that

> While the artist is able to closely observe a natural scene, he or she must still be selective when it comes to choosing which qualities of that scene, hue, tone, texture, and so on will be emphasized in the work. These qualities then become the focus of the piece, lending it coherence, and the artist lets this emerging image guide the unfolding work (a bottom-up process). The artist can also match the unfolding work against the rules that determine the choice of subject matter and style (a top-down process). (p. 356)

While the bottom-up process described by Cupchik closely matches the activity that has been designated here as *perceptual rehearsal* and the construction of inceptions, designating the top-down process as an optional alternative, as this description of artistic creativity does, leaves unspecified the criteria that guide the selection process. As pointed out earlier, Woolf— and others, such as Hooke and Faraday—did not engage in unconstrained searches of the environment to collect any and all available perceptions. These problem solvers were initially guided by an ill-defined problem and proceeded with the back-and-forth of scrupulously collecting particular perceptual information (perceptual rehearsal), refining this data (the distillation of inceptions), and constructing a mental model or emulation of the targeted phenomena. The emulations then highlighted areas of the possibility space meriting attentive, systematic search, and hence the possibility space was successively refined.

Neither the inceptual processing model described here nor Woolf's account of her writing problem solving entertain the possibility that viable problem solutions can emerge without interacting bottom-up *and* top-down processes, with the top-down process generating increasingly refined problem representations and generating additional episodes of perceptual rehearsal, inception distillation, and restructuring of the emulation as necessary.

This critique of Cupchik (1999), that the resolution of novel problems such as those tackled by artists requires *both* the bottom-up selection of data from the world and an internalized guiding emulation, is relevant to the long-standing emphasis on selection in discussions of insight and creativity. That is, discrimination between information to be included versus that which is discarded—and, hence, the importance of appropriately constraining the problem representation—is critical to the generation of innovations. For example, Henry James (quoted in Allott, 1967, p. 76) spoke of "life being all inclusion, and art being all discrimination and selection." The mathematician Poincaré said that "discovery is discernment, selection" (quoted in Westcott, 1968, p. 49). Ernest Hemingway was unsuccessful in completing his short story "The Killers" until he got the selection process right. In Hemingway's words: "I guess I left as much out of 'The Killers' as any story I ever wrote. Left out the whole city of Chicago" (quoted in Hotchner, 1966, p. 164).

It is not surprising that isolating a particular area of a vast possibility space for search requires extensive effort. Again, brute-force searches—either of problem spaces in the mind or of the world—are unlikely to yield solutions, even in the more constrained problem spaces of, say, chess. Both artists and scientists seek to reveal some aspect of reality; toward that end, they purposefully discard that which clouds or obscures, selecting that which clarifies or unifies, and the generation of a viable emulation gives direction to the selection process.

The claim, then, is that inceptual processing assists problem representation or finding. The emulation is made from real world perceptions that have been transformed into inceptions. Inceptions are discovered, and emulations are constructed, not in the real world but in the mind of the problem solver. Appearances come to have a different meaning for the problem solver after the emergence of the emulation, for this emulation now focuses the search of the possibility space as well as driving the selection of future perceptions, changing the way the individual interacts with the his or her internal world and external environment. The problem solution (what was done) then formalizes the problem representation (why one proceeded in a particular way).

It is when the problem representation refined via the heuristics of inceptual processing is described using symbols and images that the audience of a particular problem solution is able to share or understand the representation of the problem solver. On this account, the finished product (whether a novel or a mathematical proof) is a communicated problem representation that others construct in their minds. In terms of Woolf's novels, the reader was intended to share the perspective of the discoverer, so "the reader [is] relieved of the swarm and confusion of life and branded effectively with the particular aspect which the writer wishes him to see" (Woolf, 1926/1967a, pp. 132–135). Woolf (1932/1986b) wrote in a review of *Robinson Crusoe*:

All alone we must climb upon the novelist's shoulders and gaze through his eyes until we, too, understand in what order he ranges the large common objects upon which novelists are fated to gaze: man and men; behind them Nature; and above them that power which for convenience and brevity we may call God. (p. 52)

In this same vein, Woolf (1929/1967c) also wrote:

Indeed the first sign that we are reading a writer of merit is that we feel this control at work on us .... We feel that we are being compelled to accept an order and to arrange the elements of the novel—man, nature, God—in certain relations at the novelist's bidding .... From the first page we feel our minds trained upon a point which becomes more and more perceptible as the book proceeds and the writer brings his conception out of darkness. (pp. 100–101)

"The success of masterpieces seems to lie ... in the immense persuasiveness of a mind which has completely mastered its perspective" (Woolf, 1942, cited in Allott, 1959, p. 236).

Woolf's description in the passages just cited indicates that her intent was to arrange for the reader to share her problem representations. It has been argued elsewhere that this effort to arrange for the creator and audience to share the processing of information that led to a problem's resolution is characteristic of scientists such as Faraday as well (Gooding, 1990; Ippolito & Tweney, 1995). Woolf (1932/1984a) contended that the writer is a maker of meaningful worlds the reader later explores and, therefore, the reader must be similarly "capable not only of great fineness of perception ...

if you are going to make use of all that the novelist—the great artist—gives you" (p. 236). What follows is a brief account of Woolf's novelistic problem finding, primarily as to her first experimental novel, *Jacob's Room* (Woolf, 1922/1960).

WOOLF'S DEVELOPMENT OF EXPERTISE

Along with novelists Dorothy Richardson, D. H. Lawrence, William Faulkner, and James Joyce, Virginia Woolf is credited with introducing the so-called stream-of-consciousness method that revolutionized the novel (Eagle, 1970; Hussey, 1995; Wallace, 1982, 1991). The phrase *stream of consciousness* was derived from American psychologist William James's (1890/1950a, see also Hussey, 1995) characterization of the "stream of thought" in *Principles of Psychology*. The common interest of these novelists and the post-Impressionist painters was to reflect the psychological reality beneath the surface of human existence (Stansky, 1996; Woolf, 1990b). Specifically, Woolf's effort was to create a new kind of vessel to hold aspects of human psychology not previously focused on in novels.

The purpose of this section is to characterize the heuristic underpinnings of Woolf's invention of a new form of English novel and thereby elaborate her problem representation. Woolf's intention in inventing this new kind of novel was not to capture a photographic representation of human nature as in the earlier realistic novels of Austin, Hardy, and Dickens but "all the traces of the mind's passage through the world ... [to] achieve in the end some kind of whole made of shivering fragments" (Leaska, 1990, p. 393; see also Eagle, 1970).

Woolf recognized the error of the assumption underlying the realistic depiction of the physical world in works of art, that is, that perception is a direct product of a given stimulus rather than constructed "of the interactive influences of the presented stimulus *and* internal hypotheses, expectations, and knowledge" (Eysenck & Keane, 1990, p. 85). In "Mr. Bennett and Mrs. Brown," a literary essay by Woolf (1924/1984d) that centers on ruminations about a stranger on a train, she wrote that "old Mrs. Brown's character will strike you very differently according to the age and country in which you happen to be born. It would be easy enough to write three different versions of that incident in the train" (p. 199). Thus, the disconnected dabs of color or tachisme of Impressionism, and the nonlinear psychological meanderings of the stream-of-consciousness novels, displaced artistic realism, which intended a truthful depiction of a reality assumed to be universal (Cupchik, 1999; Eagle, 1970; Stansky, 1996). There is an obvious kinship between Woolf's view of art and the recognition by psychologists of the potentially constructive nature of memory and perception (Carmichael, Hogan, & Walter, 1932; Stratton, 1896). More than 50 years ago, Woolf posed a question that was subsequently the subject of extended debate among perceptual and developmental psychologists: Is perception a unidirectional process via which external physical energy impinges identically on the senses of all observers (the empirical approach), or a bidirectional process in which perception is altered by past experience and alters future interpretation of sensory events (the cognitive approach) (see Eysenck & Keane, 1990; Kail, 2001)? In Woolf's words: "Do we then know nobody? [Or do we know] only our versions of them, which, as likely as not, are emanations from ourselves" (Nicolson & Trautmann, 1977, p. 245).

The first exhibit of post-Impressionist art took place in 1910, the year England purportedly entered "the story of modernism" (Stansky, 1996, p. 1). Woolf (1924/1984d) pointed to 1910 as about the time when "all human relations ... shifted ... And when human relations change there is at the same time a change in religion, conduct, politics, and literature" (pp. 96–97).

> Whether we call it life or spirit, or truth or reality, this, the essential thing, has moved off, or on, and refuses to be contained any longer in such ill-fitting vestments as we provide .... Is it not the task of the novelist to convey this unknown and uncircumscribed spirit, whatever aberration or complexity it may display? (Woolf, 1925/1984c, pp. 287–288)

Thus, Woolf saw her role as a novelist not to continue, as realistic novels had, to accurately depict the physical world linguistically but to expand the scope of the novel to include depictions of the mental lives of human beings.

Although Woolf wrote in 1908 (when she was 26) of her intention to write a new kind of novel, she initially published two realistic novels: *The Voyage Out* (published on March 26, 1915, when Woolf was 33) and *Night and Day* (published on October 20, 1919, when Woolf was almost 38). While at work on both of these early novels, Woolf was engaged in a network of activities, including a reading program (documented from age 15 on), which involved her reading historical works as well as essays, biographies, poetry, and novels; journalistic pursuits (from 1904, when she was 22 on), including her writing essays on literature and literary reviews; informal seminars on literature, art, philosophy, history, and politics with other members of the Bloomsbury group (from 1904 on); publishing the writing of others with her husband (i.e., helping to run the Hogarth Press from 1915 when she was 33 on, the goal of which company was to publish works of literary merit); and her experimental writing (of short nonfiction and fiction pieces preserved from 1897 when she was 15 on and which were first recorded in her private journals) (Ippolito, 1999). This brief summary of the writing apprenticeship that preceded Woolf's publication of her first experimental novel, *Jacob's Room*, in late 1922 certainly is neither complete nor detailed. The purpose of this summary is to point out that Woolf was engaged in an ongoing program of activities designed to prepare her to meet her stated goal of re-forming the English novel. In essence, Woolf set out to engineer the type of paradigm shift Kuhn (1970) described of as leading to scientific breakthroughs. Her writing apprenticeship is akin to Gruber's (1989a, 1989b, 1995; see also Ippolito & Tweney, 2003) *network of enterprise*, or a group of related projects and activities designed by the problem solver to yield increases in domain-specific knowledge and germane strategies. In the case of Woolf, the domain of interest to her was not that of the extant realistic English novel but, rather, of a new kind of novel she envisioned. By engaging in a network of enterprise, "the creator participates in choosing and shaping the surroundings within which the work proceeds" (Gruber, 1989a, p. 5).

As indicated, in 1922, when Woolf published her first stream-of-consciousness novel, *Jacob's Room*, she did so after just over a decade of literary experimentation. This experimentation included several sketches, which were among Woolf's first short fiction published: "The Mark on the Wall" (1917/1989a, when she was 35), "Kew Gardens" (1919/1989b), and "An Unwritten Novel" (1920/1989c). Both "The Mark on the Wall" and "Kew Gardens" examined the effects of shifting the perspective between the material and the mental worlds of the characters. In "The Mark on the Wall," the narrator speculates what a mark on the wall might be: a nail, a crack, a rose leaf. "How readily our thoughts swarm upon a new object, lifting it a little way, as ants carry a blade of straw so feverishly, and then leave it" (Woolf, 1917/1989a, p. 83). The mark turns out to be a snail. Someone comes in and tells the narrator he is going out to buy a newspaper even though "nothing ever happens ... Damn the war!" (Woolf, 1917/1989a, p. 89). Similarly, *Jacob's Room* ends with the dilemma of what to do with the deceased Jacob's shoes, and the everyday happenings of life (e.g., a snail traversing the wall) are paramount, while the first world war is a footnote. In *Jacob's Room*—and several of Woolf's subsequent novels—the photographic or materialistic elements of life cease to be the primary focus and serve as bits of cork holding afloat a net that plumbs the depths of consciousness (Woolf, 1917/1987b). In "The Mark on the Wall," the snail is one of these bits of cork, and it serves as a real world anchor for the narrator's psychological meanderings; in *Jacob's Room*, a more complex focus—a young man named Jacob—is the center of the thoughts and feelings of the other characters.

In "Kew Gardens," the gardens are seen by the reader from the point of view of a caterpillar as a number of couples pause in nearby conversations, dislocating "the reader's accustomed perspective of the landscape" (Bishop, 1989, p. 111). E. M. Forster wrote of "Kew Gardens" and "The Mark on the Wall" that they were

> lovely little things, but they seemed to lead nowhere, they were all tiny dots and colored blobs ... Consequently when *Jacob's Room* appeared in 1922 we

were tremendously surprised. The style and sensitiveness of "Kew Gardens" remained, but they were applied to human relationships, and the structure of society. The blobs of color continue to drift past, but in their midst, interrupting their course like a closely sealed jar stands the solid figure of a young man. (quoted in Noble, 1972, pp. 188–189)

Woolf viewed "An Unwritten Novel," the third of the three stories, which served as a pilot study for the techniques employed in *Jacob's Room*, as a "great discovery" (Nicolson & Trautmann, 1978, p. 231). In this story the narrator studies fellow passengers on a train. This story contains, intertwined with the "nonevent" of a train ride, the complex flow of information in the internal mind that does not appear in realistic novels. The narrator's thoughts ultimately focus on a woman sitting nearby, and he or she creates an inner life and history for this woman. Alternatively, in *Jacob's Room*, the reader comes to know Jacob, not through the potentially biased perspective of a single individual such as the narrator of "An Unwritten Novel" (whose assessment of a fellow train passenger is inaccurate), but through a network of others' minds (Nicolson & Trautmann, 1978). It is notable that, while she is grouped with the other so-called stream-of-consciousness novelists, Woolf expressed dissatisfaction with the approaches taken by Joyce and Richardson. In 1920, when she was about to begin work on *Jacob's Room*, she wrote that she had "arrived at some idea of a new form for a novel. Suppose one thing should open out of another [but with] no scaffolding; scarcely a brick to be seen .... I see immense possibilities for the form ... I suppose the danger is the damned egotistical self; which ruins [James] Joyce & [Dorothy] Richardson, to my mind" (A. O. Bell & McNeillie, 1978, pp. 13–14).

In *Jacob's Room*, Woolf (1922/1960) dispensed entirely with the scaffolding of narrators; the reader knows Jacob not via the egotistical perspective of a single narrator but through the multiple, overlapping perspectives of Jacob's acquaintances. While the presence of a narrator provided a simple means of unifying the focus of both "The Mark on the Wall" and "An Unwritten Novel," this device also brought to these works biased, static perspectives and was analogous, at times, to the clumsy hand of a puppeteer visible above the worlds circumscribed by these short stories, giving the lie to the illusion of life simulated. Thus, the dislocation of perspective to the caterpillar, which unifies but also limits the scope of "Kew Gardens," was expanded in *Jacob's Room*. This new form of novel, which incorporated shifting, multifaceted views of Jacob, enabled a dynamic representation of human psychology more akin to the evolving mental lives of interacting human beings than the single point of-view of contemporaneous stream-of-consciousness novels or the primarily materialistic representations of earlier realistic novels (Rosenthal, 1979).

Jacob's Room relies on a structure analogous to that of perceptual figures in which figure and ground are reversible, such as the stimulus in which a white vase is sometimes perceived as the figure surrounded by a black background; at other times, two black profiles of a human face are perceived as the figure with the intervening white space ("the vase") as background. In this novel, the thoughts and feelings of the minor characters detail Jacob for the reader, with Jacob as the figure and the other characters' thoughts as background. Simultaneously, Jacob is the background against which we come to know the other characters in his "room" (via their thoughts about Jacob). Jacob emerges from the perceptions of others, and our knowledge of the other characters comes out of their interaction with and impressions of Jacob. A new form of novel emerged from several of Woolf's short experimental works just as theoretical pronouncements emerge from the findings of numerous psychological studies. Woolf wrote that she "conceive[d] '[The] Mark on the Wall,' ['Kew Gardens'], and '[An] Unwritten Novel' taking hands and dancing in unity .... I think from the ease with which I'm developing ... '[An] Unwritten Novel' there must be a path for me there" (A. O. Bell & McNeillie, 1978, p. 14). Just prior to the publication of *Jacob's Room*, Woolf wrote about this novel: "There's no doubt in my mind that I have found out how to begin (at 40) to say something in my own voice" (Nicolson & Trautmann, 1978, p. 186).

Thus, Woolf's literary problem solving clearly manifested the characteristics that research on expertise and the psychology of writing have identified as characteristic of experts: a decade or more of dedicated practice; the ability to perceive the complex, germane patterns of the flight of the mind; superior metacognitive skills; and the apportionment of substantial time to problem analysis, including—in the case of expert writers—persistent attention to the representation of even the most mundane problems across the duration of the writing task (Bereiter & Scardamalia, 1987; Carey & Flower, 1989; Chi, Glaser, & Rees, 1982; Ericsson, 1996; Ericsson & Smith, 1994; Flower & Hayes, 1980; Ippolito, 1999; Scardamalia & Bereiter, 1982, 1991). The stream-of-consciousness writers removed the scaffolding characteristic of materialism (faithful descriptions of people, places, and behavior and other long-standing novelistic conventions), but they all relied on the reader seeing the world through the mind of a single main character as the alternative. Woolf also removed the materialist scaffolding, but she went further—she saw the view of a single consciousness as inevitably biased and opted instead to adopt a chorus of perspectives to elaborate the characters and worlds of her novels. Woolf clearly used her shorter fiction as exploratory forays into the larger possibility space of her proposed re-formed novel, which was intended to devise a means of shadowing the working of the mind absent scaffolding (such as a narrator) and the static point of view (of a single character) of other stream-of-consciousness novels. These short works became the means to explore and map selected, vital regions of this problem space and turn them into something more akin to homing spaces.

As I discussed earlier, while expert writing is, in large part, indistinguishable from expertise in other domains, there are features of literature construction that are distinctive. In addition to the extended period of problem representation characteristic of writing problem solving already mentioned, "while expertise is generally considered to be domain specific, writing necessarily bridges domains; writers must be knowledgeable about writing strategies as well as the field they are writing about" (Ippolito, 2002, p. 81). Hence, Woolf developed a dual apprenticeship: (a) the one described earlier to train her in the craft of writing and the art of the novel and (b) her study of the vicissitudes of human psychology (Ippolito, 1999).

WOOLF'S INCEPTUAL PROCESSING

Woolf, as well as many other novelists

> (e.g., Guy de Maupassant, Gustave Flaubert, [Thomas] Hardy) engaged in the collection of perceptions and sought to preserve the data yielded by the effortful process of perceptual rehearsal. That is, these authors carefully observed aspects of human experience they wished to represent in their novels with some (e.g., Fyodor Dostoevsky, Woolf, [Anaïs] Nin, and Andre Gidé) carefully preserving these observations in journals. (Ippolito, 2002, p. 81; see also Allott, 1967; Briggs, 1990)

Woolf's diary and correspondence, as well as her formal essays on the novel and fiction writing, document her deployment of inceptual processing in her resolution of the ill-defined problem of reinventing the English novel (Ippolito, 1999).

Again, *perceptual rehearsal* is the "the saturation of one or more of the senses with all aspects of the phenomenon of interest to the discoverer ... [as] a means of defeating the inherent biases of our perceptual apparatus and increasing the impact of unexpected sights" (Ippolito & Tweney, 1995, p. 435). Woolf placed enormous importance on the information made available by her senses. She wrote, while at work on *Jacob's Room*, "You know my passion for sensation" (Nicolson & Trautmann, 1976, p. 477). Later, in an autobiographical essay, Woolf (1985) characterized herself as "a porous vessel afloat on sensation" (p. 133). In the midst of revising *To the Lighthouse*, Woolf's third stream-of-consciousness novel, she wrote that "otherwise [than reading] I am only an eye" (Nicolson & Trautmann, 1977, p. 309). And, at about the same time, she indicated in a letter to a friend: "That's the way I travel. Looking, looking, looking and making up phrases to match clouds. It is the passion of my life" (Nicolson & Trautmann, 1977, p. 347).

Although these quotations and those that follow favor visual sensations—and Woolf's novels, do, in fact, rely heavily on visual imagery—Woolf's insistence that rhythm was as important to prose as to poetry and her intent that the sound of the waves breaking on the shore echo throughout two of her later stream-of-consciousness novels (*To the Lighthouse* and *The Waves*) indicates that she attended to other than visual sensations, including

> those influences which play so large a part in life, ... the power of music, the stimulus of sight, the effect on us of the shape of trees or the play of color, ... the delight of movement, [and] the intoxication of wine. Every moment is the center and meeting place of an extraordinary number of perceptions. (Woolf, 1927/1967b, pp. 228–229)

Woolf would purposefully go "fishing" in the Sussex countryside and along the London streets, studying human beings living their lives. She wrote: "I pigeonhole 'fishing' thus with other momentary glimpses, like those rapid glances, for example, that I cast into basements when I walk in London streets" (Woolf, 1985, p. 135). Woolf deemed it a necessity that she gain extensive familiarity with the representative features of human existence (via perceptual rehearsal) such that viable novelistic representations of human psychology would be possible. In addition to the elements of the material world, Woolf had a self-described "passion for taking feelings to the window and looking at them" (Nicolson & Trautmann, 1976, p. 490). Woolf wrote as a young woman of 26: "I am haunted by the thought that I can never know what anyone is feeling, but I suppose at my age it can't be helped. It is like trying to jump a shadow" (Nicolson & Trautmann, 1975, p. 408). Beginning with *Jacob's Room* (written when Woolf was about 40), Woolf's novels featured the emotional lives of her characters, and she expressed a repeated interest in knowing more about psychology (e.g., A. Bell, 1977; A. Bell & McNeillie, 1980; Nicolson & Trautmann, 1975, 1978; Woolf, 1990b). She determined that, insofar as it was in her power to do so, she would even take advantage of her affliction with a mood disorder (i.e., bipolar disorder; see Q. Bell, 1972; Hussey, 1995) to discover more about human emotions and psychology, writing to a friend: "Madness is terrific I can assure you ... and in its lava I still find most of the things I write about" (Nicolson & Trautmann, 1978, p. 180). And Woolf wrote in a letter to a friend some 6 months before beginning to write *Jacob's Room*: "But isn't there some place for the theoretical heart, which is my kind—the heart which imagines what people feel, and takes an interest in it" (Nicolson & Trautmann, 1976, pp. 371–372). For Woolf, the training of the senses as to a particular class of sensory phenomena served to enhance the kind and quality of information available to her. Besides sifting through the "lava" of her mood disorder and her own and others' emotions, Woolf wrote of her effortful pursuit of every detail of commonly beheld sights and their relevance to her intent to re-form the English novel:

> I think a great deal of my future, and settle what book I am to write—how I shall re-form the novel and capture multitudes of things at present fugitive, enclose the whole, and shape infinite strange shapes. I take a good look at woods in the sunset, and fix men who are breaking stones with an intense gaze, meant to sever them from the past and the future. (Nicolson & Trautmann, 1975, p. 356)

Woolf clearly continued to resort to the process designated here as perceptual rehearsal. Some 16 years later, and about 2 years after the publication of *Jacob's Room*, she wrote that

> coming back the other evening from Charleston, again all my nerves stood upright, flushed, electrified (what's the word?) with the sheer beauty—beauty abounding and superabounding, so that one almost resents it, not being capable of catching it all, and holding it all at the moment. This progress through life is made immensely interesting by trying to grasp all these developments as one passes. I feel as if I were putting my fingers tentatively on ... either side as I grope down a tunnel, rough with odds and ends. (A. O. Bell & McNeillie, 1978, p. 311)

Erik Erikson (quoted in Briggs, 1990, p. 214) wrote of the kind of mental practice deemed here as perceptual rehearsal as analogous to child's play: "The child can throw ... [a] ball to the floor again and again, seeing it roll a hundred times ... and never be bored." Goodfield (1981) also described the similar repetition of seemingly mundane perceptual exercises by a scientist engaged in genetic research.

As noted, it was in *Jacob's Room*, which Woolf began writing in early 1920, that she was first able to devise a suitable form to hold the confluence of the material and mental lives of her characters on which she intended to focus. Thus this novel was the first fruit of Woolf's extensive apprenticeship designed to develop expertise as to a new kind of novel intended to capture the shivering fragments of the mind. However, the mere gathering of germane perceptions was insufficient to devise a representation of the problem of the re-formed novel that Woolf had set for herself. As she declared, the

> eye [merely] ... brings back the prettiest trophies ... [and] breaks off little lumps of emerald and coral as if the whole earth were made of precious stone. The thing it cannot do (one is speaking of the average unprofessional eye) is to compose these trophies in such a way as to bring out the more obscure angles and relationships. (Woolf, 1927/1984e, pp. 248–249)

"It is a mistake to think that literature can be produced from the raw. One must get out of life … one must become externalized; very, very concentrated, all at one point, not having to draw upon the scattered parts of one's character, living in the brain" (A. O. Bell & McNeillie, 1978, p. 193). Literature is not produced directly from the data collected via perceptual rehearsal.

To summarize thus far, perceptual rehearsal engendered Woolf with a specialized familiarity with the perceptual phenomena of interest, permitting her to sense more. For example, an experienced art critic is, by virtue of his or her training, exceptionally aware of the features of the work of art being scrutinized. An additional advantage of the extensive familiarity with observed phenomena that arises from perceptual rehearsal is, as mentioned earlier, the potential for decoupling perceptions from sensory input. One must sever from his or her past and future the perceptions collected during perceptual rehearsal through refining or distillation processes. Woolf examined the fragments of the mind traversing life in conjunction with the bits of physical-world cork that marked the mind's progress. However, the inceptions she constructed were severed from the unnecessary details of the real world.

Again, inceptions stand between the real and the symbolic; inceptions are veridical to the perceptions on which they are based but are reconstructed in the mind apart from the irrelevancies of their real world contexts. This decoupling serves the same purpose as partitioning numerical data into variability of interest (the constructed inceptions) and error (the nonsignificant features of the context) in order to reconstruct the workings of selected aspects of human psychology. Thus, Woolf was able to utilize inceptions to construct exemplars of aspects of the human condition to serve as characters in her novels. Even in one of Woolf's (1919/1992) realistic novels, *Night and Day*, the character of Mrs. Hilbery was not intended to be a photographic representation of Aunt Annie (Anne Isabella Thackeray, Woolf's stepaunt) but came to stand in for a larger reality beyond individual heritage and attire and utterances. "In writing one gets more and more away from the reality and Mrs. Hilbery became to me quite different from any one in the flesh" (Nicolson & Trautmann, 1976, p. 406). Mrs. Hilbery became the distilled inception of Woolf's perceptual knowledge of her aunt and, possibly, of others.

So, it happens that when a character in Woolf's *Night and Day* comes upon another character in this novel—whom he has observed intently and with whom his thoughts had been considerably preoccupied—he is unsure if she is real or, in the terminology utilized here, an *inception*. "The sudden apparition had an extraordinary effect upon him. It was as if he had thought of her so intensely that his mind had formed the shape of her, rather than that he had seen her in the flesh outside on the street" (Woolf, 1919/1992, p. 194). Woolf (1926/1967) wrote that the writer's [or innovator's] task is to take one thing [an inception in the terminology of this chapter] and let it stand for twenty; a task of danger and difficulty; ... [for] only so is the reader relieved of the swarm and confusion of life and branded effectively with the particular aspect which the writer wishes him to see. (p. 133)

So, when Woolf was writing her second novel, *Night and Day*, she spoke to her husband "not in ... [her] own character but in" the voice of one of the characters in her novel (A. O. Bell, 1977, p. 22), and in her 1939–1940 essay, "A Sketch of the Past," Woolf (1985)—in the same vein as Feynman's description of his inception of electromagnetic fields—wrote of the Stephen summer home where she spent many holidays and which she reconstructed in two novels (*To the Lighthouse* and *The Waves*):

At times I can go back to St. Ives more completely than I can this morning. I can reach a state where I seem to be watching things happen as if I were there ... I see it—the past—as an avenue lying behind: a long ribbon of scenes, emotions. There at the end of the avenue still, are the garden and the nursery. Instead of remembering here a scene and there a sound, I shall fit a plug into the wall, and listen to the past. I shall turn up August, 1890. (Woolf, 1985, p. 67)

If life has a base that it stands upon, if it is a bowl that one fills and fills and fills—then my bowl without a doubt stands upon this memory. It is of lying half asleep, half awake, in bed in the nursery at St. Ives. It is of hearing the waves breaking, one, two, one, two, and sending a splash of water over the beach.... There were passion flowers growing on the wall; they were great starry blossoms, with purple streaks, and large green buds, part empty, part full.... Then going down to the beach ... I stopped at the top to look down at the gardens. They were sunk beneath the road. The apples were on a level with one's head. The gardens gave off a murmur of bees; the apples were red and gold; there were also pink flowers; and grey and silver leaves. The buzz, the croon, the smell, all seemed to ... hum round one such a complete rapture of pleasure that I stopped, smelt; looked. (Woolf, 1985, pp. 64–66)

As indicated earlier, the solving of ill-defined problems requires the alternating of perceptual rehearsal and the distillation of inceptions as well as the construction of emulations, which transforms and solidifies the data gathered during perceptual rehearsal. These inceptions provided the raw material to construct emulations, which guided Woolf's exploration of the worlds of the newly formed novels. The characters (derived from inceptions) that peopled Woolf's novelistic worlds served the same purpose as groups of participants in psychological studies whose behavior reveals some aspect of the nature of human psychology. Woolf echoes the conception detailed here of mental emulations as akin to organisms. Specifically, Woolf (1976b) wrote that the "best novels are deposited carefully, bit by bit; and, in the end, perhaps live in all their parts" (Nicolson & Trautmann, 1975, p. 350). Works of art should be:

> Beautiful and bright ... on the surface, feathery and evanescent, one color melting into another like the colors on a butterfly's wings; but beneath the fabric must be clamped together with bolts of iron. It was to be a thing you could ruffle with your breath, and a thing you could not dislodge with a team of horses. (Woolf, 1927/1989b, p. 171)

For Woolf, emulations served the same purpose as simulations, which are intended to model a process or "imitate the conditions of a situation" (*DK Illustrated Oxford Dictionary*, 1998, p. 771). Emulations provide structural integrity and stability, capturing the interrelations of the selected inceptions in an optimal combination so that the phenomena of interest can be represented and understood. The emulation then drives the successive re-representations of the problem.

As mentioned earlier, Woolf described the emulation associated with *Jacob's Room* as three of her experimental works of shorter fiction holding hands and dancing in unison (A. O. Bell & McNeillie, 1978). In *Jacob's Room*, one thing was to open from another, and there should be an absence of scaffolding, not a brick in sight. For Woolf's next stream-of-consciousness novel, *Mrs. Dalloway* (1925/1990a), she needed a means of compressing and reflecting the emotional lifetimes of her characters in a relatively brief period of physical time. This novel, originally called *The Hours*, sought to contrast physical (Newtonian) time with psychological time (Hussey, 1995). Woolf viewed psychological time as relative in the sense that it sometimes involves the simultaneous living/reliving of the past, present, and anticipated future and is not divisible into equal-sized minutes or hours. Woolf commented on this quality of psychological time, writing at one point: "I'm sure I live more gallons to the minute walking once round the square than all the stockbrokers in London caught in the act of copulation" (Nicolson & Trautmann, 1977, p. 386).

Thus, the writing of *Mrs. Dalloway*, originally entitled *The Hours*, demanded the employment of the techniques of *Jacob's Room* and more. The means Woolf used to accomplish this was to postulate psychological caves extending into the pasts of the characters, caves that interconnected during moments of shared experiences. Woolf wrote:

> I should say a good deal about *The Hours* and my discovery, how I dig out beautiful caves behind my characters; I think that gives exactly what I want:

humanity, humor, depth. The idea is that the caves shall connect, and each comes to daylight at the present moment. (p. 263)

It took me a year's groping to discover what I call my tunneling process, by which I tell the past by installments, as I have need of it. This is my prime discovery so far .... One feels about in a state of misery—indeed I made up my mind one night to abandon the book—and then one touches the hidden spring. (Nicolson & Trautmann, 1978, p. 272)

The fluid nature of consciousness Woolf creates through the network of these caves, in which temporal boundaries and definitions lose their distinctiveness, forms a counterpoint to the inexorable tolling of Big Ben, impassively sounding out the hour. Big Ben punctuates the reveries, the pain, the pleasures with his dreary announcement that time is passing. Woolf's explorations in the timelessness of the caves is thus effectively juxtaposed to the steady progress of time on the surface, and the two express that dual sense of life—that it is at once eternal and ephemeral. (Rosenthal, 1979, pp. 89–90)

The emulation Woolf constructed that guided her exploration of the problem of *Mrs. Dalloway* was the postulated system of tunneling in the psychological caves of the characters. Thus, once Woolf had devised the emulation she called the "tunneling process," the construction of *Mrs. Dalloway* was guided by her resorting to the heuristic of tunneling and reporting on the pasts of the characters. The emulation (the tunneling *system*) was, in Woolf's words, the hidden spring—the touching of which allowed her to solve the puzzle of *Mrs. Dalloway*, a novel she had previously been about to abandon because she was unsure as to how to proceed. In this novel, Woolf contrasted the different kinds of time as well as sanity and insanity (embodied in stories with two different main characters who crossed paths throughout the novel), and tried to overcome the linearity of written language so that the novel emulated the parallelism of painting. In *The Waves*, which came later, Woolf tried to embody the sound of the waves lapping on the shore while periodically tracking the streams of consciousness of several lifelong friends in what amounted to a longitudinal novel that sought to capture the rhythms of poetry.

Echoing the definition of emulations as the arrangement of inceptions into a working model or dynamic framework that is presymbolic and postperceptual, Woolf wrote of how "a sight, an emotion, creates this wave in the mind, long before it makes words to fit it; and in writing (such is my present belief) one has to recapture this, and set this working (which apparently has nothing to do with words) and then, as it breaks and tumbles in the mind it makes words to fit it" (Nicolson & Trautmann, 1977, p. 247). The recombination of inceptions generates emulations that "serve as levels and scaffolds for building the actual creative products" (Briggs, 1990, p. 194; see also Ochse, 1990).

Like other innovators cited previously, Woolf spoke of the need to simplify, which is encompassed in the construction of emulations. In the last year of her life, Woolf (quoted in Allott, 1967, p. 236) wrote that "if there is one gift more essential to a novelist than another, it is the power of combination—the single vision." Woolf (1985) spoke, too, of her "great delight ... [in putting] the severed parts together" (p. 72). She also wrote: "This is the strongest pleasure known to me. It is the rapture I get in writing when I seem to be discovering what belongs to what" (Woolf, 1985, p. 72).

Woolf also noted the importance of the constructed emulations being runnable; that is, the characters of novels and their world must be viable independent of the author. "For, what one imagines, in a novel, is a world. Then, when one has imagined this world, suddenly people come in" (Nicolson & Trautmann, 1977, pp. 238–239). This need for a work of art to stand on its own led Woolf to advise an aspiring writer "for art's sake, ... to grasp, and not exaggerate, and put sheets of glass between him and his matter" (A. O. Bell, 1977, p. 176); she indicated that one must "provide a wall for the book from oneself without its becoming ... narrowing and restricting" (A. O. Bell & McNeillie, 1978, p. 14). Woolf intended that her writing "be judged as a chiseled block, unconnected with ... [her] hand entirely" (Nicolson & Trautmann, 1975, p. 325).

When Woolf extended her novels to include the territory the realists had explored plus the whole range of the mental lives of human beings in solitude and in interaction with others, she exponentially extended the boundaries of the space to be searched. It is thus not surprising that Woolf solving the problem of re-forming the English novel benefited from the extensive effort of devising a kind of mental model (here called an *emulation*). As pointed out, Woolf began by constructing emulations that assisted her in mapping selected areas of the possibility space of this new kind of stream-of-consciousness novel. Thus, initial exploration of areas of this ill-defined problem space yielded short fiction problem solutions in advance of Woolf tackling the larger problem of capturing the flight of the mind in novelistic form. Woolf gradually expanded and combined the emulations arising from her shorter fiction to encompass the more extensive possibility spaces of the more complex problems of writing her experimental novels and of the re-formed novel. These short fiction works (most of which were never intended for publication), as well as her journals and some of her literary essays, highlight the importance Woolf placed on her heuristic problem representation activities; she spent considerable time documenting these activities (Ippolito, 1999). The study of Woolf's writing problem solving also underscores the significance of engaging in a network of enterprises (i.e., writing, reading, publishing, studying human nature) per Gruber

(1974, 1989a, 1989b, 1995) when one's goal is to resolve an ill-defined problem and, in essence, establish a new knowledge domain.

In summary, the claim made here is that the repertoire of heuristics designated as inceptual processing assist problem finding. Appearances come to have a different meaning for the problem solver after the emergence of the emulation. This emulation focuses the search of the space of possible solutions for the targeted problem as well as contributing to the filtering of the data collected via subsequent perceptual rehearsal, altering the way the individual interacts with his or her internal world and external environment. The created invention is an emergent formalization of the details and interrelations of the problem representation.

CONCLUSIONS

I leave it to Woolf (1926/1967) to describe her use of the heuristics of inceptual processing in writing her novels. This passage demonstrates that Woolf utilized the problem-finding heuristics described earlier in tandem and that she credited them as contributing to her resolution of novelistic problems:

> The novelist—it is his distinction and his danger—is terribly exposed to life
> …. He fills his glass and lights his cigarette, he enjoys presumably all the pleasure of talk and table, but always with a sense that he is being stimulated and played upon by the subject-matter of his art…. He can no more cease to receive impressions than a fish in mid-ocean can cease to let the water rush through his gills [perceptual rehearsal]. But if this sensibility is one of the conditions of the novelist's life, it is obvious that all writers whose books survive have known how to master it and make it serve their purposes. They have finished the wine and paid the bill and gone off, alone, into some solitary room where, with toil and pause, in agony (like Flaubert), with struggle and rush, tumultuously (like Dostoyevsky) they have mastered their perceptions, hardened them, and changed them into the fabrics of art [the distillation of inceptions]. So drastic is the process of selection that in its final state we can often find no trace of the actual scene upon which the chapter was based. For in that solitary room, whose door the critics are forever trying to unlock, processes of the strangest kind are gone through. Life is subjected to a thousand disciplines and exercises. It is curbed; it is killed. It is mixed with this, stiffened with that, brought into contrast with something else; so that when we get our scene at a café a year later the surface signs by which we remembered it have disappeared. There emerges from the mist something stark, something formidable and enduring, the bone and substance upon which our rush of indiscriminating emotion was founded [the constructed emulation]. Of these two processes—the first—to receive impres-

sions—is undoubtedly easier, the simpler, and the pleasanter. And it is quite possible, provided one is gifted with a sufficiently receptive mind and a vocabulary rich enough to meet its demands, to make a book out of the preliminary emotion alone.... He can sit and watch life and make his book out of the very foam and effervescence of his emotions; or he can put his glass down, retire to his room and subject his trophy to those mysterious processes by which life becomes, like the Chinese coat, able to stand by itself—a sort of impersonal miracle. (pp. 131–135)

In this chapter I have taken advantage of Woolf's fluency and her attention to and documentation of her thinking processes as well as the extended problem representation that characterizes problem solving by expert writers and by Woolf, who was engaged in solving the ill-defined problem of altering the conventional English novel. I have attempted to describe the heuristics of perceptual rehearsal, the distillation of inceptions, and the construction of emulations in which Woolf engaged in devising solutions to novelistic problems. It should be apparent that the inceptual-processing heuristics described in this chapter are analogous to external-world techniques specifically devised to support problem solving, such as the statistical partitioning of data variability and the nomothetic approach to the study of human psychology—that is, the algorithms of statistical analysis arrange, for example, for the highlighting of particular real-world interrelations between variables; analogously, the inceptual-processing heuristics described assist the representation of aspects of real-world phenomena (in the mind and then, typically, in symbolic form) by constraining the vast possibility spaces associated with ill-defined problems.

I view the data provided here in support of the argument that the inceptual-processing heuristics are utilized in the representation of ill-defined problems as suggestive of, returning to the Simon (1981) quote with which I began this chapter, potential avenues of exploration with respect to the specific features of possibility spaces and their construction. That is, this account is akin to tachistic tiny dots and colored blobs, which suggest how problem representation works but, for now, lacks the photographic details potentially available through the conducting of nomothetic and idiographic studies. Thus, this chapter essentially describes an emulation that might guide the study of the process of problem representation.

REFERENCES

Allott, M. (Ed.). (1967). *Novelists on the novel*. London: Routledge & Kegan Paul.
Arnheim, R. (1962). *Picasso's Guernica: The genesis of a painting*. Berkeley: University of California Press.
Bell, A. O. (Ed.). (1977). *The diary of Virginia Woolf* (Vol. 1). New York: Harcourt Brace.

Bell, A. O., & McNeillie, A. (Eds.). (1978). *The diary of Virginia Woolf* (Vol. 2). New York: Harcourt Brace.

Bell, A. O., & McNeillie, A. (Eds.). (1980). *The diary of Virginia Woolf* (Vol. 3). New York: Harcourt Brace.

Bell, Q. (1972). *Virginia Woolf: A biography.* New York: Harcourt Brace.

Bereiter, C., Burtis, P. J., & Scardamalia, M. (1988). Cognitive operations in constructing main points in written composition. *Journal of Memory & Language, 27,* 261–278.

Bereiter, C., & Scardamalia, M. (1987). *The psychology of written composition.* Hillsdale, NJ: Lawrence Erlbaum Associates.

Bishop, E. L. (1989). Pursuing "it" through "Kew Gardens." In D. R. Baldwin (Ed.), *Virginia Woolf: A study of the short fiction* (pp. 109–117). Boston: Twayne.

Briggs, J. (1990). *Fire in the crucible: The self-creation of creativity and genius.* Los Angeles: Tarcher.

Bruner, J., & Potter, M. (1964, April). Interference in visual recognition. *Science, 144,* 424–425.

Bryson, M., Bereiter, C., Scardamalia, M., & Joram, E. (1991). Going beyond the problem given: Problem solving in expert and novice writers. In R. J. Sternberg & P. A. Frensch (Eds.), *Complex problem solving* (pp. 61–84). Hillsdale, NJ: Lawrence Erlbaum Associates.

Carey, L. J., & Flower, L. (1989). Foundations for creativity in the writing process. In J. A. Glover, R. R. Ronning, & C. R. Reynolds (Eds.), *The handbook of creativity* (pp. 283–303). New York: Plenum.

Carmichael, L., Hogan, H. P., & Walter, A. A. (1932). An experimental study of the effect of language on the reproduction of visually perceived form. *Journal of Experimental Psychology, 15,* 73–86.

Chase, W., & Simon, H. A. (1973). The mind's eye in chess. In A. Collins & E. E. Smith (Eds.), *Readings in cognitive science: A perspective from psychology and artificial intelligence* (pp. 461–494). San Mateo, CA: Morgan Kaufmann.

Chi, M. T. H., Glaser, R., & Farr, M. J. (Eds.). (1988). *The nature of expertise.* Hillsdale, NJ: Lawrence Erlbaum Associates.

Chi, M. T. H., Glaser, R., & Rees, E. (1982). Expertise in problem solving. In R. J. Sternberg (Ed.), *Advances in the psychology of human intelligence* (pp. 7–73). Hillsdale, NJ: Lawrence Erlbaum Associates.

Csikszentmihalyi, M. (1991). Society, culture, and person: A systems view of creativity. In R. J. Sternberg (Ed.), *The nature of creativity: Contemporary psychological perspectives* (pp. 325–339). New York: Cambridge University Press.

Cupchik, G. C. (1999). Perception and creativity. In M. A. Runco & S. R. Pritzker (Eds.), *Encyclopedia of creativity* (pp. 355–360). San Diego, CA: Academic Press.

deGroot, A. (1974). *Thought and choice and chess.* The Hague, The Netherlands: Mouton. (Original work published 1946)

DK illustrated Oxford dictionary. (1998). New York: Oxford University Press.

Duncker, K. (1945). On problem-solving (L. S. Lees, Trans.). *Psychological Monographs, 58*(5).

Eagle, D. (1970). *The concise Oxford dictionary of English literature.* New York: Oxford University Press.

Epstein, R. (1991). Skinner, creativity, and the problem of spontaneous behavior. *Psychological Science, 2,* 362–370.

Ericsson, K. A. (1996). *The road to excellence: The acquisition of expert performance in the arts and sciences, sports and games.* Mahwah, NJ: Lawrence Erlbaum Associates.

Ericsson, K. A., & Faivre, I. A. (1988). What's exceptional about exceptional abilities? In L. K. Obler & D. F. Fein (Eds.), *The exceptional brain: Neuropsychology of talent and special abilities* (pp. 436–473). New York: Guilford.

Ericsson, K. A., & Smith, J. (1994). *Toward a general theory of expertise.* New York: Cambridge University Press.

Eysenck, M. W., & Keane, M. T. (1990). *Cognitive psychology: A student's handbook.* Hillsdale, NJ: Lawrence Erlbaum Associates.

Feynman, R. P., Leighton, R. B., & Sands, M. (1964). *The Feynman lectures on physics* (Vol. 2). Reading, MA: Addison-Wesley.

Flower, L. S., & Hayes, J. R. (1980). The cognition of discovery: Defining a rhetorical problem. *College Composition and Communication, 31,* 21–32.

Flowers, J. H., & Garbin, C. P. (1989). Creativity and perception. In J. A. Glover, R. R. Ronning, & C. R. Reynolds (Eds.), *Handbook of creativity* (pp. 147–162). New York: Plenum.

Getzels, J., & Csikszentmihalyi, M. (1976). *The creative vision: A longitudinal study of problem finding in art.* New York: Wiley.

Gibson, E. J. (1969). *Principles of perceptual learning and development.* Englewood Cliffs, NJ: Appleton-Century-Crofts.

Goldenson, R. M. (1984). *Longman dictionary of psychology and psychiatry.* New York: Longman.

Goldstein, E. B. (1989). *Sensation and perception.* Belmont, CA: Wadsworth.

Goodfield, J. (1981). *An imagined world: A story of scientific discovery.* New York: Harper & Row.

Gooding, D. (1990). *Experiment and the making of meaning.* Dordrecht, The Netherlands: Kluwer Academic.

Gorman, M. E., & Carlson, W. B. (1990). Interpreting invention as a cognitive process: The case of Alexander Graham Bell, Thomas, Edison, and the telephone. *Science, Technology & Human Values, 15,* 131–164.

Gruber, H. E. (1974). *Darwin on man: A psychological study of scientific creativity.* New York: Dutton.

Gruber, H. E. (1989a). The evolving systems approach to creative work. In D. B. Wallace & H. E. Gruber (Eds.), *Creative people at work* (pp. 3–24). New York: Oxford University Press.

Gruber, H. E. (1989b). Networks of enterprise in creative science work. In B. Gholson, A. Houts, R. A. Neimayer, & W. Shadish (Eds.), *Psychology of science and metascience* (pp. 246–265). Cambridge, England: Cambridge University Press.

Gruber, H. E. (1995). Insight and affect in the history of science. In R. J. Sternberg & J. E. Davidson (Eds.), *The nature of insight* (pp. 397–432). Cambridge, MA: MIT Press.

Hayes, J. R. (1989). Cognitive processes in creativity. In J. A. Glover, R. R. Ronning, & C. R. Reynolds (Eds.), *Handbook of creativity* (pp. 135–145). New York: Plenum.

Holton, G. (1971–1972). On trying to understand scientific genius. *The American Scholar, 41,* 95–109.

Holton, G. (1978). *The scientific imagination: Case studies.* Cambridge, England: Cambridge University Press.

Hooke, R. (1987). *Micrographia.* Lincolnwood, IL: Science Heritage.

Hotchner, A. E. (1966). *Papa Hemingway.* New York: Random House.

Hussey, M. (1995). *Virginia Woolf A to Z: A comprehensive reference for students, teachers and common readers to her life, work, and critical reception.* New York: Facts on File.

Ippolito, M. F. (1994). Commentary on M. Boden's *The Creative Mind: Conscious Thought Processes and Creativity. Behavior and Brain Sciences, 17,* 546–547.

Ippolito, M. F. (1999). *Capturing the flight of the mind: A model of Virginia Woolf's novelistic problemsolving* (Doctoral dissertation, Bowling Green State University, 1998) *Dissertation Abstracts International, 60* (4-B), 1880.

Ippolito, M. F. (2002). Cognitive science. In P. Gossin (Ed.), *An encyclopedia of literature and science* (pp. 80–81). Westport, CT: Greenwood.

Ippolito, M. F., & Tweney, R. D. (1995). The inception of insight. In R. J. Sternberg & J. E. Davidson (Eds.), *The nature of insight* (pp. 433–462). Cambridge, MA: MIT Press.

Ippolito, M. F., & Tweney, R. D. (2003). Virginia Woolf and the journey to *Jacob's Room:* The "network of enterprise" of Virginia Woolf's first experimental novel. *Creativity Research Journal, 15,* 25–43.

James, W. (1950a). *Principles of psychology* (Vol. 1). New York: Dover. (Original work published 1890)

James, W. (1950b). *Principles of psychology* (Vol. 2). New York: Dover. (Original work published 1890)

Johnson-Laird, P. N. (1983). *Mental models.* Cambridge, MA: Harvard University Press.

John-Steiner, V. (1997). *Notebooks of the mind: Explorations of thinking.* New York: Oxford University Press.

Kail, R. V. (2001). *Children and their development.* Upper Saddle River, NJ: Prentice Hall.

Karmiloff-Smith, A. (1992). *Beyond modularity: A developmental perspective on cognitive science.* Cambridge, MA: MIT Press.

Kuhn, T. S. (1970). *The structure of scientific revolutions.* Chicago: University of Chicago Press.

Leaska, M. A. (Ed.). (1990). Editor's preface. In V. Woolf, *A passionate apprentice: The early journals, 1897–1909* (pp. vii–x). New York: Harcourt Brace Jovanovich.

Miller, A. I. (1989). Imagery and intuition in creative scientific thinking: Albert Einstein's invention of the special theory of relativity. In D. B. Wallace & H. E. Gruber (Eds.), *Creative people at work: Twelve cognitive case studies* (pp. 25–43). New York: Oxford University Press.

Millgate, M. (1989). *The achievement of William Faulkner.* New York: Random House.

Montgomery, H. (1988). Mental models and problem solving: Three challenges to a theory of restructuring and insight. *Scandinavian Journal of Psychology, 29,* 85–94.

Morris, W. (Ed.). (1969). *The American heritage dictionary of the English language.* Boston: Houghton Mifflin.

Newell, A., & Simon H. A. (1972). The theory of human problem solving. In A. Collins & E. E. Smith (Eds.), *Readings in cognitive science: A perspective from psychology and artificial intelligence* (pp. 33–51). San Mateo, CA: Morgan Kaufmann.

Nicolson, N., & Trautmann, J. (Eds). (1975). *The letters of Virginia Woolf* (Vol. 1). New York: Harcourt Brace Jovanovich.

Nicolson, N., & Trautmann, J. (Eds). (1976). *The letters of Virginia Woolf* (Vol. 2). New York: Harcourt Brace Jovanovich.

Nicolson, N., & Trautmann, J. (Eds). (1977). *The letters of Virginia Woolf* (Vol. 3). New York: Harcourt Brace Jovanovich.

Nicolson, N., & Trautmann, J. (Eds). (1978). *The letters of Virginia Woolf* (Vol. 4). New York: Harcourt Brace Jovanovich.

Nicolson, N., & Trautmann, J. (Eds). (1980). *The letters of Virginia Woolf* (Vol. 6). New York: Harcourt Brace Jovanovich.

Noble, J. R. (1972). *Recollections of Virginia Woolf.* New York: Morrow.

Ochse, R. (1990). *Before the gates of excellence: The determinants of creative genius.* New York: Cambridge University Press.

Perkins, D. N. (1981). *The mind's best work.* Cambridge, MA: Harvard University Press.

Perkins, D. N. (1991). The possibility of invention. In R. J. Sternberg (Ed.), *The nature of creativity: Contemporary psychological perspectives* (pp. 362–385). New York: Cambridge University Press.

Perkins, D. N. (1995). Insight in mind and genes. In R. J. Sternberg & J. E. Davidson (Eds.), *The nature of insight* (pp. 495–534). Cambridge, MA: MIT Press.

Richman, H. B., Gobet, F., Staszewski, J. J., & Simon, H. A. (1996). Perceptual and memory processes in the acquisition of expert performance: The EPAM model. In K. A. Ericsson (Ed.), *The road to excellence: The acquisition of expert performance in the arts and sciences, sports and games* (pp. 167–187). Mahwah, NJ: Lawrence Erlbaum Associates.

Rolland, R. (1929). *Beethoven the creator* (E. Newman, Trans.). New York: Harper & Brothers.

Rosenthal, M. (1979). *Virginia Woolf.* New York: Columbia University Press.

Scardamalia, M., & Bereiter, C. (1982). Assimilative processes in composition planning. *Educational Psychologist, 17,* 165–171.

Scardamalia, M., & Bereiter, C. (1991). Literate expertise. In K. A. Ericsson & J. Smith (Eds.), *Toward a general theory of expertise* (pp. 172–194). New York: Cambridge University Press.

Schooler, J. W., Fallshore, M., & Fiore, S. M. (1995). Epilogue: Putting insight into perspective. In R. J. Sternberg & J. E. Davidson (Eds.), *The nature of insight* (pp. 559–587). Cambridge, MA: MIT Press.

Shekerjian, D. (1990). *Uncommon genius: How great ideas are born.* New York: Penguin.

Shepard, R. N. (1978). Externalization of mental images and the act of creation. In B. S. Randhawa & W. E. Coffman (Eds.), *Visual learning, thinking, and communication* (pp. 133–189). New York: Academic.

Simon, H. A. (1981). *The sciences of the artificial.* Cambridge, MA: MIT Press.

Simon, H. A. (1989a). *Models of thought* (Vol. 1). New Haven, CT: Yale University Press.

Simon, H. A. (1989b). *Models of thought* (Vol. 2). New Haven, CT: Yale University Press.

Stansky, P. (1996). *On or about December 1910: Early Bloomsbury and its intimate world.* Cambridge, MA: Harvard University Press.

Stratton, G. M. (1896). Some preliminary experiments on vision without inversion of the retinal image. *Psychological Review, 3,* 611–617.

Tweney, R. D. (1990). Five questions for computationalists. In J. Shrager & P. Langley (Eds.), *Computational models of scientific discovery and theory formation* (pp. 471–484). San Mateo, CA: Morgan Kaufmann.

Tweney, R. D. (1991). Faraday's 1822 chemical hints notebook and the semantics of chemical discourse. *Bulletin for the History of Chemistry, 11,* 51–54.

Tweney, R. D. (1992a). Inventing the field: Michael Faraday and the creative "engineering" of electromagnetic field theory. In R. J. Weber & D. N. Perkins (Eds.), *Inventive minds: Creativity in technology* (pp. 31–47). New York: Oxford University Press.

Tweney, R. D. (1992b). Stopping time: Faraday and the scientific creation of perceptual order. *Physis, 29,* 149–164.

Tweney, R. D. (1994). Making waves: In which the large splash made by the hypothetico–deductive method is examined, and its claims to truth made chaotic. *BGSU Student Research in Psychology, 2,* 50–57.

Tweney, R. D. (2003). *Shifting sands: Sight and sound in Faraday's acoustical researches.* Manuscript in preparation.

Walberg, H. J. (1988). Creativity and talent as learning. In R. J. Sternberg (Ed.), *The nature of creativity: Contemporary psychological perspectives* (pp. 340–361). New York: Cambridge University Press.

Wallace, D. B. (1982). *The fabric of experience: A psychological study of Dorothy Richardson's "Pilgrimage"* (Doctoral dissertation, Rutgers University, 1982). University Microfilms International, 42(11-B), 4565.

Wallace, D. B. (1991). The genesis and microgenesis of sudden insight in the creation of literature. *Creativity Research Journal, 4*, 41–50.

Weber, R. J., & Perkins, D. N. (1989). How to invent artifacts and ideas. *New Ideas in Psychology, 7*, 49–72.

Weisberg, R. W. (1991). Problem solving and creativity. In R. J. Sternberg (Ed.), *The nature of creativity: Contemporary psychological perspectives* (pp. 148–176). New York: Cambridge University Press.

Westcott, M. R. (1968). *Toward a contemporary psychology of intuition.* New York: Holt, Rinehart & Winston.

Woolf, V. (1960). *Jacob's room.* New York: Harcourt Brace. (Original work published 1922)

Woolf, V. (1967a). Life and the novelist. In *Collected essays* (Vol. 2, pp. 131–136). New York: Harcourt Brace & World. (Original work published 1926)

Woolf, V. (1967b). The narrow bridge of art. In *Collected essays* (pp. 218–229). New York: Harcourt Brace & World. (Original work published 1927)

Woolf, V. (1967c). Phases of fiction. In *Collected essays* (Vol. 2, pp. 56–102). New York: Harcourt Brace & World. (Original work published 1929)

Woolf, V. (1984a). How should one read a book? In M. A. Leaska (Ed.), *The Virginia Woolf reader* (Vol. 1, pp. 233–245). New York: Harcourt Brace Jovanovich. (Original work published 1932)

Woolf, V. (1984c). Modern fiction. In M. A. Leaska (Ed.), *The Virginia Woolf reader* (pp. 283–291). New York: Harcourt Brace Jovanovich. (Original work published 1925)

Woolf, V. (1984d). Mr. Bennett and Mrs. Brown. In M. A. Leaska (Ed.), *The Virginia Woolf reader* (pp. 192–212). New York: Harcourt Brace Jovanovich. (Original work published 1924)

Woolf, V. (1984e). Street haunting. In M. A. Leaska (Ed.), *The Virginia Woolf reader* (pp. 248–259). New York: Harcourt Brace Jovanovich. (Original work published 1927)

Woolf, V. (1985). *Moments of being* (J. Schulkind, Ed.). New York: Harcourt Brace.

Woolf, V. (1986a). The novels of George Gissing. In A. McNeillie (Ed.), *The essays of Virginia Woolf* (Vol. 1; pp. 355–362). New York: Harcourt brace Jovanovich. (Original work published 1912)

Woolf, V. (1986b). Robinson Crusoe. In A. McNeillie (Ed.), *The second common reader* (pp. 51–58). New York: Harcourt Brace Jovanovich. (Original work published 1932)

Woolf, V. (1987a). In a library. In A. McNeillie (Ed.), *The essays of Virginia Woolf* (Vol. 2, pp. 52–54). New York: Harcourt Brace Jovanovich. (Original work published 1916)

Woolf, V. (1987b). More Dostoyevsky. In A. McNeillie (Ed.), *The essays of Virginia Woolf* (Vol. 2, pp. 83–87). New York: Harcourt Brace. (Original work published 1917)

Woolf, V. (1989a). The mark on the wall. In S. Dick (Ed.), *The complete shorter fiction of Virginia Woolf* (pp. 83–89). New York: Harcourt Brace.

Woolf, V. (1989b). Kew Gardens. In S. Dick (Ed.), *The complete shorter fiction of Virginia Woolf* (pp. 90–95). New York: Harcourt Brace. (Original work published 1919)

Woolf, V. (1989b). *To the lighthouse.* New York: Harcourt Brace. (Original work published 1927)

Woolf, V. (1989c). An unwritten novel. In S. Dick (Ed.), *The complete shorter fiction of Virginia Woolf* (pp. 112–121). New York: Harcourt Brace. (Original work published 1920)

Woolf, V. (1990a). *Mrs. Dalloway*. New York: Harcourt Brace & Company. (Original work published 1925)

Woolf, V. (1990b). *A passionate apprentice: The early journals, 1897–1909* (M. A. Leaska, Ed.). New York: Harcourt Brace Jovanovich.

Woolf, V. (1992). *Night and day*. New York: Penguin. (Original work published 1919)

11

What's So Hard About Rocket Science? Secrets the Rocket Boys Knew

Gary Bradshaw
Mississippi State University

On October 4, 1957, the Soviet launch of Sputnik astounded and dismayed the free world. Their success demonstrated a clear superiority in rocket technology over the United States. Less than a month later, Sputnik 2 was launched and carried Laika, a live dog, into orbit—further evidence of the Soviet lead.[1] These successful launches led to the development of the "space race" between the Soviet Union and the United States, with far-reaching implications for American education, technology, and the Cold War.

The Soviet Union's success generated an explosion of interest in rocketry among America's youth. Books and articles on the topic appeared in the children's literature (e.g., Yates, 1952). The enterprise rightly earned a reputation for being dangerous, although science fair projects on propulsion and rocketry were popular. Homer Hickam was among those galvanized by

---

[1]The United States was able to orbit two satellites the following spring, but U.S. launch vehicles were of low capacity compared with Soviet rockets. The Vanguard-launched TV-4 satellite weighed only 3.24 lbs (1.47 kg); the Explorer satellite weighed 31 lbs (14.1 kg). Sputnik III, launched in May 1958, weighed 2,925 lbs (1,326.8 kg).

the Soviet lead to attempt to build a rocket.[2] Hickam, a high-school student who lived in the coal-mining town of Coalwood, West Virginia, began by designing a rocket using a plastic flashlight for the rocket casement and extracted gunpowder from several cherry bombs to use as propellant. A nail punched into the bottom of the flashlight formed the opening for the nozzle of the rocket. A de-winged model airplane housed this rocket engine. True to its firework origins, the "rocket" exploded and blew off a section of the fence that had been used as a makeshift launch pad. Hickam's mother challenged the youth to take his quest seriously and build a rocket. She also cautioned him not to hurt himself or others during the process. Recognizing his lack of knowledge, Hickam enlisted the help of the scientifically oriented Quentin Wilson. Quentin was familiar with the ingredients for black powder and knew that Newton's third law was the governing principle behind rocket action.

Although they knew the ingredients of black powder fuel, neither of the boys had the chemical sophistication to compute the proper proportion of ingredients. This led to their first set of experiments to determine the ideal proportions of saltpeter, sulfur, and charcoal. They tested various mixtures by throwing samples into a coal furnace and observing the resulting flash and smoke.

After burning various mixtures, the pair decided to test the fuel as a propellant, which burns more slowly than an explosive. Quentin expressed their goal thusly: "We need to see how the powder acts under pressure. Whatever the result, we'll have a basis for modification." (Hickam, 1998, p. 83) In this statement Quentin thus expressed the idea that immediate success or failure was not at issue: Either outcome served to reduce their uncertainty about rocket fuels and so was informative. The quote also reveals a sophisticated research-and-design strategy: The pair were not searching through the space of different rocket model designs in a "generate-and-test" strategy but instead had come to the realization that they needed a body of knowledge about rockets that would permit successful design.

To test the fuels under pressure, the pair built two "rockets." These were simple craft with casements made from aluminum tubing. One end of the tubing was crimped to form a nozzle, and the other end was sealed off with a piece of wooden dowel. Cardboard fins were glued on to stabilize the rocket on the launch pad. One model "emitted a boil of nasty, stinking, yellowish smoke and then fell over, the glue on its fins melted" (Hickam, 1998, p. 84).

---

[2]This account is based on the book *Rocket Boys* (Hickam, 1998). Although the account is a retrospective memoir, several pieces of corroborative evidence support the central details of the story. All books and articles mentioned in the book have been found, and the references are accurate. Hickam (personal communication, January 31, 2003) has confirmed central details of the account.

The second model blew up. Quentin concluded "First sample was too weak, the second too strong. Now we know where we are. This is good, very good." (Hickam, 1998, p. 85). The result was considered not a failure but rather a discovery that represented an addition to their "body of knowledge" that would allow them to build a rocket.

The team grew to five members on February 1, 1958. The boys, who dubbed themselves the "Big Creek Missile Agency," continued to research proportions of black powder ingredients, again tested in the coal furnace. The boys never used chemical equations to determine ingredient proportions but instead relied on trial and error. After finding a mixture that produced a bright flash and considerable smoke, the boys considered how to use the fine powder in a rocket. Quentin noted that "I don't like this loose mix. It seems to me we ought to put some kind of combustible glue in it so it can be shaped. We could run a hole up through the center of it, get more surface area burning at once. That ought to get us more boost" (Hickam, 1998, p. 98). Hickam purchased powdered glue from the store to use as a binding agent. A small cake of powder and glue was prepared, allowed to dry for 3 days, then tested in the coal-fired water heater. It burned "vigorously" (Hickam, 1998, p. 100).

Quentin found a physics book that explained Newton's third law, using an example of a balloon. The boys recognized that a rocket is like a "hard balloon." The nozzle, analogous to the neck of the balloon, needs to be smaller than the casement. But how much smaller was not obvious, leading to a second subproblem: How should the nozzle be shaped? The boys began with a simple nozzle consisting of a washer soldered to a pipe. The Auk I had a 14-in. (36 cm) long, 1.25-in. (3.18 cm) aluminum casing. (Information about each rocket is contained in the Appendix.) One end was sealed with a metal cap covered by a wooden nose cone. A washer soldered to the opposite end of the casement formed the nozzle. Black powder with a glue binder was used for fuel, and a pencil was inserted into the wet mix to create a spindle hole that would allow a faster burn of the propellant. Cardboard fins were taped on. At testing, the rocket rose to an altitude of only 6 ft (1.8 m) before the solder holding the washer melted.

A machinist who was assisting the boys realized the nozzle would have to be welded in place, not soldered on, and suggested steel casing rather than aluminum. He also recommended using a short length of drilled steel rod in place of the washer and constructed several rocket bodies using this plan: a steel casement with a welded rod nozzle and cap.

Auks II, III, and IV were tested with the black powder propellant. Auk II and III made unsteady flights, whereas Auk IV made a smooth-but-uncontrolled flight, striking the local mine office. Before conducting any further rocket tests, the group returned to the goal of improving their propellant. Quentin designed a test stand to measure powder strengths, but it was far too complex to construct. The team then decided to pour a measured amount of powder into pop bottles and examine the size of the explosion. One finely ground mix produced a large crater, so this was selected for the next rocket tests.

The water-based glue used as a binder did not have adequate time to dry in Auk V, which reached an altitude of only 50 ft (15.2 m). Auk VI made a significant but unmeasured flight, while Auk VII made a U-turn at 50 ft (15.2 m), and Auk VIII exploded shortly after takeoff. A visiting engineer suggested that trigonometry could be used to measure the height of the rocket. This led the boys to learn trigonometry, then to fashion a theodolite to measure the altitude of their rockets.

In the fall of 1958, Miss Riley, their high school science teacher, demonstrated rapid oxidation for the chemistry class using potassium chlorate and sugar. The boys realized this would generate more propulsive gases than black powder, although Miss Riley cautioned that potassium chlorate is unstable under high temperatures and pressures. Quentin recognized that potassium nitrate (saltpeter) would have a similar chemical reaction with sugar. This formed the basis for their next rocket propellant.

Again the team had a problem with using a fine powder as a "solid" fuel. They added water to one sample, allowing it to dry for a week, but the rocket only produced a gentle thrust insufficient to lift it off the pad. A postlaunch inspection revealed that some of the fuel melted into a thick caramel-like substance. The melted fuel could still be ignited, so the boys carefully melted the explosive ingredients for their next group of rockets. A glass rod was inserted into the mixed fuel while it hardened. After it was removed, it formed a spindle hole that increased the burning surface after ignition.

This fuel proved too powerful for the nozzles used by the team, corroding the steel metal. A more durable steel was chosen for the remaining nozzles.

In January of 1959, Miss Riley gave Hickam a copy of the book *Principles of Guided Missile Design* (Merrill, 1956). This volume explained how a nozzle could be shaped to maximize thrust. Computation of the De Laval shape required calculus, which the boys had to learn, and required tests on the propellant. These matters were tackled in due course.

Later that spring, Miss Riley demonstrated another chemical reaction for a biology class in which zinc dust and sulfur were burned. Again the boys realized this would be a better fuel than saltpeter and sugar. They explored proportions of ingredients and then conducted tests on the propellant characteristics in captive rockets. Finally, they were able to solve the equations needed to determine the shape of a De Laval nozzle, which was used in their final models. The steel used in the nozzle again corroded, so a curved shape was introduced, and an ablative coating was later added to the nozzle. Using these refinements, the team produced a final series of successful rockets, one of which reached an altitude just short of 6 m (9.66 km). Hickam, representing his team, won the county, state, and national science fairs and brought home the national Gold and Silver Award.

This brief account does not cover the full range of activities carried out by the Big Creek Missile Agency as they developed their rockets. The group engaged in various activities to raise money to purchase materials and constructed a launch pad and observation blockhouse. Further details are contained in *The Rocket Boys* (Hickam, 1998).

To illuminate the activities and successes of the Big Creek Missile Agency, it is helpful to follow the approach taken by Weber and Perkins (1989; Weber, 1992) and Bradshaw (1992), whereby invention is understood as a *search* through a set of potential inventions in an effort to find effective ones. Table 11.1 identifies the 12 major design elements of rocket construction varied by the boys and shows most of the alternative choices they explored. To test the factorial combination of every possible choice on each dimension would require that nearly 2 million rockets be built, yet the young team had solved all of the important problems by their 25th model. Table 11.1 also lists more than 36 specific values on the various dimensions, so if the team had only changed *one* factor from model to model, exploring the list would have required at least 36 models. (It is difficult to identify the different levels of some parameters, such as the various fuel grinding techniques that were used, so 36 models should be considered a low estimate of the minimum number of models needed if only one change is made at a time.) Thus, their exploration of the space of model rockets was efficient. In the next section I consider factors that contributed to this inventive efficiency.

EFFICIENCIES OF INVENTION

The first efficiency of invention adopted by the youthful team was *functional decomposition* (Bradshaw, in press). Functional decomposition requires a complex invention (such as an airplane) to be divided into functional parts (wings that produce lift, propellers that produce thrust) that are refined in isolation from the whole. This "divide-and-conquer" approach produces efficiencies for two reasons. First, it is cheaper to build a component than it is to build a complete device. Perhaps more important, however, it is often easier to test a component than to test a full device. Unless most of the components of a complex invention are working correctly, the whole will fail. Changes in a component may not produce changes in the performance of the full device, even though the changes are effective. Again drawing an illustration from the world of aviation, a new and more efficient wing might be built, but if the plane is out of balance, it would fly no farther than a model with a less efficient wing. When a component is tested in isolation, the test does not rely on the performance of any number of other potentially faulty elements and so is likely to be diagnostic.

TABLE 11.1
Design Variations in Model Rockets

Fuels	Fin materials
Black powder	Cardboard
Potassium nitrate	Tin
Zinc & sulfur	Sheet aluminum
Fuel mixtures	Steel
Various proportions of reactants	**Fin attachment procedures**
Fuel preparations	Tape
Various grinding processes	Tight wire
Fuel binding agents	Strap
Dry	**Fin sizes**
Postage stamp glue	Various sizes and shapes
Water	**Nozzle geometry**
Alcohol	Washer
Casement materials	Drilled rod
Plastic tubing	Double-countersunk rod
Aluminum tubing	De Laval
Steel tubing	De Laval w/curved throat
Butt-welded steel tubing	**Nozzle treatment**
Casement diameters	None
1.25 in. (3.18 cm), 2 in. (5.08 cm), 2.25 in. (5.72 cm), 3 in. (7.62 cm)	Ablative coating
Casement lengths	
1.5 ft (45.72 cm), 2 ft (5.08 cm), 3.5 ft (106.68 cm), 4 ft (121.92 cm), 4.5 ft (137.16 cm), 6 ft (182.88 cm), 6.5 ft (198. 12 cm)	

After Hickam's initial attempt to build a complete rocket (following a "design-and-test" strategy; Bradshaw, 1992), he and his friend Quentin turned to the problem of developing a propellant for their rockets. Various mixtures of saltpeter, charcoal, and sulfur were thrown into a coal-fired furnace, and the boys examined the resulting smoke and flash. This testing continued as the team considered various binding agents (to hold the powdered solid propellant elements together) and different propellants (black powder, KNO_3 & sugar, zinc & sulfur). Propellants continued to be tested by observing the flash and smoke of samples burned in a furnace and by adding a measured amount of a mixture to a pop bottle, igniting the propellant, and observing the size of the resulting crater. Yet this appears to be the only set of variables where a functional decomposition was used by the young team. They did not explore factors such as nozzle geometry, nozzle coatings, fin materials, or casement materials in isolation from the larger rocket model. All of these factors were explored by designing and testing different rocket models. Yet we should not conclude that the team used an inefficient strategy. In the case of rockets, it was probably cheaper to build and test a model rocket than to develop test stands for nozzle shapes, wind tunnels for fins, and pressure tests for casements. Each rocket cost only a few dollars to produce, whereas corresponding instruments and testing devices would have cost much more. Testing fuels by burning samples in a coal-fired furnace was far cheaper than building and testing rockets, so the team apparently was sensitive to the relative costs and benefits of testing components versus entire rockets. Indeed, at one point Quentin described an elaborate test stand to measure the strength of propellants fashioned from "tubes and springs and pistons" (Hickam, 1998, p. 116). The boys rejected this approach, noting that their limited resources did not permit such a contraption to be built. They instead settled on the more subjective method of exploding propellant samples in pop bottles.

The Wright brothers were more aggressive in their utilization of functional decomposition as a strategy of invention (Bradshaw, in press). They decomposed the full airplane design into functional subsystems of wings (for lift), elevator (for pitch control), rudder (for yaw control), and a propulsion system (engine, propellers) for thrust. (Roll control was incorporated in the wings via wing warping.) At least 14 different functional tests were performed on isolated airplane components. Their most complex functional tests were performed using a wind tunnel and required the development of precision instruments to measure lift and drag. Other tests were more casual, such as their comparison of doped and undoped model wings to determine if the fabric was leaking air. This raises an important issue: How were the Rocket Boys able to work so efficiently without relying on an extensive decomposition?

Another factor that contributed to their efficiency is an extensive use of postlaunch fault diagnosis. After Auk I reached an altitude of only 6 ft (1.8 m), the postlaunch evaluation revealed that the propellant had melted the solder holding the washer nozzle to the casement. All subsequent models either had the nozzle welded to the casement or screwed on with machine screws. Similarly, the casement of Auk XX was made from butt-welded steel tubing, whereas all previous models had been made with solid tubing. Auk XX exploded during the test, and a postlaunch inspection revealed that the casement had burst at the weld joint. After this failure, all subsequent models were again fashioned from solid tubing. A third example occurred after the brief flight of Auk V. A postlaunch inspection revealed that considerable unburned wet propellant remained in the casing. The boys subsequently allowed their black powder fuels to dry for a much longer period.

This pattern contrasts sharply with the invention of the airplane (Bradshaw, 1992, in press). Very little useful information could be gleaned from a postlaunch attempt to fly an airplane, whereas the Rocket Boys frequently used similar information to generate improved models. To understand this paradox, it is useful to distinguish between *component failure* and *poor component performance*. The Rocket Boys could often detect component failure from a postlaunch inspection (e.g., the butt-welded casement bursting at the seam, evidence that it was not strong enough to contain the pressures created by the burning propellant). Contrast this with a situation in aviation, where a wing might not be producing sufficient lift for a craft. Component failure can be readily detected and corrected, whereas poor component performance is more subtle and likely to be masked by other flawed components.

Functional decomposition appears to be most necessary when components are not working well and less necessary when components obviously fail. Nonetheless, functional decomposition still seems to have advantages over building and testing whole systems. A factor that apparently deterred the youthful team was the cost of developing test systems for individual components: It was cheaper to build any number of model rockets than to build a single high-speed wind tunnel. Given that it was cheap to build and test a complete rocket and that component failures could be diagnosed after testing, the team relied heavily on building and testing their inventions. The Wrights were in a different situation: It was expensive to build a full-scale airplane, and it was difficult to determine what was wrong from a test, leading to their heavy reliance on functional decomposition. Comparing the two inventions reveals the strengths and limitations of each strategy.

A third important practice that contributed to the young team's efficiency in invention was the use of theory to avoid search. The final nozzle shape used by the team is known as a *De Laval nozzle*, which is designed to take low-velocity high-pressure gases in the casement, convert them to a stream moving at sonic velocity at the constriction of the nozzle, and then produce a supersonic stream at the output. Rather than testing any number of different nozzle shapes, the boys obtained a book describing the equations that could be used to compute a De Laval nozzle. Using calculus, the team identified the cross-section for the De Laval nozzle, which a machinist produced. This final calculated shape followed on a more mundane trial-and-error approach toward nozzles. The team initially used a washer. When that proved too weak, a hole was drilled into a solid rod to create a nozzle. A machinist suggested that the compression and expansion holes could be countersunk, saving weight in the rocket. These steps were not guided by theory, only the final shaping of the De Laval nozzle.

Next, the team used instruments to improve their ability to measure performance, once again leading to an efficient inventive process. Instruments were used in two different ways. One class of instruments was used to evaluate rocket fuels independently from the larger rocket design. A second class of instruments was used to measure the performance of the rocket. An example of the former class of instrument is the use of a scale to measure the thrust of the propellant. The latter class of instrument is exemplified by a theodolite used to measure the altitude of the rocket in flight. The theodolite was important in determining that later models of the rocket were not performing as well as they had been designed. This triggered another postlaunch evaluation, which revealed the De Laval nozzle was corroding.

The team also used several "scientific" strategies during their invention process. After the initial explosive failure of the plastic model, Hickam teamed up with Quentin. Hickam was aware that he did not have the knowledge to build a rocket, and he wanted to tap into Quentin's scientific knowledge. Quentin proposed that the team add to the "body of knowledge" (Hickam, 1998, p. 84) through experiment. Records were kept of experiments and tests, avoiding the need to replicate work through lost data. Quentin also argued for a strategy commonly known as VOTAT ("vary one thing at a time")—making only one change to the design of a rocket so that performance changes could be clearly attributed to the source. It is curious that the team deliberately violated this strategy to stretch their limited resources. On occasion, one major change was accompanied by several minor changes, allowing for a faster exploration of the invention space than VOTAT allows. This strategy was suggested by one of the machinists who was helping to build the rockets: "It would take you forever to find the best design if you made only one change at a time" (Hickam, 1998, p. 268). This resembles focused gambling (Bruner, Goodnow, & Austin, 1956) in concept formation tasks, where research participants vary multiple values from trial to trial. However, the version here exhibits some sophistication not possible in the concept formation task because they believed that some features of their design (such as the propellant) were more important than other features (such as the fin shape). The team usually restricted their changes to one important variable along with some others of lesser weight. In the concept formation task all variables are equally likely to be important, so no differential emphasis is possible. However, in both cases the risk arises that too many things will be changed at once, leading to failure. Quentin argued against the focused-gambling strategy: "And when this rocket blows up and you don't have a clue what caused it? What will you have learned then?" (Hickam, 1998, p. 269). Nevertheless, the team pursued the focused-gambling strategy because of financial constraints. In spite of Quentin's worries and the occasional setback, this strategy worked well for the team. A third strategy has already been mentioned: "Rockets should perform as designed." When a rocket failed to perform as it had been designed to, that was a signal that a problem still remained. The corrosion of the De Laval nozzle was detected in just this manner.

EXTERNAL EXPANSION OF THE SEARCH SPACE

On three important occasions, external events changed the course of the team. Twice the team learned about new propellants, and once they learned an effective procedure to design a rocket nozzle. As previously mentioned, Miss Riley was instrumental in all three developments. In the fall of 1958 she demonstrated rapid oxidation to the chemistry class using potassium chlorate and sugar. This led to the abandonment of black powder as a fuel and the adoption of saltpeter and sugar. In January of 1959, she gave Hickam *Principles of Guided Missile Design* (Merrill, 1956), which contained information about the construction of a De Laval nozzle. Later that year, she demonstrated the combustion of zinc dust and sulfur. Again the team adopted the new fuel and never returned to the older propellants. These developments arose not from active search but from opportunistic adoption of a demonstrably superior alternative.

One can contrast this adventitious adoption with *opportunistic reasoning* (Birnbaum, 1986). Simina and Kolodner (1995) noted that "The prerequisite for opportunistic behavior is the existence of *suspended goals* (problems), goals that cannot be pursued in the current context and are postponed" (p. 79). As far as can be determined from the account, the team never actively considered different chemicals for propellants and so did not have a corresponding suspended goal. Nevertheless, they recognized new possibilities for rocket fuel when they witnessed demonstrations of different chemical reactions. Once they learned about a new and improved propellant, they dropped their older propellant from consideration. This behavior certainly qualifies as opportunistic, but it does not appear to involve reasoning in the same way that previous accounts have formulated the processes. The presence of a suspended goal is helpful in recognizing an opportunity. This leads to the following question: What allowed the youthful team to recognize that an opportunity had arisen?

In this situation, the signal of opportunity likely stems from the similarity in behavior between their existing fuel and the burning exhibited by the chemical reactions. The team had considerable experience in burning black powder propellant and had a feel for the magnitude of the heat, brightness of the flame, and duration of the burn. When they first witnessed the potassium chlorate and sugar mix, the team at once appreciated that the chemicals burned with a bright, intense flame. This presumably reminded them of the oxidation of black powder, allowing them to spot the opportunity.

This suggests a distinction be made between *impasse-circumvention opportunities* and *design-improvement opportunities*. The former arise because a goal could not be completed and therefore had to be suspended. The latter, in contrast, can be used to improve or optimize an existing system. Recognition and use of the opportunities may be somewhat different in these two situations. One can imagine, for example, that the team might have performed some tests on the new chemical mixtures to show that they indeed represented an improvement. This would be analogous, but not identical, to the process of testing an opportunistically suggested method to achieve a goal.

A somewhat different evolution of the design space took place in the Rocket Boys' efforts to design an effective nozzle. Their initial efforts simply followed a trial-and-error methodology, but that changed once they learned about the possibility of designing a De Laval nozzle. Even though the boys had to teach themselves calculus and develop instruments and tests for their propellants, they abandoned further search on nozzle shape. Their later nozzle designs were computed algorithmically instead of being shaped experimentally.

One further point arises as we consider other accounts of the expansion of inventive search spaces. Several different researchers have emphasized turning constants into variables as a way to expand the search space (e.g., Gorman, 1992; Gorman & Carlson, 1990; Weber & Perkins, 1989). In the cases mentioned earlier, however, external agents introduced alternatives that revealed that certain constants could be considered as variables. The Rocket Boys did not exploit the new variables revealed in this way; instead, they abandoned an inferior value of a variable and adopted the new alternative. They clearly were "satisficing" (Simon, 1982) on these dimensions, either to minimize development costs or because they did not have sufficient knowledge of the relevant area to engage in an extended search.

COORDINATING MULTISPACE SEARCHES

The team searched through a space of rocket designs, another space of propellants, and a third of instruments and considered matters such as how to ignite their rockets, how to measure the altitude of their rockets, and so on. This raises an important issue about the coordination of these various searches: When should a search in one space be postponed to allow search in another? Only a partial answer can be drawn from the surviving records. The explosion of Hickam's first plastic-casement model rocket led to the exploration of rocket propellants rather than casement materials, although the latter would appear to be a more "natural" next step. Quentin Wilson's participation, along with his knowledge of rocket fuels, likely was responsible for this turn of events. On obtaining a copy of *Principles of Guided Missile Design* (Merrill, 1956) in January of 1959, the team learned about De Laval nozzles, but existing nozzles, fuels, and rocket designs continued to be tested until November 1959. Part of this delay was due to the complexity of the De Laval nozzles: The team needed to know the specific impulse of their fuel, which required a test instrument to be built. But eight rockets were designed and tested in the intervening period, so the team believed they could learn from further tests even though they knew the nozzles would later be replaced with superior models.

Although some of the searches were pursued in a serial fashion, the boys exploited opportunities for parallel searches on various occasions. In the aforementioned period from January to November 1959, the boys: (a) tested different rocket designs, (b) developed a new rocket fuel, (c) constructed two test instruments to measure the specific impulse of their fuel, and (d) solved the equations needed to fashion a De Laval nozzle. Although a serial search might have been cheaper, the team was under some time pressure to meet a science fair deadline. In any case, few clues survive as to how they chose to schedule their activities—a difficult problem in multispace searches.

MENTAL MODELS OF ROCKETS

Gorman and Carlson (1990; Gorman, 1992) have revealed the importance of mental models in the inventive process in cases such as the telephone. This was a new invention with little precedent to guide hopeful inventors. For the Rocket Boys, the situation was quite different: Rockets had been in existence for several hundreds of years prior to their work, and information about rocket design was available in books. Even before the team was formed, Quentin knew about Newton's third law and was familiar with the ingredients of black powder. Thus, one does not expect to see the same driving relationship between mental models and invention in the Rocket Boys' efforts as one sees in more original inventions. Nevertheless, there are some interesting developments in the Rocket Boys' understanding of a rocket as their research unfolded.

Homer Hickam's first attempt was motivated by an article that appeared in *Life* magazine on November 18, 1957. This issue depicted cross-sectional views of various rocket types, including a solid-fuel rocket. The drawing depicts a converging and diverging nozzle on the rocket. The early attempts made by the group had a restriction at the nozzle, but the restriction was unshaped. Hickam (1998) related the early mental model of a rocket:

> We kept trying to figure out the "why" of rockets as well as the "how." Although he hadn't found a rocket book, Quentin had finally found the physics book he'd read in the library in Welch that defined Newton's third law of action and reaction. The example given in the book was a balloon that flew around the room when its neck was opened. The air inside the balloon was under pressure, and as it flowed out of the opening (action), the balloon was propelled forward (reaction). A rocket, then, was sort of a hard balloon.
>
> Instinctively, we knew that the nozzle (the opening at the rocket's bottom), like the neck in the balloon, needed to be smaller than the casement. But how much smaller, and how the nozzle worked, and how to build one, we had no idea. All we could do was guess. (p. 100)

At this early stage, a nozzle seemed to be analogous to the neck of a balloon. It was not until they began reading a chapter from *Principles of Guided Missile Design* (Merrill, 1956) that the young team began to appreciate the subtleties of gas compression, expansion, and the fluid dynamics that governed efficient expansion for thrust. It is curious that the machinist working on their rockets decided to countersink the nozzle on both the compression and expansion sides, not to shape the gas flow but to "save weight."

Quentin did show an early appreciation for a subtle aspect of rocket design: providing a spindle hole in the propellant to allow for a greater surface area of burning propellant. This greatly increases the internal pressure and, therefore, the thrust. This innovation was adopted in the first model, Auk I. All of the rockets with solid fuel (some were loaded with powdered fuel) preserved this feature. The *Life* magazine article of November 1957 discussed leaving a "shaft" in the middle of a solid fuel rocket and described the fuel as burning from the inside out. This may have been the source of Quentin's appreciation.

Thus, in contrast to other more original inventions, the boys began with a reasonable mental model of rockets and worked largely within that original model. Only when they learn of a more sophisticated model through exposure to *Principles of Guided Missile Design* (Merrill, 1956) did their model change in any substantial way.

CONCLUSIONS

The story of the Rocket Boys is a remarkable story of a group of teenagers who embark on a systematic research effort to build a rocket. Their success was remarkable, but so was the effort that led to eventual success. Compared with other studies of creativity in invention, we observe that the team is conducting a complex search through several different spaces, including the space of designs, the space of instruments, the space of components, and the space of rocket tests. The search through the space of designs was highly efficient in part because the team exploited functional decomposition, in part because they circumvented component search through the use of theory (e.g., the De Laval nozzle), and in part because catastrophic component failure could be readily diagnosed after testing. The youthful group did not discover any significant changes to their mental model of the device but acquired a more sophisticated mental model after reading about the discoveries of others.

More detailed records are needed to understand how the team decided to switch their attention from one search space to another. Understanding the factors that trigger a switch remains a challenge for future accounts of multispace problem solving. Another challenge is to understand when VOTAT is called for and when focused gambling is appropriate. Determination of these factors would benefit from a more complete record of the hypotheses of inventors—information that is seldom a part of the archival record.

Finally, it is important to note that the youngsters maintained a tight focus in their efforts rather than broadly exploring various options. They were aware of the problem that introducing multiple changes in a design would create, but they nevertheless chose to violate this dictum in a sensible way to preserve limited resources. They also sustained their research effort through setbacks and considerable adversity, drawing on the support of friends and family. This case probably does not resemble closely the "revolutionary" inventions such as the airplane or telephone, but it may have much in common with more ordinary inventions produced in a complex, technology-rich society.

ACKNOWLEDGMENTS

I thank Ryan Tweney for early discussions on this project, Mike Gorman for his insightful suggestions on various drafts, and Isaac Shields for his assistance in tracking down sources for this article.

REFERENCES

Birnbaum, L. (1986). Integrated processing in planning and understanding. *Dissertation Abstracts International, 48*(10), 3023B.

Bradshaw, G. L. (1992). The airplane and the logic of invention. In R. N. Giere (Ed.), *Cognitive models of science* (pp. 239–250). Minneapolis: University of Minnesota Press.

Bradshaw, G. L. (in press). Heuristics and strategies of invention: Lessons from the invention of the airplane. *Cognitive Science.*

Bruner, J. S., Goodnow, J., & Austin, G. A. (1956). *A study of thinking.* New York: Wiley.

Destination, outer space—a primer on propulsion, fuels and space navigation. (1957, November 18). *Life, 43,* 42–52.

Gorman, M. E. (1992). *Simulating science heuristics, mental models, and technoscientific thinking.* Bloomington: Indiana University Press.

Gorman, M. E., & Carlson, W. B. (1990). Interpreting invention as a cognitive process: The case of Alexander Graham Bell, Thomas Edison, and the telephone. *Science, Technology, and Human Values, 15,* 131–164.

Hickam, H. H., Jr. (1998). *Rocket boys.* New York: Random House.

Merrill, G. (1956). *Principles of guided missile design (aerodynamics, propulsion, structures and design practice).* Princeton, NJ: Van Nostrand.

Simina, M. D., & Kolodner, J. L. (1995). Opportunistic reasoning: A design perspective. *Proceedings of the 17th annual conference of the Cognitive Science Society,* 78–83.

Simon, H. A. (1982). *Models of bounded rationality.* Cambridge, MA: MIT Press.

Weber, R. J. (1992). *Forks, phonographs, and hot air balloons: A field guide to inventive thinking.* New York: Oxford University Press

Weber, R. J., & Perkins, D. N. (1989). How to invent artifacts and ideas. *New Ideas in Psychology, 7,* 49–72.

Yates, R. F. (1952). *Model jets and rockets for boys.* New York: Harper.

Appendix

Model Rockets Developed by the Rocket Boys

Rocket and Test Date	Description
Unnamed; mid-November 1957	Casement: a plastic flashlight model. Nozzle: hole punched with nail. Fuel: black powder. Results: exploded and damaged supporting fence.
Auk I; spring 1958	Casement: 1.25 in. (3.18 cm) aluminum casing. Nozzle: washer soldered in place. Fuel: black powder with glue binder, spindle hole to improve burn area. Fins: cardboard fastened with electrical tape. Results: reached altitude of 6 ft (1.82 m), then solder holding washer melted.
Auks II, III, and IV; spring 1958	Casement: 1.25 in. (3.18 cm) steel. Nozzle: drilled rod welded to base. Fuel: black powder with glue binder. Results: Auk II and III had unsteady flights of about 20 ft. (6.1 m) in altitude. Auk IV has smooth, uncontrolled flight, rocket hits mine office.
Auk V; May 1958	Casement: 1.25 in. (3.18 cm) steel. Nozzle: drilled steel bar welded to base. Fuel: black powder with glue binder. Result: reached 50 ft (15.2 m) in altitude, then unsteady flight. Postlaunch examination reveals wet propellant unburned in casing.

Auks VI, VII, and VIII; July 1958	Casement: 1.25 in. (3.18 cm) steel. Nozzle: drilled steel bar welded to base. Fuel: black powder with glue binder. Fins: sheet aluminum strapped to casement with wire. Results: Auk VI rose "twice as high as mountain." Auk VII rose 50 ft (15.2 m), then made a U-turn and slammed into ground. Auk VIII exploded overhead.
Auk IX; August 1958	Casement: 1.25 in. (3.18 cm) steel. Nozzle: drilled steel bar welded to base. Fuel: granular KNO_3 & sugar. Results: rose 100 ft (30.5 m), then fell back. Postlaunch examination reveals that all propellant burned.
Auk X; August 1958	Same as Auk IX, except fuel prepared with glue binder, allowed to dry for a week. Result: did not lift off, only produced white smoke and gentle thrust. Postlaunch examination reveals considerable propellant remaining unburned inside the rocket in the form of a dark, thick liquid.
Auk XI; September 1958	Same as Auk IX. Result: leapt off the pad, then exploded. Postlaunch inspection reveals the casement burst under pressure, likely from too much propellant burning at once.
Auk XII; fall 1958	Same as Auk XI, except fuel prepared by melting together KNO_3 & sugar. Results: reached 767 ft (233.8 m) in altitude. Postlaunch inspection reveals the nozzle corroded away, reducing thrust.
Auk XIII; fall 1958	Same as Auk XII. Result: maximum altitude somewhat lower than Auk XII. Postlaunch inspection also shows corroded nozzle.
Auk XIV; fall 1958	Casement: 1.25 in. (3.18 cm) steel, 2.5 ft (76.2 cm) long. Nozzle: 1-in. (2.5 cm) piece of Society of Automotive Engineers 1020 bar stock drilled. Nozzle is screwed in place rather than welded. Fuel: melted KNO_3 & sugar. Result: reached altitude of 3,000 ft (914 m).
Auk XV; fall 1958	Casement: 1.25 in. (3.18 cm) steel, 3 ft (91 cm) long, eyebolts to guide rocket during launch. Otherwise same as Auk XIV. Result: reached altitude of 1,500 ft (457 m).
Auks XVI, XVII, XVII, and XIX; February 1959	Casement: 1.25 in. (3.18 cm) steel, three of the models are 2 ft (61 cm) long, and the final one is 3 ft (91 cm) long. Nozzle: drilled steel rod. Fuel: melted KNO_3 & sugar. Results: 2-ft (61 cm) models reach an altitude of 3,000 ft (914 m), 3-ft (91 cm) model reaches an altitude of 2 ft (61 cm). Conclusion: "Bigger is not necessarily better."
Auk XX; April 1959	Casement: 1.25 in. (3.18 cm) butt-welded steel, 3 ft (91 cm) in length. Nozzle: drilled steel rod with countersunk shape on compression and output sides. Fuel: melted KNO_3 & sugar. Fins strapped to casement. Result: explodes during test; casement burst at butt weld joint.
Auk XXI; May 1959	Casement: seamless steel tubing. Nozzle: drilled steel rod countersunk on both sides. Fuel: melted KNO_3 & sugar. Result: reaches altitude of 4,100 ft (1,250 m).
Auk XXII; September 1959	Fuel: powdered zinc and sulfur fuel. Explodes at launch.

Auk XXIIA; October 1959	Fuel: zinc and sulfur with alcohol binder. Reaches altitude of 5,766 ft (1,757 m).
Auk XXIID; October 1959	Rocket with reduced fin sizes goes astray.
Auk XXIII; November 1959	Casement: 48 in. (122 cm) long. Nozzle: De Laval shape. Fuel: zinc and sulfur with alcohol binder. Rocket was projected to reach 10,000 ft (3,048 m) but only reaches altitude of 7,056 ft (2,151 m). Postlaunch examination reveals that nozzle was eroded by fuel.
Auk XXIV; December 1959	Casement: 60 in. (152.4 cm) long. Nozzle: De Laval shape modified by curves. Fuel: zinc and sulfur with alcohol binder. Top 6 in. (15.2 cm) of casement loaded with smoke-producing fuel to track rocket. Reached altitude of 8,500 ft (2,591 m).
Auk XXV; spring 1960	Curved De Laval nozzle covered with water putty as an ablative coating. Reached altitude of 15,000 ft (4,572 m).
Auk XXVI; June 4, 1960	Simple countersunk nozzle. Reached 3,000 ft (914 m) in altitude.
Auk XXVII; June 4, 1960	Casement: 1.25 in. (3.18 cm) diameter, 3.5 ft (106.7 cm) feet long. Reached altitude of 9,000 ft (2,743 m).
Auk XXVIII; June 4, 1960	Designed for 15,000 ft (4,572 m).
Auk XXIX; June 4, 1960	Casement: 2 in. (5.08 cm) diameter, 72 in. (182.9 cm) long. Designed for 20,000 ft (6,096 m); reached altitude just under 4 miles (6.4 km).
Auk XXX; June 4, 1960	Casement: 2.25 in. (5.7 cm) diameter. Designed for 20,000 ft (6,096 m); reaches 23,000 ft (7,010 m).
Auk XXXI; June 4, 1960	Casement: 2.25 in. (5.7 cm) diameter, 78 in. (198 cm) long. Designed for 5-mile (8 km) altitude; reaches height of 31,000 ft (9,448 m).

12

A Systems-Ordered World

Thomas P. Hughes

Emeritus Professor of the History and Sociology of Science,
University of Pennsylvania, and Distinguished Visiting Professor,
Massachusetts Institute of Technology

In the late 19th century, observers believed that Western technology, especially engineering, was harnessing natural forces for the benefit of man. It was seen as creating a human-built world within a vast natural context. Great Britain, the most human-built of nations, and the United States were rapidly industrializing and urbanizing. Their citizens believed that they were improving upon nature, as canals replaced rivers; tunnels supplemented natural passes; machines substituted for labor; and steam engines replaced animal, water, and wind power. Gaslight was extending the day, improved heating was moderating the effects of climate, and the early telegraph lines were eliminating space. Nevertheless, nature in the 19th century was never more than a horizon away, even in the rapidly industrializing countries. Her great forces still determined abundance or famine, health or disease, comfort or misery.

Today, in contrast, the human-built is omnipresent, and technology is great determining force in the industrialized world. This generalization has become a truism, but we are only beginning to comprehend its ramifications. The industrialized population lives mostly in a human-built environment where nature is only a tree; a patch of blue sky; and a short exposure to excessive heat, cold, or rain. Natural rhythms have given way to human schedules. Natural substances, such as wood, are less in evidence; human-made substances pollute water and air. This widespread transformation causes considerable lament and even anxiety, but analysis, understanding, and general descriptions of the evolution and state of the human-built world are disjointed and generally unenlightening.

Historians, especially historians of technology, possess an extremely important key to understanding the modern world, for they have the potential to describe and explain the evolution of the technological systems ordering this world. Students of natural evolution have done much to enlighten us about the nature and the dynamics of natural systems. Historians, however, have as yet only scratched the surface of the highly organized and evolving human-built world. They sometimes describe, but they rarely explain, the evolution of highway, telephone, pipeline, electric transmission, television, computer, and complex production systems. There are many reasons for this failure; one is that human-built, interacting systems are as complex and as difficult to comprehend as the natural ones described by Isaac Newton, Charles Darwin, and their successors. A Newton or Darwin for the man-made world is wanting.

How do we define the systems that should be explained by the historian? What are these structures that increasingly shape history? First, it should be established that many of them are centuries old. Second, it should be stated that they tend to become larger and more complex with the passage of time. Third, it should be noted that they embody human values and, as a result, take on different styles in different cultures. Electric grids, for example, differ in detail according to whether they are located in the United States, Russia, or Japan.

How does one identify a system? Dealing with broad sweeps of time and widely separated places, historians need be wary of narrow concepts and definitions. Engineers, scientists, and social scientists focus their attention more sharply, so they can define more precisely. The historian's definition has to take into account the varied ways in which the term *system* has been used in the past, in various cultures, and by his professional peers. The lengthy entry for *system* in the *Oxford English Dictionary* is a reminder that no single definition will encompass the subject. Ludwig von Bertalanffy (1968) needed a book, not a paragraph, to define it. Members of the Engineering Systems Division of the Massachusetts Institute of Technology identified various terms associated with systems in order to define them.[1]

There are natural, or ecological, systems and human-built ones. We also encounter ecotechnological systems that involve both human-built and natural components. Some human-built ones evolve without central direc-

---

[1]Members of the Engineering Systems Division defined *systems* as have interconnected, interacting, interdependent components that function to fulfill a purpose or purposes. They decided that technical systems differ from sociotechnical systems in that the latter have both human and nonhuman components. They suggested that the environment of systems is a set of conditions external to and shaping a system (Unpublished committee notes in author's possession, spring 2001).

tion or planning, and others are planned and controlled. Early electric light and power systems in the United States, for example, were widely dispersed and evolved independently until they were consolidated regionally.

A historian might write the history of electric power systems and focus entirely on heterogeneous technical components, such as generators, transformers, and motors, but a more encompassing history could be written about electric power as a sociotechnological system having, in addition to technical components, organizational components, such as utility companies and regulatory agencies. To fulfill the goal of supplying electric power, both kinds of components are needed.

There are systems designed to create systems. For example, in the United States during the 1950s, the Air Force established a sociotechnological system involving Air Force agencies and hundreds of industrial and university contractors, including research laboratories. There were tens of thousands of administrative, scientific, blue collar, and engineering personnel. The purpose was to create the first intercontinental ballistic missile system with its heterogeneous components.

Recently the federal government, the Commonwealth of Massachusetts, numerous environmental agencies, systems engineering planners and designers, and construction firms have interacted to construct a large underground highway, the Central Artery and Tunnel, or "Big Dig," in downtown Boston. Today, several federal and state organizations, principally the Army Corps of Engineers, involving engineers and environmental scientists are interacting to restore an ecotechnological system in the Everglades of Florida.

The bewildering variety of systems creating and structuring the human-built world defies a common, universal definition. Nevertheless, there are some general characteristics that help us identify them. In general, they have a heterogeneous variety of physical and organizational components coordinated to fulfill a common goal or set of goals. The components interact, sometimes simultaneously, as in an electric power system, and other times through feedback loops with time delays, as in sociotechnological systems. They can be tightly coupled, like an electric power system, or loosely coupled, like a sociotechnological university system (Perrow, 1984).

As noted, the purpose can be unitary or complex. In the case of an operational ballistic missile system, the purpose is tightly focused. In the case of a highway, such as the Big Dig, it is designed not only to relieve traffic congestion but also to respond to a large set of environmental requirements, such as the reduction of air pollution from automobile emissions. It is also expected to stimulate economic activity in the Boston region.

The more massive and complex a system is, the more difficult the problem of controlling it. Some sociotechological systems are so complex that control cannot be centralized in a single individual or organization; control must be dispersed or diffused. In the latter case, coordination of control results from a common goal or purpose understood and accepted by the various centers of control. An aircraft carrier system for landing airplanes is such a decentralized, common purpose system (Rochlin, La Porte, & Roberts, 1987).

DEEP PLOUGHS, HARNESSED TEAMS, AND MANORS

Human-built systems date back to antiquity and beyond. The Egyptians organized systematically to build pyramids. The Roman approach to road building was also systematic. In 1962, historian Lynn White eloquently described the emergence of a system of agriculture from the 6th through the 9th centuries on the plains of northern Europe (White, 1962). In tracing its evolution, White drew on the scholarship of the eminent French historian Marc Bloch (1955/1966). According to White and Bloch, peasants worked out a balanced and efficient means of food production that culminated as an Agricultural Revolution unparalleled since the invention of tillage. Its components—the heavy plough, plough teams, open fields, three-field rotation, and horse harness—accumulated and consolidated into an interlocking and positively interacting agrarian system.

Its history can be traced in hazy outline. The heavy alluvial soil of the northern European plain necessitated the use of a different plough from the light, two-ox, scratch-plough commonly used earlier on the light soils of the Roman peninsula. Ingenious northern peasants responded with the use of a heavy plough with wheels, coulter, horizontal ploughshare, and a mouldboard that would turn the heavy soil. They found that friction of the new plough cutting through the heavy soil was so great that an eight-oxen team was needed. With the harnessing of the large team to the plough a simple, coherent system of agriculture emerged.

The use of a large plough team brought the emergence of a social, or manorial, system. Having pooled their animal resources, the peasants than pooled the land that they worked and engaged in communal decision making about the time of planting and harvesting. The manor, a peasant, communal organization, presided over a sociotechnical system.

Coherent cultivation not only spread through northern Europe and into England, but it also evolved. In the late 8th century, peasants in the region between the Loire and Rhine rivers, where the summer rains were adequate, began a three-field system of cultivation. The summer-harvested crop often included oats, a cereal not common under the old two-field system. The availability of oats allowed the peasants to keep horses, which had superior characteristics as a plough team. To fully utilize the horses, a collar harness was used. Moreover, horses were shod with iron, which increased their efficiency as a motive power. Shod horses with collars became an integral part of evolving agriculture in many parts of Europe.

To define the borders of a system is virtually impossible; to limit the analysis of dynamic interactions of the Agricultural Revolution to agriculture alone simplifies the problem of history writing but does not do justice to its complexity. Bloch, White, and others have found that the new agricultural system interacted with other momentous medieval developments, such as population growth and urbanization.

LABOR, MACHINES, AND AUTOMATONS

Evolving systems with positively interacting components were also at the core of a late 18th- and early 19th-century British Industrial Revolution. Karl Marx's (n.d.) chapter in *Capital*, entitled "Machinery and Modern Industry," is an extended analysis of the rise of a machine production. Besides explaining how the machine system contrasted with the earlier handicraft and manufacture, Marx stressed that the components of the machine system, both inanimate and human, interacted. He wrote:

> In an organized system of machinery, where one detail machine is constantly kept employed by another, a fixed relation is established between their numbers, their size, and their speed. The collective machine, now an organized system of various kinds of single machines, and of groups of single machines, becomes more and more perfect, the more the process as a whole becomes a continuous one, i.e., the less the raw material is interrupted in its passage from its first phase to its last. (Marx, n.d., p. 415)

Like Charles Babbage, author of *On the Economy of Machinery and Manufactures* (1832), Marx called the machine system an *automaton*. He wrote of "a mechanical monster whose body fills whole factories, and whose demon power, at first veiled under the slow and measured motion of his giant limbs, at length breaks out into the fast and furious whirl of his countless working organs" (Marx, n.d., p. 417). He lamented that labor had become a component subordinated to the central moving power.

Marx realized that the machine system was a component in a larger industrial system at the core of an industrial revolution. He realized that a radical change in the mode of production in one sphere of industry caused reverberations and coordinated changes in another. The introduction of a machine system in spinning, for instance, stimulated and interacted with machine weaving. Machine spinning and weaving created a demand bringing the rationalization of chemical processes for the production of bleaches and dyes. These evolving and interacting production systems stimulated the mechanization of transportation and communication networks to interconnect and inform (Marx, n.d., p. 419). Technological determinism structured his encompassing analysis.

The interactions of the British Industrial Revolution can be seen on an even higher level of abstraction. A coal, iron, and steam system functioned at the core of the revolution. The introduction of pumps driven by steam engines increased the production of coal, which, when transformed into coke, fueled furnaces producing iron in large quantities, used in the making of steam engines. These not only pumped water from the mines but also drove machines pumping air into iron furnaces to increase their yield. Steam engines also powered locomotives that carried coal, iron, and other products. The coal, iron, and steam interactions involved numerous positive feedback loops.

SECOND INDUSTRIAL REVOLUTION

Evolving production systems were also at the heart of the Second Industrial Revolution. Occurring in the United States and Germany between 1880 and 1930, this revolution involved both technological and organizational changes. Independent inventors and, later, scientists in industrial research laboratories, spurred invention, research, development, and innovation, resulting in the telephone, electric light and power, wireless radio, the automobile, and the airplane.[2] New materials included steel, aluminum, plastics, and reinforced concrete. Countless workers involved in mass production and managers applying scientific management dramatically increased the production of goods by giant industrial corporations. Engineering and scientific education flourished in universities and colleges. Industrialized metropolitan centers mushroomed.

At the technical core of the Second Industrial Revolution were internal combustion engine and electric power transmission systems. The light, high-rpm internal combustion engine made possible the automobile production and use system. It involved the mass production of automobiles best exemplified by Henry Ford's Highland Park and River Rouge plants. Automobile production interacted positively with petroleum refining in a feedback loop. One could not expand without the expansion of the other. Gasoline-fueled autos interacted with the growth of a street and highway system. Utilities, including service stations, were also necessary. The hardware of the automobile production and use system interacted with the corporations that presided over it, thereby constituting a sociotechnical system.

Electric power provides an outstanding example of a tightly coupled system with near-instantaneous feedback loops. The flow of electrons ap-

---

[2]Simultaneity of invention was common, because inventors recognized common problems in expanding systems. On simultaneity, see Gorman, Mehalik, Carlson, and Oblon (1933).

proached the speed of light. Supply and demand were so linked that change in one brought near-instantaneous change in the other. An excess of supply from generators would increase the frequency of the system's electric current, which triggered automatic, negative feedback control devices that reduced generator output. If, on the other hand, the load on the system increased, this was indicated by a decrease in frequency that automatically initiated an increase of generator output until the standard frequency was obtained.

Like automobile production and use, the tightly coupled electric power system involved positive, interactive feedback loops. For instance, electric power distribution and transmission lines required copper, and copper was refined using electric power. Electric motors in machine shops drove machine tools used in making steam and water turbines that in turn were used to generate electricity.

Large transmission systems required the introduction of complex management that articulated and implemented such concepts as diversity and load factor. Electric light and power managers measured their success by the value of the *load factor*, the shape of the load curve that graphically represents it. A high and flat load curve showed that the installed generator capacity was extensively used around the clock. As a result, the cost of capital investment in equipment would be spread over a large number of production units and lower the cost of supply.

To achieve a high and flat load curve, utility managers exploited load diversity. To do this, the utility needed to supply electricity over a large area with a variety of load types. A diverse mix of light, motor, electrochemical, and traction loads usually leveled the load curve. The diversity ensured that many consumers would not demand a peak load simultaneously and result in a poor load curve with peaks of consumption.

There was a positive feedback relationship between management and supply. If the manager achieved a good load factor, this reduced the cost of supply, which, when passed on to the consumers, increased their consumption or increased the number of consumers.

CONCLUSION

Only a few systems of technology could be discussed in this chapter, but the range of these suggests how widely the concept applies and how deeply systems shape technological and social change. An agricultural revolution driven by a system of land cultivation occurred in a preindustrial society. Marx (n.d.) wrote of a 19th-century Industrial Revolution driven by a coal, iron, and steam system. The United States and Germany experienced a Second Industrial Revolution with a technical core involving internal combustion engines and electric power transmission.

The history of these three revolutions raises questions about the role systems play in bringing about major technological revolutions. All three involved positive feedback systems centered on a major power source. Multi-animal teams drawing, and positively interacting with, deep plows were at the heart of the medieval Agricultural Revolution. Steam engines interacting in a positive feedback way with coal and iron drove the British Industrial Revolution. The spread of automobile production and use during the Second Industrial Revolution is an excellent example of positive feedback.[3]

Introduction of new power sources often brought technological revolutions, because power is omnipresent in the technological world. As a result, a change in power source brings about changes in numerous systems dependent on power. For example, the steam engine replaced water, wind, and animal—especially horse—power in countless systems. Similarly, the internal combustion engine replaced steam power, and the light high-revolution engine found entirely new applications, especially in the field of transport. Electric power transmission also widely displaced existing power transmission by leather belt and found entirely new uses. It should be stressed that changing the power component in technological systems necessitated changes in the other components in the affected system.

A question also arises concerning systems and the *Information Revolution*. Like the Second Industrial Revolution, the present-day Information Revolution involves the spread of new pervasive technology, in this case, an information-transmission technology rather than a power technology. Like power in the past, information is omnipresent today in the technological world. Numerous technological systems depend on information, especially for communication and control. Introducing a new means of transmitting information also has cascading effects as digital information displaces older forms of information in various systems. Information and electric energy have similar effects and generate comparable developments, because both are means of transmission and distribution, in one case energy and in the other information.

Finally, a question arises about the capacity of individuals and society to foster technological revolutions and control those already underway. Judging from consideration of the ones discussed earlier, I conclude that they result primarily from unintended and unanticipated confluences. Individuals and societies can help shape them only if they observe the unfolding events and give gentle nudges at critical junctions. The federal government's funding in the 1950s and 1960s of evolving computer developments and the resulting stimulus for an Information Revolution is a case in point.

---

[3]The classic example of positive feedback is an amplifier tube in a wireless communication system.

ACKNOWLEDGMENT

This chapter is an extensive revision of my 1980 article, "The Order of the Technological World" (Hughes, 1980).

REFERENCES

Babbage, C. (1832). *On the economy of machinery and manufactures*. London: [s.n.]

Bertalanffy, L. von. (1968). *General systems theory: Foundations, development, applications*. New York: Braziller.

Bloch, M. (1966). *French rural history: An essay on its basic characteristics*. Berkeley: University of California Press. (Original work published 1955)

Gorman, M. E., Mehalik, M. M., Carlson, W. B., & Oblon, M. (1933). Alexander Graham Bell, Elisha Gray and the speaking telegraph: A cognitive comparison. *History of Technology, 15*, 1–56.

Hughes, T. P. (1980). The order of the technological world. In A. B. Hall & N. Smith (Eds.), *History of Technology* (pp. 3–16). London: Mansell.

Marx, K. (n.d.). *Capital: A critique of political economy* (F. Engels, Ed.). New York.

Perrow, C. (1984). *Normal accidents: Living with high-risk technology*. New York: Basic Books.

Rochlin, G., La Porte, T., & Roberts, K. H. (1987, Autumn). The self-designing high-reliability organization: Aircraft carrier flight operations at sea. *Naval War College Review*, 76–90.

White, L. Jr. (1962). *Medieval technology and social change*. Oxford, England: Clarendon.

13

Levels of Expertise and Trading Zones: Combining Cognitive and Social Approaches to Technology Studies

Michael E. Gorman
University of Virginia

Nancy Nersessian's chapter (chap. 2) at the beginning of this volume, highlights the historic tensions between cognitive and sociological approaches to the study of scientific thinking. She looks for a synthesis in the area of situated cognition, where thinking is considered an activity that occurs in the environment as well as in the brain (Gorman, 1997).

Recently, Harry Collins, one of the originators of the social constructivist position in science studies, has advocated a new approach that holds promise for an integration between cognitive and social approaches. He and his colleague Robert Evans discuss two waves, or movements, in science studies, and propose a third.[1] During the first wave, social scientists "generally aimed at understanding, explaining and effectively reinforcing the success of the sciences" and "a good scientific training was seen to put a person in a position to speak with authority and decisiveness in their own field, and of-

---

[1]Collins and Evans (2002) recognized that their description of the previous waves is simplified almost to the point of caricature.

ten in other fields, too" (Collins & Evans, 2002, p. 239). The second wave is said to begin roughly in the 1970s, shortly after publication of Kuhn's *Structure of Scientific Revolutions*. The term *social constructivism* is often applied to this wave, which held that science is like other forms of social activity and therefore that scientific knowledge is the product of social interaction. (In her chapter, Nersessian documents how an anticognitive bias came in with constructivism.)

Collins and Evans (2002) proposed that the third wave in science and technology studies should focus on the study of experience and expertise (SEE). This move ought to bring sociological studies of science closer to cognitive approaches, because there is a cognitive literature on scientific expertise. Much of this literature is concerned with differences between experts and novices on the sorts of normal science problems encountered in textbooks. For example, in a classic study, Chi and her colleagues found that "that experts tended to categorize problems into types that are defined by the major physics principles that will be used in solution, whereas novices tend to categorize them into types as defined by the entities contained in the problem statement" (Chi, Feltovich, & Glaser, 1981, p. 150). Similarly, in another classic article, Larkin, McDermott, Simon, and Simon (1980) found that expert physicists working on familiar problems worked forward from the information given, and once a problem was appropriately categorized the rest of the problem-solving process consisted of relatively automatic procedures. Novice physics students, in contrast, had to struggle backward from the unknown solution, trying to find the right equations and quantities (Larkin et al., 1980).[2] For experts working in familiar domains, using pattern recognition skills to classify a problem is the first step toward solving it (Schunn & Anderson, 2001).

In these and other expert–novice studies (Clement, 1991; Green & Gilhooly, 1992), cognitive scientists were more concerned with how students get turned into normal scientists than with how scientific expertise was negotiated in a variety of settings. Nersessian in suggests in chapter 2 how expertise is negotiated in interdisciplines where disciplinary expertises are combined in novel ways. Negotiation need not be limited to social contexts, of course; Ippolito and Tweney (1995) discussed how Faraday "negotiated" a new expertise in a novel area of optics, and Gooding (e.g., 1990) has studied similar issues in other areas of Faraday's expertise.

---

[2]Consider an example: Suppose experts and novices have to find the friction coefficient for a block resting on an inclined plane. The weight of the block, the angle of the plane, and the force pushing against the block are given. Experts will work forward from the givens, generating the necessary equations, which in turn can be solved by familiar algorithms. The novice, in contrast, will typically start from the end point, trying to find values for the variables in the equation he or she must use at the end of the problem by generating other equations that use the givens (Anzai, 1991).

SEE has great potential for shedding light one the kinds of multidisciplinary interactions that are becoming prevalent on the cutting edge of technology, but only if combined with other concepts from the literature on science–technology studies. In this chapter I outline a framework that links SEE to other important concepts in the science and technology studies (STS) literature and apply this new SEE framework to two examples: (a) a converging-technologies initiative sponsored by the National Science Foundation and (b) an attempt to create a new kind of expert on human–environmental systems.

THREE TYPES OF EXPERTISE

Collins and Evans (2002) distinguished among three levels of shared expertise that can occur when more than one discipline is involved in scientific or technological problem solving.

1. None

Here two or more disciplines communicate only in the most superficial way, without any exchange of knowledge. An example would be a sociologist or anthropologist who studies a scientific laboratory without making any effort to understand the content, which was the approach taken by Latour and Woolgar (1986) in their classic study of laboratory life. A similar lack of understanding initially prevailed between AIDs activists who did not want to be in a placebo control group and scientists who insisted that was the only way to conduct appropriate trials of new medications (Epstein, 1996).

2. Interactional

Here the different disciplines have to be able to partially share domain-specific knowledge at a level of generality sufficient to complete a task. To continue our example, a sociologist or anthropologist studying a laboratory in this situation would learn enough about a scientific domain to permit thoughtful conversation with experts over matters of disciplinary content and method. Similarly, some AIDs activists educated themselves in scientific terminology to the point where they were able to interact with researchers. For example, Epstein (1995) described one activist who came into meetings with "seven earrings in one ear and a mohawk and my ratty old jacket on" and was dismissed initially as one of those street activists who had no expertise worth attending to. However, when she showed her growing command of the technical language, she won respect.

3. Contributing

Here practitioners from one discipline learn enough to make a real contribution to another. Pamplin and Collins (1975) conducted a sociological study of paranormal spoon bending and ended up coauthoring an article that made an important contribution to the psychokinesis literature, documenting instances of fraud and recommending improved research methods.

In the AIDs case, activists allied themselves with biostatisticians and others who succeeded in broadening the research protocols to include a more representative HIV positive population and to permit the use of other medicines taken in combination with the one on trial. The net effect of these modifications was to make the trials less rigorous from an experimental standpoint, but more ecologically valid from the standpoint of AIDs treatment. A small number of activists therefore contributed to AIDs research by becoming more like scientists—behavior that angered other AIDs activists.

Another example is the way in which Luis Alvarez combined his background in physics with his son Walter's expertise in geology to come up with a new hypothesis for the extinction of the dinosaurs and led a search for evidence (Alvarez, 1997).

These three types of expertise can be viewed as a continuum, shifting not only as an individual learns more about a domain of expertise but also as the nature of the problem shifts. Consider an example from one of the current projects at the University of Virginia. A group of us, supported by the National Science Foundation, are studying societal dimensions of nanotechnology.[3] Most of the funding goes to support a new nanotechnology student, who will serve as a participant–observer of her own learning process. In addition, she has been asked by her thesis advisors to look at the potential impact of her research on "world ills." Our goal is to make her a contributing expert in nanotechnology, at the level typical for Masters students. She might also gain contributing expertise in cognitive psychology of science, as she collaborates with Michael E. Gorman and Jeff Shrager on an analysis of her own cognitive processes.

In terms of ethical issues and global challenges, the initial goal is to make her an interactional expert, relying on the work of others to identify problems—but if her research involves a possible technological solution to any aspect of one of these problems, she will become a contributing expert. As one of her advisors, my goal right now is to become an interactional expert in nanotechnology and in her specific research area, but I want our team to

---

[3]The project is titled "Social and Ethical Dimensions of Nanotechnology" and is funded by NSF award SES-0210452.

make a significant contribution to an understanding of the societal dimensions of nanotechnology.

In contrast, Collins and Evans (2002) demarcated technical expertise from what they call the *public* or *political domain*—in which technical expertise carries no more weight than the expertise or views of any other stakeholder group. By extension, this argument implies that ethicists and policymakers do not need to interact with the expertise of a domain, just the implications of the research. The problem with this view is that one can only deal with social impacts after they are known, that is, when it is often too late. I think it is worth trying to see what happens if someone sensitized to societal dimensions is also capable of contributing to the creation of new technologies. This person would then blur the boundaries between the esoteric scientific domain and the public domain.

In both my case and the students', we begin with little expertise and move as quickly as possible to the interacting stage. The extent to which we get to the contributing stage will depend partly on the nature of the problems that emerge as we push into this new area.

TRADING ZONES

These three kinds of expertise can be linked to another important concept from the STS literature: trading zones:

> Two groups can agree on rules of exchange even if they ascribe utterly different significance to the objects being exchanged; they may even disagree on the meaning of the exchange process itself. Nonetheless, the trading partners can hammer out a *local* coordination, despite vast *global* differences. In an even more sophisticated way, cultures in interaction frequently establish contact languages, systems of discourse that can vary from the most function-specific jargons, through semispecific pidgins, to full-fledged creoles rich enough to support activities as complex as poetry and metalinguistic reflection. (Galison, 1997, p. 783)

Galison used the trading zone metaphor to describe the interactions among scientists and engineers developing radar and particle detectors.

Baird and Cohen (1999) applied the trading zone metaphor to an analysis of problems with the application of magnetic resonance imaging. Between 1987 and 1990,

> it became fashionable for physicians to reduce the rather long MR (magnetic resonance) imaging times by using anisotropically shaped (i.e., nonsquare) imaging pixels in studies of the spine. As it turned out, this resulted in a prominent dark line appearing within the spinal cord. The dark line was a Gibbs

ringing artifact. Unfortunately, clinicians, not aware of this kind of artifact—for not being conversant with the mathematics used to transform the
instrument signal into an image—at times interpreted this artifact as a disease
process: a fluid filled lesion known as a "syrinx" requiring aggressive medical
treatment. (Baird & Cohen, 1999, p. 238)

An individual who bridged medicine and physics detected the problem—too late for many patients. The creole among medicine, physics, and
engineering was insufficient in this and other instances and needed constant updating.

THREE NETWORK STATES

Gorman and Mehalik (2002) proposed three states in networks that couple
people with technology and link them with trading zones. The first state is a
network controlled by an elite in which there really is no trade: Those not in
the elite either obey, or they are ignored. The elite can be a group of experts
who use their specialized knowledge to dictate how a sociotechnical system
will function. The expertise of such an elite is black-boxed for other participants in the network. Communications are top-down, consisting primarily
of directives that must be obeyed. Stalinist agricultural and manufacturing
schemes used in the Soviet Union are examples of these elite control networks (Graham, 1993; Scott, 1998). State control overrode the experiential expertise of farmers, workers, and engineers. Similarly, AIDs activists
felt that they were being forced to conform to a research protocol that did
not honor their experiential knowledge of what it would take to recruit a
representative sample of participants (Epstein, 1996).

The second is a more equal trading zone, in which experts from different
fields interact around the development of a technological system, such as
radar or magnetic resonance imaging. Such systems can serve as boundary
objects, occupying a space at the focal point of several expertises, each of
which has a somewhat different view of the emerging technology.

According to Bowker and Star (1999), boundary objects are "plastic
enough to adapt to local needs and constraints of the several parties em-
ploying them, yet robust enough to maintain a common identity across
sites" (p. 298). Boundary objects are "most useful in analyzing cooperative
and relatively equal situations; issues of imperialist imposition of standards,
force, and deception have a somewhat different structure" (p. 297). Therefore, boundary objects will be absent from trading zones that are dominated
by an elite group but are likely to exist in trading zones where the partners
are relatively equal.

In this kind of a trading zone, a system often serves as a boundary object
that links the participants in the network (Hughes, 1998), where experts

from different fields have different mental models of the system. One could adopt the term *boundary system* to refer to these cases (I consider the Everglades as an example later). A mental model, in this case, is a representation of the boundary system that is incomplete because each expert views the system in terms with which she or he is familiar. A good analogy is the "blind men and elephant problem," where the one feels the tail, another the trunk, another the leg, and each constructs a different beast consistent with his perspective.

The idea of a mental model would be anathema to sociologists of science, several of whom—as Nersessian noted in chapter 2—endorsed a moratorium on cognitive approaches to the study of science. A sociological reductionist might view the mind as epiphenomenal, as simply the by-product of social interactions—although even some sociologists reject this extreme position (Woolgar, 1987). In trading zones, however, people spend time trying to imagine each others' intentions, and they create and sustain different views of the evolving technological system, as Bijker (1995) showed in his excellent work on technological frames. Some parts of these imaginings are tacit—virtually impossible to articulate—and therefore cannot be completely captured in inscriptions or conversation. Tweney et al.'s efforts to replicate Faraday's experiments in chapter 7 are partly an effort to recover his tacit knowledge; see also Gooding's (1990) attempts to re-create a different set of experiments by Faraday. These tacit representations are frequent sources of misunderstandings in multidisciplinary trading zones (Gorman, 2002b). I use the term *mental model* to designate both individual and shared representations that are partly tacit and typically have visual and/or kinesthetic components.

For example, I have shown how Alexander Graham Bell based his telephone on a visual and kinesthetic analogy to the human ear that was unique to his experience and expertise (Gorman, 1997). In chapter 2, Nersessian suggests that mental models can include external devices as well as mental representations. Similarly, Carlson and I showed that these mental models could be partly embodied in devices, creating mechanical representations (Carlson & Gorman, 1990).

Human representations are themselves a boundary object between different disciplines, with sociologists studying people's inscriptions and conversations about reality and psychologists trying to study how people represent the world mentally and neurophysiologically. The term *mental model*, therefore, is not meant to be a step toward cognitive reductionism. Instead, I hope it will remind readers that as much as some sociologists, philosophers, behaviorists, and neuroscientists would like to wish it away, the age-old problem of how people represent their experience is still central to understanding human affairs.

AIDs activists and scientists from different disciplines learned to interact over a research protocol that served as a way of regulating a trading zone, in which those with AIDs would agree to certain guidelines for being enrolled in research and, in return, the guidelines were modified to make the protocol more inclusive. The creole that evolved involved AIDs activists learning scientific terms and scientists incorporating the arguments of activists in publications (Epstein, 1995).

Contributing expertise brings us to the third kind of trading zone, in which the participants share both a common understanding of a goal and a continually evolving representation (Tweney, 2001) of a technosocial system. In a State 2 trading zone, a technosocial system serves as a boundary object; in a State 3 network, the object or system moves from the boundary to the center of a growing, shared expertise. It is possible that a small number of AIDs activists and scientists achieved this kind of shared expertise for brief periods of time, but because Epstein (1995) did not use this framework, it is impossible to be certain. The free, collaborative exchange characteristic of the group that developed the Advanced Research Project Agency (ARPANET) is an example (Hughes, 1998).

Table 13.1 shows the relationship among the three levels of expertise, the three types of trading zone and three levels of intergroup communication. The three states, like the three types of expertise, are on a continuum. At State 1 on one extreme, there is a top-down, "my way or the highway" network dominated by one group. Communications are in the form of instructions from the elite to the other participants, and the elite is solely concerned with evidence of obedience. State 2 covers the broadest range on the continuum, because it encompasses different kinds of negotiations among groups who begin with different areas of expertise, but where no one group is able to dominate the others. On the left, or State 1 end, would be networks in which different groups of experts try to "throw their parts of a technology over the wall" to each other. If a group cannot dominate the network, at least it can try to black-box its own part of a multidisciplinary project, demanding that other network participants just accept their contribution as given. Here the different groups resist developing a creole and try either to pull the boundary object under their control or shove it

TABLE 13.1

The Three Types of Trading Zone and Their Respective Levels
of Expertise and Communication

	State 1	State 2	State 3
Trading zone	Elite control	Approximate parity	Shared mental model
Shared expertise	None	Interactional	Contributing
Communication	Orders	Creole	Shared meanings

into another group's responsibility. Farther to the right are State 2 networks that involve interactional expertise. Here participants have developed a creole that allows them to exchange important knowledge around a boundary object, partially opening each others' black boxes when necessary. A variety of hierarchical relationships may facilitate or hinder these State 2 networks, but if one group comes to dominate the hierarchy, the network shifts to a State 1.

Finally, on the State 3 end, there is a network that has to evolve a shared mental model to address what is typically a cutting-edge problem that stretches beyond anyone's disciplinary expertise. Here all participants are committed to a common goal and are often engaged in developing a new language to describe what they are doing. In this kind of a "skunkworks" environment, hierarchy is virtually ignored.

States in a network shift over time. AIDs research protocols initially served as a boundary object for activists, medical researchers, and statisticians, each of whom saw the protocols in a different way. Groups of activists initially demanded changes in the protocols, and researchers made some concessions—a primitive trading zone, more adversarial than cooperative. However, as activists grew in sophistication and attracted allies among statisticians, the trading zone became more collaborative, and the activists moved from reacting to interacting to, in some cases, contributing.

States also shift on the basis of how much of a system one includes in the network. Those inside the emerging ARPANET group operated mainly in a State 3, with experts from different disciplines working equally with managers to come up with novel solutions. ARPA's goal was to "find people that they thought were sufficiently smart and sufficiently motivated, give them a ball and let them run with it" (Hughes, 1998, p. 287). When asked how he knew what people and universities to fund, Joseph Licklider at ARPA said it depended on "a kind of networking. You learn to trust certain people, and they expand their acquaintance. And the best people are at the best universities, which one knows by reputation" (Hughes, 1998, p. 264). In other words, those not in a small, elite group of institutions could not get ARPA funding. For those on the outside, the ARPANET group constituted an elite into which they could not break; they were in an unequal trading zone with respect to what emerged.

I now apply this framework to two emerging areas of multidisciplinary collaboration.

CONVERGING TECHNOLOGIES
FOR HUMAN PERFORMANCE

The NSF recently held a conference on "Converging Technologies (NBIC) for Human Performance," where the N stands for nanotechnology, the B for biotechnology, the I for information technology, and the C for cognitive science.[4] The model for collaboration that emerged was disciplinary depth combined with the ability to share expertise. The framework outlined previously allows us to discuss possible levels of sharing, on a continuum:

1. None, in which each discipline tries to dominate the trading zone or threatens to exit.
2. Interactional, in which disciplinary experts create creoles around boundary objects representing technological possibilities.
3. Contributing, in which experts from these four technologies engage each other deeply, learning enough to contribute jointly to development of a new technological system.

Consider, for example, how converging technologies could be used to create a "super soldier," [5] featuring the following:

- Information technology to link the soldier into a command network that includes information on threats and support'
- Nanosensors to provide information about the immediate environment, including biological and chemical threats;
- Training on how to integrate all of this information into a cognitive system that can adapt rapidly;
- Genetic modifications that include improvements to the soldier's physique and nervous system.

Each of these capabilities is coupled with the other ones, so at least a close interaction among expertises is required—and just as AIDs activists should have been involved in shaping research protocols involving themselves as participants, so too soldiers should be involved in these new technologies from the earliest design phases. Because this super soldier has the potential to become a Golem (Collins & Pinch, 1998) or even a Frankenstein,[6] another kind of expertise obviously needs to be added: ethics.

The question is whether this ethics expertise should belong to the political sphere that Collins and Evans (2002) set up as a separate trading zone. The risk is the same as that with the expert's regress: Ethicists, politicians,

---

[4]For a pdf version of the report, see http://www.wtec.org/convergingtechnologies/ .

[5]See http://www.technologyreview.com/articles/talbot1002.asp for Massachusetts Institute of Technology's ideas on how to use nanotechnology to create a super soldier.

[6]The Golem is a powerful, bumbling giant from Jewish mythology who does unintentional harm. Frankenstein's monster, in contrast, took deliberate revenge on his creator. If the supersoldier technology becomes a Golem, it will have unintended side effects that will be harmful. If it becomes a Frankenstein, it might be turned against us by those entrusted with its use, or who obtain it.

and other stakeholders in a separate sphere will be reacting to the Golem after it has already been created, instead of being present at the moment a breakthrough occurs. An alternative would be to have practical ethicists join this super-soldier project as interacting experts, working closely with the researchers to explore societal implications as the technology produces new breakthroughs. Indeed, we might go a step farther and have ethicists and social scientists contributing to decisions about which set of experiments or potential designs a research team ought to conduct next—contributing in the sense that the ethicists are not philosopher-kings who dictate the research direction but simply members of the team, adding their expertise to the matrix of considerations.

Here I use *expertise* in the broadest sense, to include wisdom (Gorman, 2002a). An example of such wisdom is the ability to engage in moral imagination, which involves going "out toward people to inhabit their worlds, not just by rational calculations, but also in imagination, feeling, and expression" (Johnson, 1993, p. 200). Ethicists and social scientists need to help scientists and engineers imagine the potential impact of new technologies on a wide variety of stakeholders, especially at the point where new research directions are under serious consideration.

EARTH SYSTEMS ENGINEERING MANAGEMENT

Brad Allenby, a prominent industrial ecologist, has called for this kind of contribution from social scientists and ethicists in the development of Earth Systems Engineering Management (ESEM), a new approach to human–environmental systems (Allenby, 2001). ESEM is discussed in more detail in chapter 14, by Allenby. For now, it will suffice to say that ESEM begins with the premise that no corner of the globe is unaffected by human beings, given our technological "advances," and therefore we have a responsibility to manage our planet intelligently. Human beings, nature, and technology are all closely coupled in a dynamic system whose interactions are hard to predict.

Therefore, ESEM is what Collins and Evans (2002) referred to as a *reflexive historical science*, or one in which "the long-term outcomes are affected by humans themselves." [7] The global environment is a complex dynamic system that is affected by human activities, including our attempts to understand and manage it; therefore, it is a reflexive historical science. This kind of system requires continuous monitoring and technologies whose impacts

---

[7]Collins and Evans's (2002) other types of science include *normal science*, in which consensus in the community of experts has been achieved; *Golem science*, in which an eventual consensus is likely, but none exists at present; and *historical sciences*, in which consensus will take a long time because the system itself is complex and needs to be studied over an extended duration.

can be reversed if the monitoring indicates an unexpected change in the system state. Most environmental regulation involves a trading zone among multiple agencies with interacting expertise (Jasanoff, 1992), but ESEM implies moving beyond this kind of careful, often-adversarial trading zone to a continuous dialogue with the complex system that will produce shared representations—a dialogue that will include knowledge scientists and ethicists as contributing experts.

Consider the Everglades, home to 68 endangered species, multiple ecosystems, and a rapidly growing urban landscape that depends on the "river of grass" for water.[8] Management of this system is currently a complex trading zone among multiple stakeholders, linked by a multibillion dollar restoration plan that incorporates different mental models. The U.S. Park Service wants to restore the Everglades to its natural state, including the preservation of indigenous species and elimination of invasives.

The South Florida Water Management District (SFWMD) sees the park and surrounding lands as a huge reservoir. Their mission is to provide water to a growing urban area and control flooding, although they are committed to working on restoration where it does not interfere with those goals. The Army Corps of Engineers has a restoration mandate from Congress but also flood control and drinking-supply mandates from its history and from its work with the SFWMD. The corps is locked by legislation into a 50:50 trading zone with the SFWMD.

For agricultural interests, particularly sugar, the Everglades is a giant waste treatment facility. Sugar plantations are built right on the compost left over from centuries of sawgrass and other organics. The sugar industry claims that the Everglades could tolerate phosphorus levels of up to 20 parts per billion, whereas the SFWMD and the Florida Department of Environmental Protection established the level at 10 parts per billion. The SFWMD has designed artificial wetlands near the sugar plantations to treat phosphorus and other agricultural runoff, but current models cannot ensure that this strategy will sufficiently mitigate downstream impacts (Sklar, Fitz, Wu, Zee, & McVoy, 2001).

These examples only begin to suggest the complexity of the trading zone that continues to evolve around the Everglades as a boundary system. To appreciate the tensions, consider a boundary object within this system: the Tamiami canal, which keeps water from the north from entering the Everglades Park, except through a system of small dams controlled by the Army corps and the SFWMD. If the natural flow of the Everglades is to be restored, the entire levy and dam system will have to be removed and the

---

[8]Much of the material in this Everglades section is based on interviews and observations I and one of my students (Charles Jansen) conducted on a field trip, supported by ESEM funding from AT&T.

existing road replaced by an overpass. Once the overpass is built and the levy is gone, it is hard to dam the water again—and the flow effects may not be restorative, given how other parts of the system are managed. For example, if phosphorus levels continue to increase in the northern part of the system, then greater flow will spread the pollution more rapidly and widely. More of the natural floods would occur, with catastrophic consequences for development. If greenhouse warming continues, then the salt water could move up from the southern part of the system without any dams to check it.

This example illustrates how a boundary object can be embedded in a boundary system and how the possibilities for the object are linked to the mental model of the system. If one's goal is to restore the Everglades to an earlier system state, then one sees the canal as an obstruction. If one sees the Everglades as a reservoir, one sees the canal as an essential component in flood control.

The current trading zone is managed by volumes of regulations, codes, and plans (SFWMD, 2002). All of these agreements are valuable, but if participants in the trading zone still see each other as having competing goals, then no documents will suffice for genuine cooperation across a system as complex as this.

The Everglades is an example of the kind of system an earth systems engineer (ESE) should be able to help us manage. An ESEM expert would have to create at least occasional State 3 moments in this State 2 network, to engage a core of participants in the kind of dialogue that might shift the Everglades from a boundary object to a shared mental model. What kind of a "natural" system are we trying to restore? The Everglades would continue to be a changing system, anyway, but human technology has greatly accelerated the changes and taken them in new and unexpected directions. Now we have to choose system toward which goals to work, collect information, and create models that allow us to adjust quickly to unanticipated perturbations. These goals need to be synergistic: One cannot have unlimited development, unlimited access to water, no flooding, and restoration of the Everglades as it existed centuries ago.

The Everglades are, of course, part of the larger global environmental system and cannot be managed in isolation. For example, the ESE will have to consider the impact of global climate change.

New technologies and scientific methods would emerge from this kind of collaboration, as in the case of other multidisciplinary collaborations, such as converging technologies. How, for example, could a system be created to allow extended flow across the Tamiami in a way that could be reversed quickly if unanticipated negative consequences occurred? Perhaps new materials made available by advances in nanotechnology might help, along with intelligent agents that can give feedback on minute changes in the flow and respond after alerting the ESE.

DISCUSSION

To understand and improve multidisciplinary collaboration, cognitive and social approaches to STS have to be integrated. The framework outlined in this chapter integrates SEE, trading zones, boundary objects, and cognitive representations. The three network states proposed in this chapter make us aware of significant shifts in relationships among practitioners—especially experts—that are both social and cognitive. For example, a boundary system is represented in unique ways by different disciplinary participants in a State 2 trading zone; in a State 3 network, all participants need to share a mental model of the system they are creating.

This framework will, of course, be modified and elaborated by future research and perhaps even abandoned—although old frameworks rarely die, they just get absorbed into new ones. In particular, this framework makes the prediction that State 3 networks are essential to breakthroughs in multidisciplinary science and technology, especially breakthroughs that truly benefit society. This prediction should be explored in both current and historical case studies.

ACKNOWLEDGMENTS

Portions of this chapter were drawn from my 2002 article, "Levels of Expertise and Trading Zones: A Framework for Multidisciplinary Collaboration," *Social Studies of Science, 32* (5–6), although the argument here is greatly expanded and revised. I thank Ryan Tweney, Brad Allenby, and Jeff Shrager for their comments.

REFERENCES

Allenby, B. (2001). Earth systems engineering and management. *IEEE Technology and Society Magazine, 19*(4), 10–21.

Alvarez, W. (1997). *T.rex and the crater of doom*. Princeton, NJ: Princeton University Press.

Anzai, Y. (1991). Learning and use of representations for physics expertise. In K. A. Ericsson & J. Smith (Eds.), *Toward a general theory of expertise* (pp. 64–92). Cambridge, England: Cambridge University Press.

Baird, D., & Cohen, M. (1999). Why trade? *Perspectives on Science, 7*, 231–254.

Bijker, W. E. (1995). *Of bicycles, bakelites and bulbs: Toward a theory of sociotechnical change*. Cambridge, MA: MIT Press.

Bowker, G. C., & Star, S. L. (1999). *Sorting things out: Classification and its consequences*. Cambridge, MA: MIT Press.

Carlson, W. B., & Gorman, M. E. (1990). Understanding invention as a cognitive process: The case of Thomas Edison and early motion pictures, 1888–1891. *Social Studies of Science, 20*, 387–430.

Chi, M. T. H., Feltovich, P. J., & Glaser, R. (1981). Categorization and representation of physics problems by experts and novices. *Cognitive Science, 5*, 121–152.

Clement, J. (1991). Nonformal reasoning in experts and science students: The use of analogies, extreme cases, and physical intuition. In J. F. Voss, D. N. Perkins, & S. J. W. (Eds.), *Informal reasoning and education* (pp. 345–362). Hillsdale, NJ: Lawrence Erlbaum Associates.

Collins, H. M., & Evans, R. (2002). The third wave of science studies. *Social Studies of Science, 32*, 235–296.

Collins, H., & Pinch, T. (1998). *The Golem at large: What you should know about technology*. Cambridge, England: Cambridge University Press.

Epstein, S. (1995). The construction of lay expertise: AIDs activism and the forging of credibility in the reform of clinical trials. *Science, Technology & Human Values, 20*, 408–437.

Epstein, S. (1996). *Impure science: AIDs, activism, and the politics of knowledge*. Berkeley: University of California Press.

Galison, P. (1997). *Image & logic: A material culture of microphysics*. Chicago: University of Chicago Press.

Gooding, D. (1990). *Experiment and the making of meaning: Human agency in scientific observation and experiment*. Dordrecht, The Netherlands: Kluwer Academic.

Gorman, M. E. (1997). Mind in the world: Cognition and practice in the invention of the telephone. *Social Studies of Science, 27*, 583–624.

Gorman, M. E. (2002a). Turning students into professionals: Types of knowledge and ABET engineering criteria. *Journal of Engineering Education, 91*, 339–344.

Gorman, M. E. (2002b). Types of knowledge and their roles in technology transfer. *Journal of Technology Transfer, 27*, 219–231.

Gorman, M. E., & Mehalik, M. M. (2002). Turning good into gold: A comparative study of two environmental invention Networks. *Science, Technology & Human Values, 27*, 499–529.

Graham, L. R. (1993). *The ghost of the executed engineer: Technology and the fall of the Soviet Union*. Cambridge, MA: Harvard University Press.

Green, A. J. K., & Gilhooly, K. J. (1992). Empirical advances in expertise research. In M. T. Keane & K. J. Gilhooly (Eds.), *Advances in the psychology of thinking* (Vol. I, pp. 45–70). London: Harvester Wheatsheaf.

Hughes, T. P. (1998). *Rescuing Prometheus*. New York: Pantheon.

Ippolito, M. F., & Tweney, R. D. (1995). The inception of insight. In R. J. Sternberg & J. E. Davidson (Eds.), *The nature of insight* (pp. 433–462). Cambridge, MA: MIT Press.

Jasanoff, S. (1992). Science, politics and the renegotiation of expertise at EPA. *Osiris, 7*, 195–217.

Johnson, M. (1993). *Moral imagination*. Chicago: University of Chicago Press.

Kuhn, T. S. (1962). *The structure of scientific revolutions*. Chicago: University of Chicago Press.

Larkin, J. H., McDermott, J., Simon, D. P., & Simon, H. A. (1980, June 20). Expert and novice performance in solving physics problems. *Science, 208*, 1335–1342.

Latour, B., & Woolgar, S. (1986). *Laboratory life: The construction of scientific facts*. Princeton, NJ: Princeton University Press.

Pamplin, B. R., & Collins, H. (1975). Spoon bending: An experimental approach. *Nature, 257*, 8.

Schunn, C. D., & Anderson, J. R. (2001). Acquiring expertise in science: Explorations of what, when and how. In K. Crowley, C. D. Schunn, & T. Okada (Eds.), *Designing for science: Implications for everyday, classroom, and professional settings* (pp. 83–114). Mahwah, NJ: Lawrence Erlbaum Associates.

Scott, J. C. (1998). *Seeing like a state: How certain schemes to improve the human condition have failed.* New Haven, CT: Yale University Press.

Sklar, F. H., Fitz, H. C., Wu, Y., Zee, R. V., & McVoy, C. (2001). The design of ecological landscape models for Everglades restoration. *Ecological Economics, 37,* 379–401.

South Florida Water Management District. (2002). *2002 Everglades Consolidated Report.* West Palm Beach, FL: Author.

Tweney, R. D. (2001). Scientific thinking: A cognitive–historical approach. In K. Crowley, C. D. Schunn, & T. Okada (Eds.), *Designing for science: Implications for everyday, classroom, and professional settings* (pp. 141–173). Mahwah, NJ: Lawrence Erlbaum Associates.

Woolgar, S. (1987). Reconstructing man and machine: A note on sociological critiques of cognitivism. In W. E. Bjiker, T. Hughes, & T. Pinch (Eds.), *The social construction of technological systems* (pp. 311–328). Cambridge, MA: MIT Press.

14

Technology at the Global Scale: Integrative Cognitivism and Earth Systems Engineering and Management

Brad Allenby

Environment, Health and Safety Vice President, AT&T

It is usual to deal with technology and technological systems at many scales, but it is perhaps too seldom that we stand back and evaluate what humans, as toolmakers, have really created: the anthropogenic Earth, a planet where technology systems and their associated cultural, economic—and, indeed, theological—dimensions increasingly determine the evolution of both human and natural systems. A principal result of the Industrial Revolution and associated changes in human demographics, technology systems, cultures, and economic systems has been the evolution of a planet increasingly defined by the activity of one species: ours. Although it is apparent that in many cases we do not even perceive this to be the case—indeed, powerful discourses such as environmentalism are predicated on mental models that preclude such perception—and we certainly don't know how to think about such a planet, it is equally clear that the evolution of complex patterns of technology and society, which have been going on for centuries, have called forth something that is indeed new: the human Earth. Once that is recognized, it becomes increasingly dysfunctional, and arguably unethical, not to strive toward a higher level of rationality and morality in managing the Earth than we have yet achieved. Continued resiliency of both human and natural systems will require development of the ability to rationally and ethically engineer and manage coupled human–natural systems in a highly integrated fashion—an Earth Systems Engineering and Management (ESEM) capability.[1]

In this chapter I discuss the scientific and historical context of ESEM and then define it, including a few brief illustrative examples. I then turn to the principal focus of this discussion: the need to establish a philosophical basis on which to begin thinking about ESEM. To do so, I present the concept of *integrative cognitivism*. I apologize immediately for introducing yet another esoteric term, but although I draw from the recent philosophic school of externalism, as well as the idea of "actor networks," I go beyond them, especially in the important need to understand the role of intentionality in an ESEM world—and it had to be called something.

Also, it is important to emphasize from the beginning that *earth systems engineering and management* is not intended to imply a technocratic bias. In particular, it is apparent that the design challenges and processes required by earth systems cannot be met by traditional engineering approaches or by relying only on the physical and biological sciences. Thus, "earth systems" include such human systems such as economies, political and cultural institutions, and religions. Moreover, the engineering involved is not a bounded activity leading to an artifact; rather, it is a dialogue with the complex natural and human components of the earth systems that will necessarily continue for the foreseeable future—for, absent economic, cultural, and demographic collapse, the anthropogenic earth will continue for the foreseeable future. ESEM is perhaps best thought of as the necessary competency for living rationally and ethically in an anthropogenic earth, not as an "engineering" or "management" discipline.[2]

EARTH SYSTEMS ENGINEERING AND MANAGEMENT

The scientific foundation of the ESEM discourse is the increasing recognition that the critical dynamics of most fundamental natural systems are increasingly dominated by human activity. The physics and chemistry of every cubic meter of air and of water on or above the Earth's surface has been affected in one way or another by human activity. Critical dynamics

---

[1]Whether resiliency, stability, or some other system behavior is the desired goal is at base a decision about values and requires a dialogue that has not yet started, as well as considerable more objective knowledge about what likely systems paths are and the costs, benefits, and distributional equity patterns involved with each.

[2]The inadequacy of *earth systems engineering and management* as a title for this set of activities has been noted by many, but so far I have not been able to think of a better, equally comprehensive and admonitory term. Suggestions are encouraged.

of elemental cycles through the human and natural spheres—those of nitrogen, carbon, sulfur, phosphorus, and the heavy metals—as well as the dynamics at all scales of atmospheric, oceanic, and hydrological systems, are increasingly affected by the technological activities of our species ("Human-Dominated Ecosystems," 1997; McNeill, 2000; Socolow, Andrews, Berkhout, & Thomas, 1994; Turner et al., 1990; Vitousek, Mooney, Lubchenco, & Melillo, 1997). The biosphere itself, at levels from the genetic to the landscape, is increasingly a human product. At the genomic level, the human genome has been mapped, as have those of selected bacteria, yeast, plants, and other mammals ("Genome Prospecting," 1999; "The Plant Revolution," 1999). At the organism level, species are being genetically engineered to, among other things, increase agricultural yields, reduce pesticide consumption, reduce demand for land for agriculture, enable plant growth under saline conditions and thereby conserve fresh water resources, produce new drugs, reduce disease, increase hardiness, and support a healthier human diet. Few biological communities can be found that do not reflect human predation, management, or consumption. As Gallagher and Carpenter (1997, p. 485) remarked in introducing a special issue of *Science* on human-dominated ecosystems, the concept of a pristine ecosystem, untouched by human activity, "is collapsing in the wake of scientists' realization that there are *no places left on Earth that don't fall under humanity's shadow* [italics added]." Even those considered "natural" almost inevitably contain invasive species, frequently in dominant roles. The terraforming of Mars has long been the stuff of science fiction and the occasional visionary: It is ironic that we have all the while been terraforming earth.[3]

This becomes apparent after even a cursory inspection of the evolutionary patterns of our species. Population growth over time, linked to technology state, has trended steadily upward (allowing for the usual perturbations; Cohen, 1995; Graedel & Allenby, 1995). However it is modeled technically, human population growth clearly accelerated strongly with the advent of the Industrial Revolution, which essentially created unlimited resources for the expansion of the human population (not unlike yeast in a new growth medium; Cohen, 1995; Allenby, 1999). This is mirrored by economic data, which illustrate similar growth patterns on both a global gross

---

[3]Science fiction has long dealt with the subject of terraforming planets other than Earth, especially Mars; this has also been explored by scientists and engineers in NASA and elsewhere (Oberg, 1999). It is interesting that Keith (2000, p. 254) commented that "the terraforming community has generated a more robust debate about ethical concerns than exists for geoengineering," perhaps implicitly confirming the point that it is easier to evaluate and judge hypotheticals than our own (terrestrial) behavior, even when the latter is in front of our eyes. One is tempted to argue, in fact, that the interest in terraforming other planets is a sublimation of the "illicit" knowledge that we are doing precisely that to our own planet.

domestic product and a per capita basis (see Table 14.1). This overall growth is not homogeneous, of course: It consists of different technology clusters (see Table 14.2), with different patterns of interaction with co-evolving natural systems. For example, early industrialization depended heavily on mining, whereas the latest technology cluster, bio- and information technology, depends much more heavily on intellectual capital and knowledge (Grubler, 1998; McNeill, 2000).

Although it may be true that at the individual, conscious level humans did not deliberately set out to dominate the natural world, it is apparent that the growth and increasing complexity of technological, economic, and human population systems characteristic of the Industrial Revolution have had precisely that effect. Moreover, although the scale of these effects, and the concomitant emergent behaviors (e.g., global climate change) are new with the 20th century (McNeill, 2000; Turner, Clark, Kates, Richards, Mathews, & Meyer, 1990), the increasingly tight coupling between predominantly human systems and predominantly natural systems is not a sudden phenomenon. Greenland ice deposits reflect copper production during the Sung Dynasty in ancient China (ca. 1000 BC) as well as by the ancient Greeks and Romans; spikes in lead concentrations in the sediments of Swedish lakes reflect production of that metal in ancient Athens, Rome, and medieval Europe (Hong, Candelone, Patterson, & Boutron, 1996; Renberg, Persson, & Emteryd, 1994). Anthropogenic carbon dioxide buildup in the atmosphere began not with the Industrial Revolution, with its reliance on fossil fuel, but with the deforestation of Eurasia and Africa over the past millennia (Grubler, 1998; Jager & Barry, 1990). Human impacts on ecosystems have similarly been going on for centuries, from the probable role of humans in eliminating megafauna in Australia and North

TABLE 14.1
Global Economic History: 1500–1992

Date	World GDP (indexed to 1500 = 100)	Per Capita World GDP (1990 dollars)	Per Capita (indexed to 1500 – 100)
1500	100	565	100
1820	290	651	117
1900	823	1,263	224
1950	2,238	2,138	378
1992	11,664	5,145	942

Source: Based on J. R. McNeill, 2000, Something New Under the Sun (New York: W. W. Norton & Company). Tables 1.1 and 1.2, pp. 6–7, and sources cited therein.

TABLE 14.2
Technology Clusters

Cluster	Major Technology	Geographic Center of Activity	Date	Nature of Technology	Nature of Environmental Impacts
Textiles	Cotton mills/ coal and iron production	British midlands/ Lancashire	1750– 1820	Physical infrastructure (materials)	Significant but localized (e.g., British forests)
Steam	Steam engine (pumping to machinery to railroads)	Europe	1800– 1870s	Enabling physical infrastructure (energy)	Diffuse; local air impacts
Heavy Engineering	Steel and railways	Europe, U.S., Japan	1850– 1940	Physical infrastructure (advanced energy and materials sectors)	Significant; less localized (use and disposal throughout developed regions)
Mass Production & Consumption	Internal combustion engine, automobile	Europe, U.S., Japan	1920s– present	Application of physical infrastructure, mass production	Important contributor to global impacts (scale issues)
Information	Electronics, services, and biotechnology	U.S., Pacific Rim	1990s– present	Development of information (non-physical infrastructure)	Reduction in environmental impact per unit quality of life?

America (Alroy, 2001; Roberts et al., 2001) to the clear role of human transportation systems in supporting invasive species around the world (Jabolonski, 1991). As Kaiser (1999, p. 1836) noted, "The world's ecosystems will never revert to the pristine state they enjoyed before humans began to routinely crisscross the globe." In fact, the long evolution of agriculture as an important factor in increasing human dominance of natural systems is both apparent, and frequently overlooked, because it is so mundane and common to virtually all societies (Redman, 1999). Modern technological systems, however, from air travel to medicine, have dramatically increased these impacts: Writing in *Science,* Palumbi (2001) noted that "human ecological impact has enormous evolutionary consequences ... and can greatly accelerate evolutionary change in the species around us ...

technological impact has increased so markedly over the past few decades that humans may be the world's dominant evolutionary force" (p. 1786).

Thus, the need for an ESEM capability arises from the increasingly integrated, complex, and global behaviors that characterize coupled natural and human systems. ESEM can accordingly be defined as the capability to rationally and ethically engineer and manage human technology systems and related elements of natural systems in such a way as to provide the requisite functionality while facilitating the active management of strongly coupled natural systems (Allenby, 2000–2001). As an obvious corollary, ESEM should aim to minimize the risk and scale of unplanned or undesirable perturbations in coupled human or natural systems. Moreover, it is apparent that such complex systems cannot be "controlled" in the usual sense. Accordingly, ESEM is a design and engineering activity predicated on continued learning and dialogue with the systems of which the engineer is an integral part, rather than the traditional engineering approach, which is predicated on a predictable, controllable system that is external to the engineer and his or her actions. It is obvious that ESEM in many cases will deal with large and complex systems, with complicated biological, physical, governance, ethical, scientific, technological, cultural, and religious dimensions and uncertainties. In fact, in most cases, the scientific and technical knowledge necessary to support ESEM approaches is weak or nonexistent, and the evolution of the institutional and ethical capacity necessary to complement ESEM is, if anything, in an even more primitive state. Accordingly, ESEM is best thought of as a capability that must be developed over a period of decades, rather than something to be implemented in the short term.

Operationally, however, ESEM is as much evolutionary as discontinuous, for it builds on practices and activities that are already being explored, some newer and less developed than others. From a technical perspective, these would include industrial ecology methodologies such as *life cycle assessment; design for environment;* and *materials flow analysis,* as well as experience gained in systems engineering (Graedel & Allenby, 1995; Hughes, 1998; Pool, 1997; Socolow et al., 1994). From a managerial perspective, it draws on the literature about managing complexity and "learning organizations" (e.g., Senge, 1990) as well as that on "adaptive management," which has been developed primarily by scientists and ecologists in the context of resource management (Allenby, 2000–2001; Berkes & Folke, 1998; Gunderson, Holling, & Light, 1995). A few examples of ESEM might help illustrate the concept and some of its implications.

Consider as one example the Florida Everglades, a unique ecosystem that has been altered most recently by an 1,800 mile network of canals built over the last 50 years to support agricultural and settlement activity, which has diverted some 1.7 billion gallons a day of water flow. Invasive species, especially the melalu, Brazilian pepper, and Australian pine, are outcompeting native species, aided by human disturbances such as draining marshes and more frequent fires. The nesting success of birds, a predominant animal form in the Everglades, has declined at least 95% since the mid-1930s (Kloor, 2000). Changes in the Everglades have, in turn, affected Florida Bay, through perturbations in salinity levels, nitrogen cycling, and other mechanisms, leading to die-offs in important species such as the seagrasses (Schrope, 2001). The natural cycles that once defined and supported the Everglades in a stable condition, including rainfall and water distribution patterns and nutrient cycles, have been profoundly affected not just by human settlement patterns, agriculture, tourism, industry, and transportation systems but also by the various management regimes attempted over the past 100 years. Moreover, the increasingly powerful coupling between primarily natural and primarily human systems is well illustrated in this case, for perhaps the most important single contributor to the present biological structure of the Everglades is not competition, not predator–prey relationships, but the sugar subsidy, which supports the predominant form of agriculture in the area (Light, Gunderson, & Holling, 1995).

In response to the imminent danger of collapse of the Everglades as a functioning ecosystem, a $7.8 billion Everglades "restoration" project has begun. Its intent is to restore waterflow in nonhuman systems to functional levels while continuing to support industrial, agricultural, settlement, and other human activity. The hope—in ESEM terms, one would say the *design objectives and constraints*—is to preserve and rebuild a desired ecological community while maintaining human activity at a politically acceptable level; it is not clear that this extraordinary challenge can yet be successfully accomplished. An important part of the plan is its emphasis on continuing dialogue with both the human and the natural systems that, integrated together, compose the project, including continuing scientific review of both plans and the natural systems' reactions to what is already being done (Kaiser, 2001).

What are the implications of this case study? To begin, it is obvious that the Everglades is now, and will continue to be for the foreseeable future, a product of human design and human choices. As Schrope (2001, p. 128) commented, "Returning the Everglades to their natural state is not an option. The project's goal is instead to restore flow as much as possible while also considering the needs of local residents and farms." Human modification of the fundamental dynamics of the Everglades has already occurred: There is no "pristine" system to which to return, and the Everglades will never be "natural" again. It is not "restoration" but the creation of a new engineered system, that will display those characteristics—including preservation of flora and fauna, if that is a design objective, and the system can be engineered to do so—that humans choose. The challenge is not to restore a hypothetical past state but to exercise ethical and rational choice based on projections of systems evolution over time.

It cannot be overemphasized that, although ESEM does not imply an artifactual world, it does require, as in this case, that humans consciously accept design responsibility for their actions, even (perhaps especially) when part of the design objective is maintenance of "wild" or "natural" areas. Moreover, the Everglades will continue to be a complex evolving system, and its human designers will continue to be active in managing it in light of the objectives and constraints society imposes on their project. This case thus illustrates another important point: ESEM is not traditional engineering that seeks to create a defined end point (e.g., a car design); rather, it is an ongoing dialogue between human designers and human–natural systems—a dialogue that, absent collapse of the human or natural systems involved, will continue indefinitely. There is no escaping our responsibility for our designs.

This point is emphasized in a second example of nascent ESEM, the Aral Sea debacle, an important cautionary warning against the high modernist linkage of the modern scientific and technological apparatus with the political and cultural power of the state (McNeill, 2000; Scott, 1998). Like many large hydrologic projects—especially regional irrigation systems, canal and dam construction, and the like—what amounts to an engineered destruction of the Aral Sea resulted from efforts to support higher levels of agricultural activity. Actually a lake, the Aral Sea, which only decades ago was the fourth largest lake in the world, has in a few short years lost about half its area and some three fourths of its volume because of the diversion of 94% of the flow of two rivers, the Amu Dar'ya and the Syr Dar'ya, primarily to grow cotton. It is ironic that some estimates are that only 30% of the diverted water, carried in unlined canals through sandy desert soils, reaches its destination; moreover, the resulting cotton is of such a low grade that it cannot even be sold on competitive world markets (Feshbach, 1998; McNeill, 2000). The unintended results are staggering: desertification of the region (the resulting Ak-kum desert, expected to reach 3 million hectares in the year 2000, did not even exist 35 years ago); generation of some 40 to 150 millions of toxic dust per year, with substantial detrimental impacts on regional agriculture and human health; potential impacts on the climate regimes of China, India, and Southeastern Europe; increased salinization of the Aral Sea with concomitant loss of 20 out of 24 indigenous fish species, and a drop in the fish catch from 44,000 tons in the 1950s to 0 today, with a concomitant loss of 60,000 jobs; a reduction in nesting bird species in the area from 173 to 38; and possibly even the release of biological warfare agents previously contained because they were worked with on an island in the Aral Sea (Voskreseniye Island) that is now becoming a peninsula (Feshbach, 1998; Postel, 1996).

The case can be summed up in one word: hubris. Such projects, frequently associated with state-sponsored high modernism, can be extremely damaging to both human and natural systems (and it is interesting that those states that thus cavalierly destroy natural systems tend to treat their human populations brutally as well; Shapiro, 2001). The failure to assume responsibility for the human Earth that has been created over the past centuries certainly can lead to a quasi-Utopian blindness to moral responsibility (this tends to be the environmentalist response, because of the theological and ideological positioning of "Nature" as sacred Other, which precludes perceiving it as dominated by the human). On the other hand, however, the high modernist ideological predisposition to large engineering projects that neglect both the human and environmental contexts within which they inevitably unfold, combined with an ideology of human dominance of nature, can lead to equally unethical—and economically and environmentally costly—outcomes. Both ideological blindness and high modernist hubris are failure modes in the age of the anthropogenic Earth.

Other examples will quickly suggest themselves to the reader, with perhaps the most obvious being the complex international response to global climate change. This clearly is ESEM, even if it is not recognized as such, for what is being determined is not only whether and how to stabilize "natural" systems—in this case, the carbon cycle and the climate system, and associated systems, such as oceanic circulation patterns—but also what evolutionary paths will be permitted for human economies and societies. This is far too complex an example to be quickly summarized, but one aspect clearly stands out for the technologist: the technological naiveté of the dialogue so far. To choose just a simple example, it is clear that the negotiations have tended to take as a given that production and use of fossil fuels must be stringently limited and that changes in lifestyle must be imposed. To anyone versed in technological evolution, however, this is a questionable assumption, for it now appears that carbon dioxide resulting from fossil fuel combustion in power plants can be captured and sequestered for centuries to millennia in the ocean, deep aquifers, geologic formations, or other long term sinks at relatively low risk and cost compared with the alternatives. Although there are some economic and energy penalties, the technologies for accomplishing this are available and feasible (Socolow, 1997; U.S. Department of Energy, 1999). With appropriate engineering and feedstock selection, the power plant can be designed not as an emitter or as a nonemitter but as a variable-release control node in a managed carbon cycle system (Fig. 14.1). Moreover, the plant can produce a mixture of secondary energy sources, especially hydrogen and electricity, which can support most energy use throughout the economy.

Although the carbon sequestration technology is critical given highly probable energy generation scenarios that contemplate significant reliance

Fossil Fuel Energy System with Sequestration

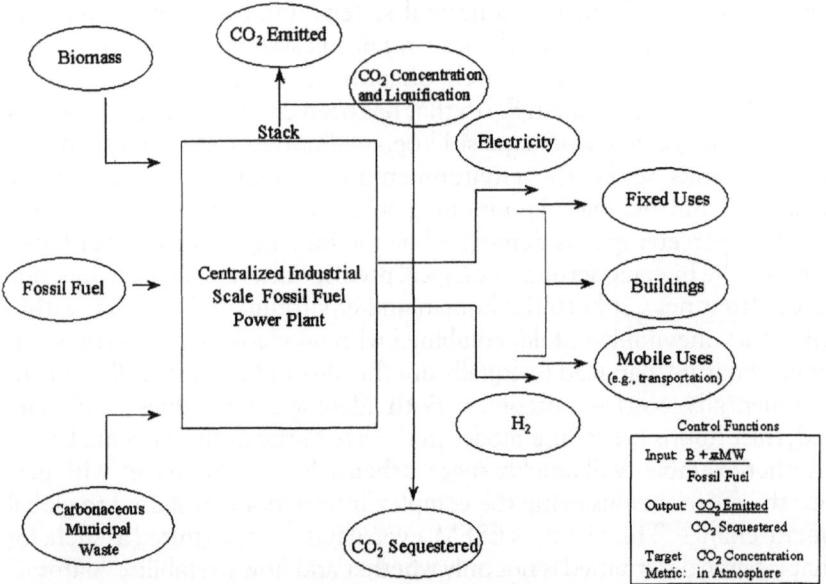

FIG. 14.1.   Fossil fuel power plant as a carbon cycle governance component. Current technologies enable the isolation and sequestration in geologic formations of carbon dioxide produced as a result of energy generation in fossil fuel power plants. Although not yet implemented, such technologies, by varying the inputs and amount of carbon dioxide sequestered, could enable the use of fossil fuel power plants as mechanisms to help regulate atmospheric concentrations of carbon dioxide.

on fossil fuels, at least over the next 50 years,[4] many other developing technologies offer means to manage the carbon cycle and, thereby, to stabilize the climate system. Biomass plantations, ocean fertilization, energy efficiency, evolving technology mixes, different patterns of living and built infrastructure—all will have some impact (and all are dynamically coupled, if for no other reason than because of their linkages to the carbon cycle and, as a result of the Kyoto negotiation process, to human economic, cultural, and political patterns). They, in turn, are coupled to other systems, such as the nitrogen and hydrologic cycles, which must be considered as part of a sys-

---

[4]There is little possibility that traditional fossil fuels will run out over at least the next century (U.S. Department of Energy, 1999). In any event, there are more than 10 trillion tons of methane in oceanic hydrates—twice the amount of carbon as in all known coal, oil, and gas deposits (Suess, Bohrmann, Greinert, & Lausch, 1999; Zimmer, 2001).

temic, ESEM approach to climate change (Fig. 14.2).[5] These "natural" systems in turn are tightly coupled to human cultural, economic, and technological systems, for it is clearly the case that any meaningful effort to address global climate change must significantly affect these human systems as well. In short, the global climate change negotiation process is attempt-

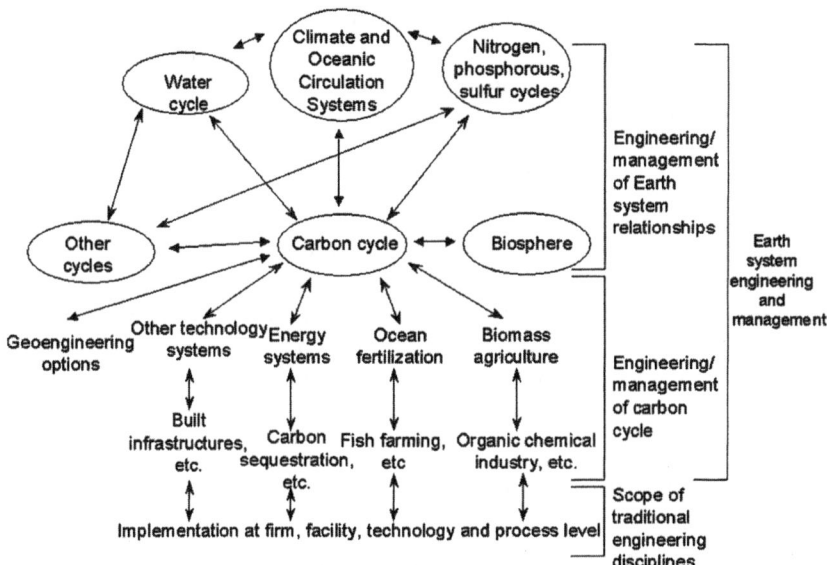

FIG. 14.2. Earth systems engineering and management: Climate and carbon cycle schematic. Management of global climate change is today usually thought of as an isolatable end in itself. This initial naïve understanding is slowly being replaced by the understanding that the climate system, and underlying carbon, nitrogen, hydrologic, and other cycles, must be understood as integrated systems—and inclusive of human technological, economic, and cultural systems as well. Understanding, much less designing and managing, these complex coupled systems is currently beyond the state of the art, but a critical first step is the recognition of the real nature of the challenge.

---

[5]There are also additional technological means by which global climate change can be managed, which come under the general rubric of geoengineering. For example, a number of proposals that would change the reflectivity of the planet have been made, including injecting a fleet of thin aluminum balloons into space between the earth and the sun, which would reduce the amount of energy striking the earth's surface (Keith, 2000). A particularly intriguing possibility is the removal of carbon dioxide from ambient air by chemical extraction, a technology which, if operationalized economically at scale, would essentially negate any need for emissions controls (Elliott et al., 2001). Whether the risks involved in these options (especially the geoengineering alternatives) are worth the benefits has not been comprehensively evaluated, although in many cases their estimated costs (continued on next page)

ing to engineer the world but pretending it is merely addressing a difficult single issue, and this inability to perceive, much less understand, what it is really about is one reason for its political failure in jurisdictions such as the United States.

This point is clarified by considering only one of the technological options that might be available to mitigate global climate change forcing: industrial-scale atmospheric engineering. The technology would involve large-scale systems that would scrub carbon dioxide from ambient air. Although unproven, such a technology not only negates the demand that fossil fuel use be reduced but also requires that we actively decide on the level of global climate change forcing we think is appropriate—in other words, it demands that we design the atmosphere to an explicit human standard, based on a set of values that must also be made explicit. It thus also forces us to recognize that the global climate change negotiating process has failed to address the question around which the entire process revolves: What kind of world do you want? And, indeed, for ESEM as a whole this is the critical question, at this point in our history both inescapable and unanswerable. And yet we must begin to develop the capability to do so, for we are busy creating the anthropogenic Earth, and if we do so without asking or attempting to answer that question we are failing as sentient and ethical beings.

Integrative Cognitivism

These examples demonstrate perhaps the critical dynamic characterizing the anthropogenic Earth: It is a planet where the sum total of human behavior and cognition, including emergent behaviors and characteristics at various levels, is reified and integrated, in a process that is neither predictable nor teleological but clearly reflects human activities and, in a complicated way, human intentionality. Increasingly, technological systems integrate at global scales to create a monocultural planet, one that an alien, seeing the Earth for the first time, would take from the objective evidence before it as representing the conscious intent of the species taken as a whole—including, of course, subsystems ranging from the individual human entity, to hu-

---

[5](continued)   (which admittedly are fairly speculative at this point) are minor compared with the currently favored alternative of substantial emissions reductions. For example, Keith (2000, pp. 270–271) presented estimates that several albedo modification technologies could be implemented at less than a dollar per ton of carbon dioxide mitigated, compared with estimates ranging from $10 to $100 per ton for carbon sequestration in managed forests and proposed tax rates of $50 per ton. Many of these options have not been part of the negotiating process, at least in part because of the aversion to technology that generally characterizes the environmentalist discourse.

man institutions at different scales, to the global scale. The Earth now clearly exhibits at a planetary scale the appearance of planning and intelligence and is thus the physical expression of the "diversity in unity" of human activity. At its most fundamental level, the anthropogenic Earth is the final resolution of the Cartesian dichotomy, for it marks the reintegration of mind and matter at the global scale: human Mind becomes the world. The analytical framework is one of complexity and self-catalysis, marked by accelerating networking within and among systems at differing temporal and spatial scales and the unpredictable but significant integration of vast new information structures—notably, but not only, biological genomes—into human systems.

In this sense, then, the anthropogenic Earth is a product of human activity but one that, until recently, has not been perceived as a result of human design.[6] This lack of perception has facilitated a failure to assume moral and rational responsibility for that design. Necessarily, as a predicate to ESEM we must then address the question of how we may begin both to perceive the reality of that which we have wrought as well as to assume the burden of design of earth systems—a burden that, in particular, requires a new understanding of the critical importance of intentionality in design at this scale, for design necessarily implies choice and intent, and out of choice and intent flow moral responsibility. Conversely, without intentionality, choice

---

[6]Many familiar systems are the product of human activity but, because of their complex and autocatalytic nature, not considered to be designed by humans. One example familiar to all is the city: All elements, from building to infrastructure, are designed by individuals, yet the city itself is subject to dynamics and behaviors that emerge only at levels above those designed by humans, as many a mayor can attest. Another prime example is the Internet. As Barabasi (2002, pp. 145, 148–150) noted:

Though human made, the Internet is not centrally designed. Structurally, the Internet is closer to an ecosystem than to a Swiss watch ... no single company or person controls more than a negligible fraction of the whole Internet. The underlying network has become so distributed, decentralized, and locally guarded that even such an ordinary task as getting a central map of it has become virtually impossible ... while entirely of human design, the Internet now lives a life of its own. It has all the characteristics of a complex evolving system, making it more similar to a cell than to a computer chip .... What neither computer scientists nor biologists know is how the large-scale structure emerges, once we put the pieces together.

More broadly, one cannot help but think of Vico's words regarding human history in this regard (quoted in Thompson, 1978, p. 291):

It is true that men have themselves made this world of nations, although not in full cognizance of the outcomes of their activities, for this world without doubt has issued from a mind often diverse, at times quite contrary, and always superior to the particular ends that men had proposed to themselves .... That which did all this was mind, for men did it with intelligence; it was not fate, for they did it by choice; not chance, for the results of their always so acting are perpetually the same.

and intent are meaningless. Thus, to establish a firm conceptual basis for ESEM, we must consider not just cognition but also intentionality.

The astute reader will have noted the use of the term *intentionality* rather than the perhaps more common term *free will*. The two concepts are obviously related, a topic I revisit in more detail later. But the idea of free will carries at this point considerable baggage unnecessary to our discussion, as well as the difficult implication that free will is something freestanding and independent, a primary characteristic of, for example, mature human beings. To the contrary, I argue that to the extent to which free will is expressed at all, it is within the context of internal and external system states that significantly determine how it can be expressed, the way it is expressed, or, indeed, whether it can even exist at certain points. Thus, an operating definition of *intentionality* is that it is free will as actually expressible in an environment of complex and dynamic internal and external networks, a contingent rather than absolute property.

To understand the importance of intentionality and design in the anthropogenic earth, let us begin by turning first to the familiar concept of "actor networks," defined by Callon (1997, p. 93) for technological systems: "An actor network is simultaneously an actor whose activity is networking heterogeneous elements and a network that is able to redefine and transform what is it made of …. The actor network is reducible neither to an actor alone nor to a network." Actor networks are complicated by the fact that they operate at many different levels in human systems; moreover, every entity almost always is part of many different actor networks simultaneously. Each network acts *as if* it were an actor seeking certain goals; a common one, important for most actor networks, is continued existence. Additional goals—ethical behavior for humans, economic performance for firms, support of family members for families or clans—characterize and particularize each as well. Each contains components that are "natural" as well as human. Moreover, each simultaneously interacts with exogenous elements and systems in its external environment and reconfigures itself internally; as Callon (1997, p. 93) wrote:

> The actor network should not … be confused with a network linking in some predictable fashion elements that are perfectly well defined and stable, for the entities it is composed of, whether natural or social, could at any moment redefine their identity and mutual relationships in some new way and bring new elements into the network.

To complicate things even more, one entity—say, one human individual—in practice is composed of a number of different actor networks. Thus, an individual may be simultaneously a citizen, an environmentalist, a family member, and a representative of a larger subjectivity (e.g., a manager in a firm, an active member of a local church; Sagoff, 1996). An individual may thus be conceptualized as a dynamic and constantly changing network of networks, which is in turn a component of networks at many different scales.[7]

An actor network may thus be seen as one way to conceptualize an active agent in a complex system. The concept is similar in some interesting ways to that of a *representational system*, which Hofstadter and Dennett (1988, pp. 192–193) defined as:

> an active, self-updating collection of structures organized to "mirror" the world as it evolves ... a representational system is built on categories; it sifts incoming data into those categories; when necessary refining or enlarging its network of internal categories; its representations or symbols interact among themselves according to their own internal logic; this logic, although it runs without ever consulting the external world nevertheless creates a faithful enough model of the way the world works that it manages to keep the symbols pretty much "in phase" with the world they are supposed to be mirroring.

Comparing the two models is instructive. Both are bounded: There are operations internal to the system and an external system that they mirror and with which they interact. Both are based on networks, and these networks reflect, and change in response to, changes in the environment external to the bounds of the system. In both cases, changes in network structure reflect changes in information content of the system. In both cases, it is not clear that there is any such concept as an "optimum" state; rather, being good enough to continue to operate over time is the measure of success.

However, neither of these formulations appears to deal robustly with the critical element of intentionality. In the case of the actor network, there appears to be an implied necessity for intentionality inasmuch as *actor* implies action based on intentionality directed beyond the boundaries of the network when necessary, whereas in the case of the representational system this does not appear to be a necessary property. Thus, for example, it would seem to be entirely feasible for a salt marsh to be a representational system—but a salt

---

[7]To this extent, one might be inclined, by analogy to the gas laws, to argue that a significant challenge for future research is to develop the statistical mechanics of dynamic networks of networks. This is complicated enough when dealing with nonintentional systems, but when the existence of intentionality, and thus contingency, is admitted, it becomes daunting indeed. Parenthetically, this illustrates that the common argument that free will is an unnecessary "ghost in the machine," and can thus be ignored philosophically, is at least in part flawed, for contingency greatly increases complexity and, especially in human systems, arises at least in part from the existence of intentionality. Thus, contingency arises from intentionality, which in turn is free will as expressed in complex dynamic systems.

marsh, unlike a human, does not exhibit intentionality. This difference is profound: If there are only representational systems, and not actor networks, there need not be intentionality, and without intentionality there is not the contingency that otherwise characterizes human systems and is such an inherent part of their complexity.[8] Moreover, without intentionality there is not design but deterministic evolution[9]—and thus the discourse of ESEM becomes marginally relevant, at best.

In beginning this section, I noted that an alien looking at the Earth might well conclude from the evidence before it that this planet was in the advanced stages of being terraformed by a "conscious" entity to reflect the interests of one species: humans. It is now time to reflect more closely on that formulation—and, in particular, to move beyond the use of the term *conscious*, which I want to avoid for three reasons. First, there is no agreement about what consciousness is (Harnad, 2001). Second, it is unnecessary for our purposes, for we can substitute for *consciousness* the concept of *cognition* and define that term more concisely without the baggage that *consciousness* carries with it. Third, *consciousness* suffers from overuse as disguised analogy and can mislead seriously in this guise. It is an intuitive concept to individuals yet has been widely applied to systems higher in the hierarchy—states (e.g., Kant & Hegel), classes (e.g., Marx), firms and organizations (e.g., Senge), and the like. There, it carries with it the faint odor of analogy but not the rigorous assumption that such systems are "conscious" in the same way that an individual human is. However, by identifying a set of functions associated with cognition, and determining whether such systems exhibit those functions, one can fairly credibly determine whether they are cognitive systems. We thus have the hope of moving beyond analogy.

More rigorously, then, I define cognitive systems as having five characteristics: (a) they are dynamic, and distributed and networked across temporal, spatial, informational, and complexity scales; (b) they act as active perceptual filters, in that there will always be some information in the ambient environment that could be perceived but, by system design or choice, is not; (c) they process and manipulate information; (d) they display intentionality, and the exercise of judgment making choices; and (e) they create new information, which in turn is both internalized to the creating system and feeds into other networked cognitive systems. This, for example, leads to the reflexivity of human systems so often noted by sociologists

---

[8]The interrelatedness of elements of complex systems is again illustrated here: The structure of the system determines the extent and form through which latent free will can be expressed as intentionality, and that intentionality in turn introduces contingency, and thus a higher degree of complexity, into the system.

[9]Deterministic evolution does not necessarily imply predictability, as chaos theory reminds us.

(Giddens, 1990) and the auto-catalytic nature of technology systems (Grubler, 1998).

This definition does not preclude recognizing that system levels above the individual (firms, nation–states, nongovernmental organizations, etc.) are cognitive systems as well. It is necessary, however, to differentiate between the *appearance* of intentionality such higher level systems display, which has undoubtedly been one reason they are so frequently analogized to human individuals, and the *source* of intentionality. I argue that all intentionality—and, consequently, moral responsibility—derives ultimately from the exercise of free will by individuals, albeit distorted, complicated, combined, and confused by the complex systems within which they are expressed.

This concept of cognitive systems is the core of the framework concept of *integrative cognitivism,* which in turn is an extension of existing concepts, including "actor networks," "externalism," and "philosophic environmentalism" to the scale of technology at the level of the anthropogenic Earth.

To set the stage for understanding the implications of Mind in the anthropogenic world, it is important to recognize, and move beyond, the most dominant underlying paradigm of "Mind" in the Western tradition. Descartes, in his efforts to foundationally ground modern philosophy, not only famously created the mind–body duality but also located mind, and cognition, internally to the thinking subject—indeed, it had to be internal, because its action was the only means by which the existence of the subject could be established (Descartes, 1637–1641/1998; Rowlands, 1999). This "internalist" approach, although continued through Kant and Hegel to the present, has become increasingly questionable. In part this has occurred because of the recognition of the increasing focus on language, and its determinative effects on perception and cognition (Habermas, 1975; Rorty, 1989), and in part because of the broader recognition of the importance of social context and cooperation in determining what individuals can and do think (Burge, 1998; Giddens, 1990).

As will become more apparent shortly, the degree to which knowledge exists external to the individual, and is both accessed as needed, and powers the reflexivity of social evolution, is an important part of this re-integration of the internal and external environments (Giddens, 1990).[10] This approach is the basis for a relatively new school of philoso-

---

[10]As Giddens (1990, pp. 53–54) noted: "The production of systematic knowledge about social life becomes integral to system reproduction, rolling social life away from the fixities of tradition .... Knowledge reflexively applied to the conditions of system reproduction intrinsically alters the circumstances to which it originally referred."

phy called *externalism,* which holds that (Ludlow & Martin, 1998, p. 3) "the social character of content pervades virtually every expression of our language, and virtually every aspect of our thoughts." However, externalism by itself does not go far enough to explain the dynamics of Mind in an anthropogenic earth. This is because, although it may attack the first Cartesian principle, foundational mind–body dualism, it still allows for the second Cartesian principle, internalism. What I assert, however, is that cognitive function is, in fact, not internal any more in any meaningful way: that the defining characteristic of the anthropogenic world is that Mind, and cognition, increasingly extend throughout it in fact and not just in some vague, "ghost in the machine" sense.

In doing so, I begin by following Rowlands's (1999) argument for a position he called *environmentalism.* Unfortunately, I have to coin another term for his approach: First, to avoid confusion with the completely different use of *environmentalism* in ESEM (as the discourse of environmentalists), and second, because I extend his arguments in speculative ways and would not want him blamed for that.[11] Given that this line of argument undermines both elements of the Cartesian position, I call it *integrative cognitivism:* It integrates the internal and the external in the act of cognition, and it focuses on cognition rather than consciousness. In doing so, it provides, I hope, a more powerful basis for understanding some of the philosophic implications of the anthropogenic Earth.

The gist of integrative cognitivism as presented by Rowlands (1999, pp. 22, 26) is relatively straightforward: Cognitive processes are not exclusively internal to cognizing organisms and thus cannot be understood by focusing only on internal processes. Rather, cognitive processes are necessarily composed of internal processes and external structures that function as loci of information. In cognition, the organism identifies and appropriates the necessary information by manipulating such external structures. As Rowlands (1999, p. 29) put it:

---

[11]Rowlands was very careful to indicate throughout that he was not trying to *refute* the prevailing internalist paradigm but to make its nature as mythology clear, and to *unseat* it, to show that it "is not how we have to think about cognition" (p. 12). Thus, he does not seek to rigorously disprove internalism but to demonstrate that "there is, in other words, no theoretically respectable reason for thinking of cognitive processes as purely and exclusively internal items" (p. 12). I, on the other hand, believe his approach, when reasonably extended, is far more powerful than merely challenging a particularly stubborn mental model and that "integrative cognitivism" provides a necessary tool for understanding and exploring the increasing scale and complexity of cognitive systems.

Minds are not purely internal things; they are, in part, *worldly* in character. That is, minds are hybrid entities, made up in part of what is going on inside the skin of creatures who have them, but also made up in part of what is going on in the environment of those creatures.

Clark (2001) reinforced the critical insight of integrative cognitivism, that cognition—thinking—is a function that no longer occurs within the human organism but in an intricate, dynamic network of symbols, information technologies, language, and culture, emphasizing:

the distinctive way human brains repeatedly create and exploit various species of cognitive technology so as to expand and reshape the space of human reason ... creating extended cognitive systems whose computational and problem-solving abilities are quite different from those of the naked brain. Human brains maintain an intricate cognitive dance with an ecologically novel, and immensely empowering, environment: the world of symbols, media, formalisms, texts, speech, instruments and culture. The computational circuitry of human cognition thus flows both within and beyond the head.

A few examples may help clarify and support this fundamental assertion of integrative cognitivism. Thus, for example, the development of writing and external information storage devices enabled the development of "collective memory hardware":

Individuals in possession of reading, writing, and other relevant skills have access to the "code" that allows them to plug into this external [information storage] network. They thus become somewhat like computers with networking capabilities; they are equipped to interface, to plug into whatever network becomes available. And, once plugged in, their memory capacities are determined both by the network and by their own biological endowments. Those who possess the "code," the means of access to the external system, share a common memory system. And, as the content of this system expands far beyond the scope of any single individual, the system becomes by far the most significant factor in the memory of individuals. (Rowlands, 1999, p. 144)

To take a simple example, my memory of Shakespeare's plays exists not in my mind but in the system composed of my book containing the complete works of Shakespeare, an index system that I understand that allows me to search that book, and my internal capabilities to process the information. Moreover, memory structured in this way is far more powerful, and capable of storing far more information, than the "episodic" memory characteristic of oral traditions.[12]

Lest these be thought trivial observations, remember that cognition is power: A more cognitive entity is a more powerful entity. In this light, Diamond's (1997, pp. 67–81) discussion of the battle at Cajamarca, in the Peruvian highlands, between the Spanish conquistador Francisco Pizarro and the Inca emperor Atahuallpa, is instructive. Pizarro had 168 Spanish soldiers (62 on horseback, 102 on foot), was fighting in an unfamiliar cultural and physical environment, and had no support available. Atahuallpa, on the other hand, was the absolute ruler of the largest and most advanced state in the Americas. He was in the middle of his empire, surrounded by 80,000 soldiers, when he walked into a Spanish trap at Cajamarca, was captured, held for ransom and then, when that was paid, executed by the Spanish. Indian losses were in the thousands; not a single Spaniard was killed. This was not unusual: At four similar battles—Jauja, Vilcashuaman, Vilcaconga, and Cuzco—the same pattern repeated.

In any such extraordinary clash of cultures, a number of factors come into play—such as, for example, the Spanish superiority in armament and use of horses. However, that alone cannot account for the fact that not just the battle of Cajamarca, but the clash of cultures, ended decisively in favor of the Spanish. Here, Diamond's (1997) comments about the disparity in cultural memory are instructive. Atahuallpa had only the information

---

[12]A number of more complex examples have been researched by those studying technological evolution. For example, Gorman (1997) took an integrative-cognitivism approach to the invention of the telephone (although, of course, he did not call it that, because I just made up the term). In evaluating how two innovators—Alexander Graham Bell and Elisha Gray—invented devices capable of transmitting tones, he illustrated a process of dialogue involving different individual mental models, mechanical representations, and heuristics that anchored the cognitive process of invention of the telephone not solely in the individual, or in the external context within which the individual worked, but firmly in both:

Cognition clearly is both embodied in brain, hands and eyes, and also distributed among various technologies and shared across groups ... in order to compare two inventors, one needs to understand their mental models and mechanical representations. This kind of understanding is possible only because so much of cognition is in the world. (Gorman, 1997, pp. 589, 612)

In other words, Bell and Gray functioned within two different cognitive systems, and the results of one proved superior to the other in enabling an economically and socially relevant function: the transmission of voice over distance in real time. Other examples were provided by Bijker, Hughes, and Pinch (1997): Technological innovation is a fruitful area for case studies in integrative cognitivism precisely because the complexity of developing artifacts and practices which by definition must be functional in a much broader social, economic, and cultural context illustrates quite plainly the coupling between internal and external elements of cognitive systems.

about the Spanish that could be transmitted by word of mouth; the Spanish, on the other hand, belonged to a literate tradition. The amount of information to which they had access, and which they could therefore process, was far greater. Indeed, "Pizarro explicitly modeled his ambush of Atahuallpa on the successful strategy of Cortez" (Diamond, 1997, at 80):

> In short, literacy made the Spaniards heirs to a huge body of knowledge about human behavior and history. By contrast, not only did Atahuallpa have no conception of the Spaniards themselves, and no personal experience of any other invaders from overseas, but he also had not even heard (or read) of similar threats to anyone else, anywhere else, anytime previously in history. That gulf of experience encouraged Pizarro to set his trap and Atahuallpa to walk into it.

From the perspective of integrative cognitivism, what happened in Peru was that two systems with unequal cognitive ability came into conflict, and that with the greater cognitive ability won. This is not an argument for some form of cognitive determinism, for certainly it is conceivable that the battle of Cajamarca could have gone the other way, and the Spanish eventually ejected from the New World. Moreover, it is evident that, as with virtually any other important historical event, there are no simple or linear causalities: The real world is not kind to such assertions. In general, however, the system with the greater cognitive ability will outcompete those with which it comes into conflict. In part, of course, this is because cultures with greater cognitive ability have a suite of associated characteristics—higher technological development, more developed economies, more complex social organization—that, taken together, also may provide advantages in any competition with other cultures.[13]

Rowlands' (1999) arguments are phrased in terms of individual organisms and their functions, as befits the undermining of an essentially Cartesian worldview based on *cogito, ergo sum*. It is significant that an important element of his argument is evolutionary: that in general it is more efficient for an organism to offload cognitive functions into its external environment than to attempt to internalize it all. (Why should one have to remember all of Shakespeare internally when one can just as easily remember the much

---

[13]This analysis also does not support in any way the quasi-racist argument that the Incas were, in individual cognitive ability or in other qualities, such as courage, somehow inherently inferior to the Spanish: There is no evidence for this whatsoever (indeed, this is a major point of Diamond's work). However, as integrative cognitivism would hold, it is not the intelligence internal to the individual human organism that is relevant but the larger cognitive system—a structure heavily dependent on culture—that determines the overall cognitive ability to which each individual has access. Internal Spanish cognition was multiplied by the cultural structures, such as writing, available to it; the same was not true for the Incas.

shorter algorithm for deriving such information from books? And then one can remember many more books and access their information and wisdom on an as-needed basis. Indeed, similar structures are built into efficient software and communications systems everywhere.) Also, given this efficiency, it follows that evolution would favor those mechanisms, and organisms, that distributed cognition across internal and external components in such a way. However, Diamond's (1997) example, and the fact that the world today is predominantly a product of the European experience,[14] extends that argument, for it is not just individuals interacting with their world but the structure of cultures and societies, and the mechanisms and pathways by which they magnify and extend internal processes to far greater temporal and spatial scales, that determine evolutionary success. Whether we wish it or not, cultures compete (Harrison & Huntington, 2000), and it is an important insight that cultures with a greater systemic cognitive capability— in some meaningful sense, a better collective ability to think—have a greater chance of success.

Is it possible to be more explicit? Perhaps. A critical determinant of the cognitive capabilities of a culture is its technological sophistication and the degree to which the culture does not unnecessarily constrain the self-catalyzing dynamic of technology.[15] As Castells (2000, p. 7) wrote:

---

[14]As Giddens (1990, pp. 136, 174; see also Russell, 1945/1972; Diamond, 1997; and Landes, 1998) commented:

[Modernity] is institutionally based on the nation-state and systemic capitalist production .... Both have their roots in specific characteristics of European history and have few parallels in prior periods or in other cultural settings. If, in close conjunction with each other, they have since swept across the world, this is above all because of the power they have generated. No other, more traditional, social forms have been able to contest this power in respect of maintaining complete autonomy outside the trends of global development. Is modernity distinctly a Western project in terms of the ways of life fostered by these two great transformative agencies? To this query, the blunt answer must be "yes" ... The modes of life brought into being by modernity have swept us away from *all* traditional types of social order, in quite unprecedented fashion.

[15]Again, it must be emphasized that the technological sophistication of a culture, which is correlated to a significant degree with the cognitive capabilities of that culture, is not deterministic of success in itself: Indeed, technological sophistication and evolution within a culture is an integral part of a much more complex context that may include, *inter alia*, educational processes, tax and economic structures, cultural approbation of innovation, individualism and entrepreneurial activity, and governance structures of all kinds. Thus, Castells (2000, p. 5) remarked that:

The technological blossoming that took place in the early 1970s [in Silicon Valley, California] can be somehow related to the culture of freedom, individual innovation, and entrepreneurialism that grew out of the 1960's culture of American campuses. Not so much in terms of its politics, since Silicon Valley was, and is, a solid bastion of the conservative vote, and most innovators were meta-political, but with regard to social values of breaking away from established patterns of behavior, both in society at large and in the business world.   (*continued on next page*)

The ability or inability of societies to master technologies that are strategically decisive in each historical period, largely shapes their destiny, to the point where we could say that while technology *per se* does not determine historical evolution and social change, technology (or the lack of it) embodies the capacity of societies to transform themselves, as well as the uses to which societies, always in a conflictive process, decide to put their technological potential.

Integrative cognitivism strongly suggests that a fundamental role of technology is to increase the cognitive capability of the culture or society within which it exists. To a first-order approximation, the evolution of technology is the evolution of cognition—and the evolutions of human biology, culture, and technology become an entwined evolution of cognition. Moreover, technology is auto-catalyzing: The more technologically adept a culture already is, the more technology it is capable of evolving (Grubler, 1998). This implies an important corollary: that cognition, which is correlated with technology, exhibits similar dynamics—the more cognitive a culture already is, the more likely, and easier, it is for the culture to continue to broaden and deepen its cognitive function.[16]

In doing so, the society will become more complex, more fragmented, and, simultaneously, more integrated, creating deeper and more networked cognitive systems. Simple belief systems will be challenged and become increasingly dysfunctional. The collapse of space and time boundaries, the ahistorical pastiche so beloved of postmodernists, is, in fact, a reflection of

---

[15] *(continued)*   More fundamentally, and perhaps somewhat controversially, technological sophistication appears to be associated with capitalism, individual liberty, and democratic institutions (Harrison & Huntington, 2000), which are in turn associated with Eurocentric culture and Christianity (the most famous work along these lines perhaps being *The Protestant Ethic and the Spirit of Capitalism*, by Max Weber, 1904/1998; see also Noble, 1998).

[16]The ability of cultures to support technological evolution, however, changes over time. Castells (2000, pp. 7–8) noted the example of China, which around 1400 is considered by many to have been the most advanced technological civilization in the world but by 1800 had evolved little further and had clearly been surpassed technologically and scientifically by Europe. The reasons for this are complex and debated, but at least one seems to have been "the rulers' fears of the potentially disrupting impacts of technological change on social stability" (Castells, 2000, p. 9). A similar explanation has been suggested for the differential rates of success of various cultures within Europe, where the dynamic northern European Protestant countries, such as Great Britain and the Netherlands, outcompeted the more static and rigid societies of Catholic southern Europe (Landes, 1998). The obvious question raised by such analyses is whether the ideological environmentalist opposition to new foundational technologies, such as biotechnology, especially in Europe, will have a similar negative impact on the ability of European cultures to compete with potentially less constrained ones such as China, and perhaps India and Brazil, over time.

such a culture but, contrary to pessimistic prediction, need not result in fundamental alienation, either of individuals or of institutions. This is because even if the overall system is highly fragmented, complex, and contingent, it is also characterized by patterns of order that are local within time and space and thus offer safe foundations for both individuals and higher entities. Local hierarchy coexists with networked evolutionary complexity: Local truth suffices where attempts to establish foundational Truth become dysfunctional, increasingly atavistic, and finally untenable.[17] This structure, of local patterns of order embedded in changing and undefined larger systems, seems to stabilize systems. Thus, Kitano (2002, p. 207) noted, in referring to the lamda phage decision circuit, "the archetypal genetic switch" and cellular systems in general that "structurally stable network configurations increase insensitivity to parameter changes, noise, and minor mutations ... stable [genetic] switching action arises from the structure of its network, rather than the specific affinities of its binding sites."

This analysis strongly suggests that the human condition is perhaps more fundamentally fragmented, yet paradoxically more dynamically ordered, than traditional postmodern analysis suggests. In exploring this, we can use the State 1, 2, and 3 model suggested by Gorman (chap. 13, this volume). The critical realization is that not just discourses, but ontologies, are not totalizing: Each faithfully captures reality, but none captures reality completely. The analogy to quantum particles is apparent: neither wave, nor particle, but both. Thus, we can view State 1 as the default assumption that most people make about their world: There is a single, totalizing ontology, and a single worldview prevails. State 2 is a cautious and bounded coexistence of totalizing ontologies, in the sense that the communities involved do not give up their belief in the absolute authority of their ontology but are willing to negotiate compromises at the language level. This explains why State 2 issues tend to remain in State 2 and cannot evolve into State 3: The communities in State 2 are perfectly willing to discuss the immediate problems until the boundary of their ontology is reached, and then they are unable to compromise, because they have not put the validity of their ontology into play. State 3, on the other hand, requires accepting the contingency of one's ontology and becoming comfortable with understanding the world as too complex to be captured in any single ontol-

---

[17]In any such discussion it is apparent that one is speaking of truth in the human and not religious sense. That a multicultural world is complex and confusing, and tends to corrode local truth posturing as global Truth does not logically imply that the kind of transcendental Truth characteristic of religion does not exist, merely that human instantiations of it in historical and cultural contexts are contingent and partial.

ogy. Thus, by analogy to cognitive dissonance, one learns to live with "ontological dissonance." So I, as everyman, become comfortable with the fact that engineers, using an ontology based on the existence of an absolute exterior reality and its structuring through external laws, can build a jet plane that flies, and English critics cannot—but then, English critics, using an ontology of relativism and thus personal structuring of reality, can write a brilliant critique of a new art film, but engineers cannot. Thus, in the State 3 system I have to learn to suspend my personal ontological belief system (which, for almost everyone, is absolute at the personal level) in evaluating the broader social issues, accepting the conflicting ontologies as both contingent and, for their adherents, totalizing. The pattern is familiar: Ontologies are indeed ordered and hierarchical, but they are localized; they are the cognitive tools by which an inquisitive species has learned multiple ways to deal with a complex reality. They form one set of local order on which the evolution of the anthropogenic world is predicated.

It is in the light of integrative cognitivism that the profound implications of new foundational technologies can be made explicit. Consider the three technological systems that are likely to have perhaps the greatest impacts on the anthropogenic world going forward: (a) nanotechnology; (b) biotechnology (including genomics); and (c) the many dimensions of information technology, especially the creation of ubiquitous "intelligent networks." Each in its own way is an extension of human design and function into a new realm: In the case of nanotechnology, the very small—indeed, in some cases, molecular—structure of the physical world; in the case of biotechnology, into biological functions from the genetic and molecular to the regional and global (as genetic design changes organisms, and, accordingly, restructures biological communities); in the case of information technology, the coupling into networks of previously disparate artifactual and structural systems. In other words, each extends cognitive structure into realms that previously were independent of it—human cognition extending into the molecular and, through many couplings, to the global.

This extension of cognition is, obviously, the extension and reification of human Mind in the anthropogenic earth. As Rowlands (1999, pp. 135–136) noted, "on this planet, at least, only *homo sapiens* has been able to develop external representational structures." The importance of this development is significant:

> The great schism in the development of the modern mind was the incorporation of external means of representation. The significant development was not representation *per se*. This can be found in many non-human animals. All episodic experience is representational, at least in the minimal sense that it is

about other things. What is distinctive about the modern human mind is not representation as such, but the development of *external* means of representation. (Rowlands, 1999, p. 129)

This is the core of the dynamics behind the anthropogenic earth and the reason that it is uniquely humans who are creating it, for it is human culture and technology that form the framework for the unprecedented expansion of cognition to the level of global systems that is the defining characteristic of the anthropogenic earth. It also expresses the challenge of ESEM, for we are not used to thinking in terms of integrative cognitivism: of our thinking process as extended out into the nonhuman, cultural, economic, and technological systems that surround us. Accordingly, the observation that ESEM requires a new way of thinking is more than hyperbole: It requires that we begin at the very beginning, with an exploration of how we think, and how we know—not just individually but at the level of increasingly broad and complex cognitive systems.

Integrative cognitivism also makes clear that saying that the earth increasingly reflects human Mind is more than a figure of speech. Rather, it is a recognition that elements of systems previously considered "natural" increasingly become integrated into cognition—cognition springing from humans and their institutions but now spread far beyond the individual, with consequences and implications that are difficult to predict, for integrative cognitivism does not hold that internal human cognition simply becomes more powerful as it is externalized in increasingly complex ways in systems that are themselves evolving. Instead, the cognitive process, although still emanating from the individual, becomes different in kind.[18] Cognition and cognitive power become coupled to the particular elements of cultural and technological evolution that characterize different societies; the unpredict-

---

[18]Some readers will, perhaps, object to this realization because it appears to them to simply be a restatement of Hegelian *Geist* or perhaps Heideggerian Being, come to consciousness. There are, however, major and important differences. Unlike either concept, the process of expanding cognition over time is not teleological; that is, it ends neither at Hegel's desk with the coming into consciousness of *Geist* nor with the teleology of the (semi-Taoist Christian?) Being. Complex systems not only continue to evolve but also do so in ways that are not teleological (although for complex evolving systems with significant human components, teleologies, by affecting the exercise of intentionality, affect the evolutionary paths actually taken by the system—although usually not in ways expected by those exercising the intentionality). Second, the evolution of cognitive systems, and human Mind as reified in the anthropogenic Earth, is not a description of an endpoint, or a deterministic system; rather, it is an observed process with certain implications. Third, cognitive systems, no matter how widespread, emanate from and remain based in the will of individual human beings: Remove humans from the world, and the technologies, built environments, and systems that compose today's cognitive systems collapse. We have not yet—so far as we know—created an independent will.

able contingency that history so clearly demonstrates is characteristic of human systems; and the dynamics of networks of artifacts, information systems, individuals, technologies, and other elements that now form the operative platform for cognition.

ESEM requires that, to the extent possible, these cognitive processes are elevated from the implicit and unperceived to the explicit, for only by doing so can two important cognitive functions be integrated into the evolution of the anthropogenic earth: ethics and design. Thus, ESEM itself is dependent on the existence of intentionality, for it rests on the assumption that design is possible, and design in turn implies the ability to choose among options that make a difference to the designed system—in this case, the evolutionary path of the anthropogenic Earth. If there is no intentionality, no actual choices to be made, then ESEM is illusory, for determinism by any other name is determinism. Moreover, intentionality, freedom and free will have traditionally been associated with what Kane (1998, p. 4) called "moral responsibility in the deep sense," that entity that wills, that originates purposes, also creates for itself fundamental, nondelegable, moral responsibility. The assumption of moral responsibility is also foundational for ESEM. Accordingly, both the rationality and the moral responsibility demanded by ESEM require intentionality, which in turn requires the potentiality for freedom. Is this a reasonable assumption?

This is not an area to be entered lightly, for the issues of freedom and free will have been disputed for millennia. It is daunting that Jalalu'ddin Rumi, the Persian poet, commented in the 12th century that "There is a disputation [that will continue] till mankind are raised from the dead between the Necessitarians and the partisans of Free Will" (quoted in Kane, 1998, p. 3).[19] Accordingly, I do not pretend to do more here than sketch an outline of the relationship between free will, a potential whose expression depends on systems conditions and which arises from the human individual, and intentionality, which is free will as actually expressed in the behavior of the appropriate system. I explore these ideas from the perspective of integrative cognitivism and ESEM

To begin, it is apparent that "freedom" and "free will" are not unitary concepts. One clear demarcation line is usually drawn between external (political) freedom and internal (spiritual) freedom. There is thus an intuitive and obvious difference between freedom from restrictions imposed by the exter-

---

[19]Kane (1998) noted that modern philosophy has tended to move away from the very concept of "will" on the grounds that it, like many concepts, such as phlogiston, has no content but rather reflects a state of ignorance that has passed. Kane's book takes the opposite position, as, indeed, do many theologians and some ethicists: Although they may not agree on what "free will" is, they recognize a deep connection between the ability to will and moral responsibility.

nal environment on a cognitive system and the freedom of the cognitive system to control its internal system state and potential responses to changes in the external environment. Regarding the latter, there are two formulations worth mentioning in the ESEM context. The first is that of Sartre (and, more broadly, existentialism): the radical freedom of choice. Sartre argued that humans always have a choice as to how they will respond to events, and to life itself—after all, one can always kill oneself rather than accept whatever option is presented. This choice is constant and ongoing and the means by which an individual creates his or her personal world—and, importantly, not to recognize and embrace this freedom, painful as it might be, is inauthentic: a flight from both freedom and responsibility.

The second form of personal freedom can be called *perverse freedom* (Palmer, 1996). This is the freedom of Dostoyevsky's antihero in *Notes From the Underground;* knowing he is sick, he refuses to go to a doctor, for reason and reality itself are intrusions on his freedom.

> One's own free unfettered choice, one's own caprice—however wild it may be, one's own fancy worked up at times to a frenzy—is that very "most advantageous advantage" which we have overlooked, which comes under no classification and against which all systems and theories are continually being shattered to atoms. And how do these wiseacres know that man wants a normal, a virtuous choice? What has made them conceive that many must want a rationally advantageous choice? What man wants is simply *independent* choice, whatever that independence may cost and wherever it may lead. (Dostoyevsky, quoted in Palmer, 1996, p. 226)

These freedoms are not absolute, however, for a cognitive system inevitably contains elements external to the individual organism (indeed, this is central to integrative cognitivism). Thus, one may be existentially free to commit suicide, but one is not free to purchase a product that is nowhere made, and indeed there are many instances where the system state precludes the exercise of free will. In other words, the system state mediates the creation and expression of intentionality from inchoate free will. However, such a concept of freedom creates an important insight into the dynamics of the anthropogenic world, for the contingency and (to a lesser extent) reflexivity that characterize human systems and thus, increasingly, the dynamics of natural systems is the contingency and reflexivity that, to a large extent, reflects the radical freedom of Dostoyevsky, Sartre, and Heiddeger—a radical freedom that springs uniquely from human intentionality. The argument is not that personal freedom is always and everywhere extant: Indeed, it follows from an integrative cognitivism analysis that freedom reflects a complex and constantly changing system state rather than an inherent characteristic of individuals under all circumstances. Within that system state, however, the perverse and radical freedom of the human individual can, and does, make decisions that alter the future paths of evolution of natural and human systems. It is this freedom, in other words, that is the ground from which intentionality arises, and intentionality in turn is the source of contingency in human systems, a contingency that is not found in nonhuman systems.

It is here that the concept of "will" is critical, for it is intentionality that powers cognitive systems and that has created over millennia the anthropogenic earth. Although, like *freedom*, the term *will* has been used in many ways over the centuries, I will follow Kane (1998), who defines "will" as

> a *set of conceptually interrelated powers or capacities*, including the powers to deliberate, or to reason practically, to choose or decide, to make practical judgments, to form intentions or purposes, to critically evaluate reasons for action, and so forth. (p. 22)

Will is a teleological phenomenon: It "signifies an orientation or inclination toward some objective or end that is desired, chosen, or striven for ... different senses of *will* and *willing* represent different ways in which agents may be directed toward or tend toward *ends* or *purposes*" (Kane, 1998, p. 27) "Will" for our purposes is thus the intentionality expressed in, among other things, designs of systems. This has a critical implication, for intentionality is a necessary predicate for moral responsibility. Thus, Kant noted (quoted in Kane, 1998, p. 34) that a "man ... must make or have made himself into whatever in the moral sense, whether good or evil, he is or is to become. Either [good or evil character] must be the effect of his free choice for otherwise he could not be held responsible for it and could therefore be neither morally good nor evil." This principle, of course, is enshrined in the insanity defense against criminal charges in the Anglo-American legal tradition: A person may not be held morally, and thus legally, culpable for his or her actions where he or she is demonstrably incapable of intentionality by reason of insanity (Black's, 1990).

At this point we enter deeper waters, for these formulations of will and intentionality are focused on the individual. The systems of concern in ESEM, and indeed much of the complexities of the modern, arise instead from the operation of systems (e.g., markets and cultures) and institutions that, although composed of individuals, exhibit far different characteristics, and operate at far different spatial and temporal scales, than the individual. This question is better understood by returning to the understanding of cognitive systems that underlies integrative cognitivism. In particular, consider three of the requirements for the existence of a cognitive system: (a) perception, (b) manipulation of information, and (c) intentionality. It is clear that, in most cases, entities such as cultures, firms, and classes perceive and process information; moreover, they do so in ways that serve as self-definitional and, by including previous output in ongoing information process, are auto-catalytic. Accordingly, in most cases such institutions clearly meet the first two criteria.

Despite the language of subjectivity often used, however, there is little evidence at this point that firms, nations, or institutions exhibit either "will" or intentionality as a characteristic endogenous to themselves as potential subjects, rather than expressing, in some form or other, the intentionality of the individuals, past and present, who have constituted them. In other words, so far as we can ascertain, human individuals alone have will and ultimate intentionality; institutions exhibit derivative intentionality.[20] We thus come to a model where, regardless of how complex a cognitive system is, the will that energizes it derives, directly or indirectly, in ways not always clearly understood, from individual humans. Thus, the *ultimate source* of will in a cognitive system is the human individual(s), but the *expression* of will in a complex system context may be—in fact, almost always is—something very different. The grounding is free will; the expression is intentionality.

With intentionality, then, as with many other characteristics of complex systems, we see the well-known phenomenon of emergent behavior. In this case, just as it would be impossible to predict the actions of a human individual on February 4, 2009, from knowledge of the atoms that constitute that individual, it is impossible to predict the dynamics of the anthropogenic carbon cycle from knowledge of the individual wills that provide the intentionality behind that cognitive system. However, the existence of this difficult-to-predict behavior at higher levels of the system does not mean that the system is not composed of more basic elements: The body is indeed composed of atoms, and the intentionality of the carbon cycle cognitive system is indeed composed of different individual wills. However, to follow this analogy, the disciplines and intellectual approaches required to understand a complex system must be targeted at the particular level of the hierarchy of concern: A chemist studies atoms and molecules, a cellular biologist studies how these elements operate at a cellular level (including the dynamics emergent at the cellular level), a doctor or physiologist studies the system at the level of the organism, a psychologist studies the behavior that accompanies the organization of a

---

[20]This formulation should not be read to imply that individual will is fixed or clear—a sort of "rational man" of the will. Indeed, there is much evidence that human intentionality is not only highly subjective but also changes depending on external circumstances—mobs, for example, will do things that the individuals, not under their influence, would not do individually.

human organism and its expression of mind, and a sociologist studies the emergent properties of complex groups of humans. There certainly are interrelationships among the levels: A psychologist, for example, may be very interested in the behavioral effects of introducing certain molecules into human organisms. However, at each level different emergent characteristics require different disciplinary approaches.[21]

This is the core dilemma of the anthropogenic world: As a species, we are now operating at levels of the system where we have yet to develop the understanding—indeed, even the language—that allows us to comprehend and manage the expression of intentionality at the level of highly complex and broadly scaled cognitive systems. Intentionality as expressed in the human carbon cycle is very different from summed individual wills—indeed, we do not yet know how to address that issue. Rather, the reactions seem to be (a) denial (there are no perturbations); (b) retreat (if we just do a little less, all will be well); (c) advocacy of unrealizable utopias (just assume a world with only a few hundred million people, and all will be "natural" and "wild"); or, most problematically, (d) creating ideologies that define categories and belief structures such that others can be identified as evil, blamed, then attacked. It is now apparent, however, that these options must be rejected, for two reasons: (a) They deny the ability to ethically and rationally design where the species through institutions and as a whole is already engineering and, perhaps most unacceptable, (b) they facilitate the evasion of moral responsibility. For if we allow ourselves the illegitimate claim that there is no individual intentionality expressed in these large and complex cognitive systems, we also implicitly reject moral responsibility for the outcomes. This is both profoundly inauthentic and moral cowardice.

---

[21]In fact, each level may require different ontologies. Many physical scientists, for example, assume without question the existence of an external physical world, which they attempt to characterize and understand; at the other extreme, many social scientists assume the world is significantly structured by humans and human intentionality, the extreme being those who hold that the world is entirely constructed by human intellect. The need for different ontologies for working with different parts of the system is quite clear: One would not want, for example, a literary critic who was convinced that the world was as he decided it should be to design an airplane; conversely, most engineers would not be good at deconstructing a contemporary novel to understand its relevancy in building new social constituencies. Extending the postmodernist critique to conclude that there are not only no absolute, foundational discourses but also no absolute, foundational ontologies is not only appropriate but also necessary at this stage of the intellectual development of the species. Reality is always too complex for capture in our languages, our philosophies, and our ontologies—sometimes light is a particle, and sometimes it's a wave; sometimes reality is physically absolute, and sometimes it is constructed. The challenge, particularly highlighted by ESEM, is to evolve philosophies, ontologies, and understanding that enable continued evolution in the context of an anthropogenic world too complex for existing approaches.

Following the analogy of atoms and the human body, it is not that we need to stop trying to understand these cognitive systems at the level of the individual as we are now trying to do. Rather, it is that *in addition* we need to develop new approaches, new ways of thinking, that define these systems differently and are scaled to the appropriate levels of the hierarchy and the concomitant emergent behavior. Thus, we should continue the study of, for example, various genetic engineering technologies, and their impacts—but we should also learn how to study the anthropogenic carbon cycle, in all its human and physical complexity, which will require different tools and different approaches. Moreover, in doing so we cannot only look at the science, the technology, or the existing policies, most of which are appropriate for their scale but almost by definition fail at the level of the carbon cycle itself, or just rely on heavily ideological discourses, for they also fail at this scale. We also need to understand how intentionality, design, and moral responsibility function on such a scale: How, for example, should theological systems that have heretofore focused on the individual evolve to provide guidance at the level of the carbon cycle?

Perhaps an example will help illustrate this dynamic. Consider a European shipbuilder who is building a unique design of sailing ship, one that will enable Europeans to sail around the world; colonize much of it; and, at the end of a long process, create a globalized culture derived from the Enlightenment. His will is expressed directly in his design, the ship, which is part of one relatively limited cognitive system (the ship design, relevant ship manufacturing and sailmaking capabilities, known characteristics of the oceanic and weather systems, navigation capabilities, sailors' skill sets, etc.; Law, 1997). He directly intends the ship to sail open oceans and to do so as well as his design capabilities and the state of technology allow. Moreover, should the ship sink in calm weather, we would feel justified in criticizing his design, and possibly in moral reproach as well, should the design have been obviously and negligently flawed. In other words, at this scale and within this cognitive system the shipbuilder's intentionality, based on his will, is expressed simply and directly and implies not only free choice but also moral responsibility.

With our centuries of hindsight, however, we can see that this particular technology carried within it implications unknowable to our ship designer. To take only one, once the European nations, in competition with each other, began their period of global exploration and migration, virtually every island biology was changed. Species such as the dodo were exterminated; rats were transmitted to other islands; European landscapes traveled with new settlers. In this instance, there is obviously a gap between the will of the shipbuilder—to build an oceangoing vessel—and the results of the implementation of that technology. Even if the technology worked as planned, some might criticize it on the grounds of the unpredicted effects it generated, but most would hesitate to hold the ship-builder morally responsible for the results, for at least two reasons. First, there are intervening wills, contingencies, and the like that greatly complicate the ability of the shipbuilder to perceive, or predict, these more general effects (e.g., the sailors who choose to kill the last dodos). Second, it is unrealistic to impute to the shipbuilder knowledge, and thus the ability to will, the scale and complexity of the cognitive system that encompasses the expansion of European culture and presence around the world.

At this largest stage of cognitive systems, then, which is characteristic of ESEM, does the model of human intentionality, arising from will, combined with technological and cultural systems that vastly extend the cognitive activity of the individual mind, break down? The model, I think, does not; rather, the issue is that the complexity of the cognitive systems, at heart energized by the intentionality of individual humans, has become so great that the emergent behaviors at higher levels appear completely disconnected to the individuals constituting the system. Put another way, the dynamics of these complex systems are such that it appears that single individuals are unable to display any intentionality at all within them. The collective intentionality expressed in the functioning of the technological systems and globalized cultural context as a whole is of a different type, and certainly of a different scale, than the individual intentionality on which it is based. If there is no linkage between these scales—if intentionality at the level of large systems arises somehow from their inherent subjectivity, rather than as an integrated (and often systemically reinterpreted) expression of individual intentionality—it would negate much effort to "improve" the world (always depending on how one interprets that term). What is much more probable, however, is that we are confusing very different levels of intentionality, that of the individual versus the individual as a component of a large, complex cognitive system. In some senses, this is analogous to mistaking the design activities involved in creating a switch for the Internet with the "design" of the Internet itself: The emergent characteristics of the latter mean that effects of router design are not additive but contribute in ways we don't yet know how to think about to the evolution of an autocatalytic system whose design is emergent only at the level of the system taken as a whole.

It is not just complexity, however, that complicates the understanding of intentionality in ESEM systems, but also perception. Focused perception is not just common; it is a *raison d'être* of cognitive systems. Moreover, control of perception is a fundamental mechanism for the exercise of human power and authoritarianism, which is a common barrier to the exercise of free will at the individual level, which requires at least some knowledge of options that are actually available. Concomitantly, if the option space is significantly limited before I become aware of it, my exercise of free will as well as my freedom to participate in a nondominated discourse (Habermas, 1975) have been denied. One could argue that such a situation is created, for example, when the scientists involved in global climate change implicitly remove sets of technologies, such as geoengineering possibilities, from the negotiating arena (Keith, 2000), thus presenting the broader public, which will not be able to perceive such options on its own, with an already-attenuated option space.

This reconfirms the integrative cognitivist concept of intentionality as not an inherent characteristic of an individual but as a function of systems state, with the systems including elements both internal and external to the organism. Concomitantly, the responsibility for rational and ethical design becomes more complex as well, for it also is a property not of the individual but the cognitive system(s) within which that individual functions.

This interplay between systems constraints and intentionality is a complex one. Determinism in some cases need not preclude options, and there is always the question of how proximate to the act in question intentionality must have been exercised to make that act one of free will. An example frequently discussed is the case of Luther who, when he broke with the Church in Rome, famously proclaimed "Here I stand. I can do no other." Assume that he was simply being factual, and that he, in fact, at that point could "do no other." His act could still be one of intentionality, and thus moral accountability, if he were responsible for making himself the person that he was and that, as a consequence of the choices that went into personal creation, at that time and place had to act as he did. In discussing this case, Kane (1998) comments:

> Those who know something about Luther's biography know about the long period of inner turmoil and struggle he endured in the years leading up to that fateful "Here I stand." By numerous difficult choices and actions during that period, Luther was gradually building and shaping the character and motives that issues in his act. If we have no hesitation in saying that he was responsible for the final affirmation, I think it is because we believe that he was responsible through many past choices and actions for making himself into the kind of man he then was. And, if this is so, the question of whether Luther could have done otherwise shifts backwards from the present act to the earlier choices and actions by which he formed his character and motives. If he is ultimately accountable for his present act, then at least some of these earlier choices or actions must have been such that he could have done otherwise with respect to them. If this were not the case, then what he was would have never truly

been "up to him" because *nothing he could have ever done would have made any difference to what he was*. (pp. 39–40)[22]

The importance of this example is that it emphasizes in another dimension that intentionality is based on free will, but the latter is inchoate until expressed in a particular context, which includes previous decisions that have altered systems states in ways that are now immutable. The particular fact situation that exists may be entirely deterministic—"I can do no other"—but the choices that created the evolutionary path to that fact situation may be such that one is justified in allocating moral responsibility in the event. Again, intentionality properly understood arises from the cognitive system(s) within which the individual functions and not from any capacity internal to the individual alone.

The implication of this discussion for ESEM is clear. We now know that we have created an anthropogenic world and that knowledge creates an overarching moral responsibility to exercise intentionality—design and management—in ways that are ethical and moral. However, the more specific knowledge that would enable moral choices in particular areas, such as designing the Everglades and the Baltic, or managing the global carbon cycle and climate system, is lacking. That, combined with refuge in perceptual systems (e.g., ideologies) that filter out knowledge of the anthropogenic world, are at this point enabling denial of moral responsibility. But this is a temporary and increasingly unethical state for, like Luther, humanity cannot deny the moral obligations that arise from choices made in the past, even where they were not made with full knowledge of the future. We are, as Luther, responsible for creating ourselves as the dominant species of the Earth, and although we may wish, like Luther, to proclaim "here we stand— we can do no other," that does not evade moral responsibility, for we—our wills as individuals, expressed as intentionality collectively operating through increasingly complex cognitive systems—have created this current situation through choice. Rather, it reinforces it. Thus, the study of ESEM is not discretionary but, at this point in human history, is a moral imperative.

---

[22]The question of "congealed intentionality" is an interesting one. This arises quite simply as choices that initially are allowed by the system over time become institutionalized and therefore no longer subject to change. Institutions without question are an important filter between the ground of free will and its actual expression as intentionality, for they significantly define (and often reduce) option space. However, institutions arise from previous exercises of intentionality, which become in their turn constraints on the future exercise of intentionality. See Berger and Lockmann (1967) for a wonderful discussion from a sociological viewpoint of the evolution of institutions from human freedom. The whole question of how we can understand this systemic dimension of intentionality, and free will, and its interplay across time, remains to be addressed.

Integrative cognitivism thus implies that cognitive systems both extend and restrict intentionality, which, like any system parameter, scales depending on system structure. The system state within which individuals are embedded determines the existence of intentionality in any particular context; it is best viewed as an emergent characteristic of the particular state and dynamics of complex systems reflecting the transitory conditions of the particular system. There is, for example, no question that the existence of complex cognitive systems allows individuals—a Hitler, a Mao, a Stalin—to affect events in ways that are profoundly contingent (because dependent on their exercise of intentionality), but also have far more impact as a result of the scale of the cognitive systems within which they operated. A Hitler without Germany, or a Mao without China, might have still had intentionality, but its exercise would have had far less impact on other systems than was in fact the case.

The cognitive systems of which an individual is a part also bound intentionality in quite obvious ways, especially in a world of large institutional powers—nation–states, transnational corporations, nongovernmental organizations. Moreover, it is more subtly bounded in complex systems: Models have demonstrated that where there is a moderate level of connectivity among system components, favorable outcomes can be obtained by choice among options, but where there is a high degree of connectivity it is much harder to identify favorable outcomes and, in fact, they may not exist (Kauffman, 1993). This, of course, is static complexity, but an analogous principle appears to hold with dynamic complexity: Over time, many complex systems, including social systems, can be modeled as cycling through phases, some of which are very stable, making the exercise of intentionality difficult if not impossible, and some of which are chaotic and fluid, and the exercise of intentionality can make discontinuous differences.[23] Thus, periods of chaos can facilitate revolutions in governments, and chance and contingency may throw up a Hitler, Lenin, or Mao who significantly change the course of histories. In other words, intentionality is in large part a system function, and will vary over temporal and spatial scales as other systems parameters do: It is not inherent in the individual subjectivity, but an emergent characteristic of the systems

---

[23]Such models are quite rough, of course, and are perhaps best thought of as schematics at this point. Westly (1995) offered a social cycle of change that includes (a) creation of a new social order; (b) encoding and institutionalizing of the new order into organizational structure; (c) increasing rigidity of the social order, leading to short-term efficiency but-long term loss of adaptability; thus paving the way for (d) revitalization and reorganization, a chaotic period of innovation, learning, and experimentation; leading to (e) reintegration leading to the generation of new myths and creation of a new social order. Expression of intentionality is least effective in (b) and especially (c), and is most effective in (d) and, to some extent, in (e).

themselves.[24] Thus, although ultimate intentionality derives from individual human will, its potential expression, and the form of that expression, are systems dependent. By implication, then, one must understand the nature of the complex systems within which one is embedded before one can act with rationality and ethical responsibility to help evolve them.

The anthropocentric Earth, and the integrative cognitivism analysis, make it clear that authenticity—the continuing difficult effort to perceive with integrity and vision—and to act—to design—on the basis of that knowledge—is a moral mandate. Following the existentialist formulation (and, for that matter, going back to Socrates's injunction to "know thyself"), an authenticity necessary for our times will require as a first element a recognition and acceptance of the world as it is, not as various ideologies would wish it to be. With knowledge of the anthropogenic Earth comes an existential crisis, as the honest perception demanded by authenticity reveals a chaotic, unpredictable, highly problematic planet in the throes of anthropogenic change, with a complexity that neither existing intellectual tools nor language itself is adequate to address. Each individual is profoundly ignorant and strives hard to remain ignorant even of his or her ignorance; naiveté and willful perceptual and intellectual blindness become comfortable characteristics of discourse. The result is a fleeing into ideology, random myths, and stories, the creation of mental models that simplify reality into manageable fantasy and reduce perception until it no longer threatens. This is understandable, but it is cowardice; it is bad faith; it is profoundly inauthentic. It is a flight from freedom, from responsibility, from integrity. As Sartre said in the context of the individual, "Man is condemned to be free," and this is a far more daunting challenge in the context of an anthropogenic world that, having created, we now want to pretend not to see. For now this freedom, from whence rises moral obligation, is neither comfortable, nor, sometimes, even bearable—but it is the freedom demanded by the historical moment, and it is nondelegable.

He, only, merits freedom and existence

Who wins them every day anew.

—Goethe (1833/1984, *Faust*, lines 11, 575–76)

---

[24]For example, complex systems such as technology "can undergo a sequence of evolutionary instabilities at the micro level of economic agents while still maintaining structure and evolutionary stability at the macro levels of industrial sectors and the overall economy" (Grubler, 1998, p. 114). Other human systems, such as urban centers (Dear, 2000; Jacobs, 1961/1992), and economies (Arthur et al., 1997), are also becoming much more complex over time, and thence displaying similar characteristics. Thus, one could have choice, and intentionality, at a subsystem level but not at a systems level, at a particular point in system evolution: a locally contingent but globally deterministic state (the opposite, a globally contingent state with local determinism, is also possible).

ACKNOWLEDGMENTS

The views expressed herein are mine and not necessarily those of any institution with which I am associated. Much of this chapter is drawn from my Batten Institute Working Paper, "Observations on the Philosophic Implications of Earth Systems Engineering and Management" (Allenby, 2002), and is used by permission.

REFERENCES

Allenby, B. R. (1999). *Industrial Ecology: Policy framework and implementation.* Upper Saddle River, NJ: Prentice Hall.

Allenby, B. R. (2000–2001). Earth systems engineering and management. *Technology and Society, 19,* 10–24.

Allenby, B. R. (2002). *Observations on the philosophic implications of earth systems engineering and management.* Batten Institute Working Paper, Darden Graduate School of Business, University of Virginia.

Alroy, J. (2001). A multispecies overkill simulation of the end-Pleistocene megafaunal mass extinction. *Science, 292,* 1893–1896.

Arthur, W. B., Durlauf, S. N., & Lane, D. A. (1997). *The economy as an evolving complex system II.* Reading, MA: Addison-Wesley.

Barabasi, A. (2002). *Linked: The new science of networks.* Cambridge, MA: Perseus.

Berger, P. L., & Lockmann, T. (1967). *The social construction of reality.* New York: Anchor.

Berkes, F., & Folke, C. (Eds.). (1998). *Linking social and ecological systems: Management practices and social mechanisms for building resilience.* Cambridge, England: Cambridge University Press.

Bijker, W. E., Hughes, T. P., & Pinch, T. (Eds.). (1997). *The social construction of technological systems.* Cambridge, MA: MIT Press.

Black's law dictionary. (1990). St. Paul, MN: West Corporation.

Burge, T. (1998). Individualism and the mental. In P. Ludlow & N. Martin (Eds.), *Externalism and self-knowledge* (pp. 21–84)). Stanford, CA: Center for the Study of Language and Information.

Callon, M. (1997). Society in the making: The study of technology as a tool for sociological analysis. In W. E. Bijker, T. P. Hughes, & T. Pinch (Eds.), *The social construction of technological systems* (pp. 83–106). Cambridge, MA: MIT Press.

Castells, M. (2000). *The Rise of the Network Society* (2nd ed.). Oxford, England: Blackwell.

Clark, A. (2001). Natural born cyborgs. Retrieved January 10, 2001, from http://www.edge.org/3rd_culture/clark/clark_index.html

Cohen, J. E. (1995). *How many people can the earth support?* New York: Norton.

Dear, M. J. (2000). *The postmodern urban condition.* Oxford, England: Blackwell.

Descartes, R. (1998). *Discourse on method and meditations on first philosophy* (4th ed., D. A. Cress, Trans.). Indianapolis, IN: Hackett. (Original work published 1637–1641)

Diamond, J. (1997). *Guns, germs and steel.* New York: Norton.

Elliott, S. K., Lackner, K. S., Ziock, H. J., Dubey, M. K., Hanson, H. P., & Barr, S. (2001). Compensation of atmospheric CO_2 buildup through engineered chemical sinkage. *Geophysical Research Letters, 28,* 1235–1238.

Feshbach, M. (1998). Former Soviet Union's environmental and health problems. In B. R. Allenby, T. J. Gilmartin, & R. F. Lehman, III (Eds.), *Environmental threats and national security: An international challenge to science and technology* (pp. 173–188). Livermore, CA: Lawrence Livermore National Laboratory.

Gallagher, R., & Carpenter, B. (1997). Human-dominated ecosystems: Introduction. *Science, 277,* 485.

Genome prospecting. (1999). *Science, 286,* 443–491.

Giddens, A. (1990). *The consequences of modernity.* Stanford, CA: Stanford University Press.

Goethe, J. W. von. (1984). *Faust parts I and II.* (S. Atkins, Trans.). Princeton, NJ: Princeton University Press. (Original work published 1833)

Gorman, M. (1997). Mind in the world: Cognition and practice in the invention of the telephone. *Social Studies of Science, 27,* 583–624.

Graedel, T. E., & Allenby, B. R. (1995). *Industrial ecology.* Upper Saddle River, NJ: Prentice Hall.

Grubler, A. (1998). *Technology and global change.* Cambridge, England: Cambridge University Press.

Gunderson, L. H., Holling, C. S., & Light, S. S. (Eds.). (1995). *Barriers & bridges to the renewal of ecosystems and institutions.* New York: Columbia University Press.

Habermas, J. (1975). *Legitimation crisis* (T. McCarthy, Trans.). Boston: Beacon.

Harnad, S. (2001). No easy way out. *The Sciences, 41*(2), 36–42.

Harrison, L. E., & Huntington, S. P. (Eds.). (2000). *Culture matters: How values shape human progress.* New York: Basic Books.

Hofstadter, D. R., & Dennett, D. C. (Eds.). (1988). *The mind's I.* New York: Bantam Books.

Hong, S., Candelone, J., Patterson, C. C., &. Boutron, C. F. (1996). History of ancient copper smelting pollution during Roman and medieval times recorded in Greenland ice. *Science, 272,* 246–249.

Hughes, T. P. (1998). *Rescuing Prometheus.* New York: Pantheon.

Human-dominated ecosystems. (1997). *Science, 277,* 485–525.

Jablonski, D. (1991). Extinctions: A paleontological perspective. *Science, 253,* 754–757.

Jacobs, J. (1992). *The death and life of great American cities.* New York: Vintage Books. (Original work published 1961)

Jager, J., & Barry, R. G. (1990). Climate. In B. L. Turner, W. C. Clark, R. W. Kates, J. F. Richards, J. T. Mathews, & W. B. Meyer (Eds.), *The earth as transformed by human action* (pp. 000–000). Cambridge, England: Cambridge University Press.

Kaiser, J. (1999). Stemming the tide of invading species. *Science, 285,* 1836–1841.

Kaiser, J. (2001). NRC panel pokes holes in Everglades scheme. *Science, 291,* 959–961.

Kane, R. H. (1998). *The significance of free will.* Oxford, England: Oxford University Press.

Kauffman, S. A. (1993). *Origins of order: Self-organization and selection in evolution.* Oxford, England: Oxford University Press.

Keith, D. W. (2000). Geoengineering the climate: History and prospect. *Annual Review of Energy and the Environment, 25,* 245–284.

Kitano, N. (2002). Computational systems biology. *Nature, 420,* 206–210.

Kloor, K. (2000). Everglades restoration plan hits rough waters. *Science, 288,* 1166–1167.

Landes, D. S. (1998). *The wealth and poverty of nations.* New York: Norton.

Law, J. (1997). Technology and heterogeneous engineering: The case of Portuguese expansion. In W. E. Bijker, T. P. Hughes, & T. Pinch (Eds.), *The social construction of technological systems* (pp. 111–134). Cambridge, MA: MIT Press.

Light, S. S., Gunderson, L. H., & Holling, C. S. (1995). The Everglades: Evolution of management in a turbulent ecosystem. In L. H. Gunderson, C. S. Holling, & S. S. Light (Eds.), *Barriers & bridges to the renewal of ecosystems and institutions* (pp. 103–168). New York: Columbia University Press.

Ludlow, P., & Martin, N. (Eds.). (1998). *Externalism and self-knowledge.* Stanford, CA: CSLI.

McNeill, J. R. (2000). *Something new under the sun.* New York: Norton.

Noble, D. F. (1998). *The religion of technology.* New York: Knopf.

Oberg, J. (1999). Missionaries to Mars. *Technology Review, 102,* 54–59

Palmer, D. (1996). *Does the center hold?* (2nd ed.). London: Mayfield.

Palumbi, S. R. (2001). Humans as the world's greatest evolutionary force. *Science, 293,* 1786–1790.

The plant revolution. (1999). *Science, 285,* 367–389.

Pool, R. (1997). *Beyond engineering: How society shapes technology.* Oxford, England: Oxford University Press.

Postel, S. (1996). Forging a sustainable water strategy. In L. Starke (Ed.), *State of the world 1996* (pp. 40–59). New York: Norton.

Redman, C. L. (1999). *Human impact on ancient environments.* Tucson: University of Arizona Press.

Renberg, I., Persson, M. W., & Emteryd, O. (1994). Pre-industrial atmospheric lead contamination in Swedish lake sediments. *Nature, 363,* 323–326.

Roberts, R. G., Flannery, T. E., Ayliffe, L. K., Yoshida, H., Olley, J. M., Prideaux, G. J., et al. (2001). New ages for the last Australian megafauna: Continent-wide extinction about 46,000 years ago. *Science, 292,* 1888–1892.

Rorty, R. (1989). *Contingency, irony, and solidarity.* Cambridge, England: Cambridge University Press.

Rowlands, M. (1999). *The body in mind.* Cambridge, England: Cambridge University Press.

Russell, B. (1972). *A history of western philosophy.* New York: Simon & Schuster. (Original work published 1945)

Sagoff, M. (1996). *The economy of the earth.* Cambridge, England: Cambridge University Press. (Original work published 1988)

Schrope, M. (2001). Save our swamp. *Nature, 409,* 128–130.

Scott, J. C. (1998). *Seeing like a state: How certain schemes to improve the human condition have failed.* New Haven, CT: Yale University Press.

Senge, P. M. (1990). *The fifth discipline.* New York: Doubleday.

Shapiro, J. (2001). *Mao's war against nature: Politics and the environment in revolutionary China.* Cambridge, England: Cambridge University Press.

Socolow, R. C. (Ed.). (1997). *Fuel decarbonization and carbon sequestration: Report of a workshop* (Princeton University Center for Energy and Environmental Studies Report No. 302). Princeton, NJ: Princeton University Press.

Socolow, R., Andrews, C., Berkhout, F., & Thomas, V. (1994). *Industrial ecology and global change.* Cambridge, England: Cambridge University Press.

Suess, E., Bohrmann, G., Greinert, J., & Lausch, E. (1999). Flammable ice. *Scientific American, 281,* 76–83.

Thompson, E. P. (1978). *The poverty of theory.* London: Merlin.

Turner, B. L., Clark, W. C., Kates, R. W., Richards, J. F., Mathews, J. T., & Meyer, W. B. (Eds.). (1990). *The earth as transformed by human action.* Cambridge, England: Cambridge University Press.

U.S. Department of Energy. (1999). *Carbon sequestration: Research and development.* Washington, DC: Author.

Vitousek, P. M., Mooney, H. A., Lubchenco, J., & Melillo, J. M. (1997). Human domination of Earth's ecosystems. *Science, 277,* 494–499.

Weber, M. (1998). *The protestant ethic and the spirit of capitalism* (T. Parsons, Trans.). Los Angeles: Roxbury. (Original work published 1904)

Westly, F. (1995). Governing design: The management of social systems. In L. H. Gunderson, C. S. Holling, & S. S. Light (Eds.), *Barriers & Bridges to the renewal of ecosystems and institutions* (pp. 391–427). New York: Columbia University Press.

Zimmer, C. (2001). "Inconceivable" bugs eat methane on the ocean floor. *Science, 293,* 418–19.

15

The Future of Cognitive Studies of Science and Technology

Michael E. Gorman
University of Virginia

Ryan D. Tweney
Bowling Green State University

David C. Gooding
University of Bath

Alexandra P. Kincannon
University of Virginia

This volume illustrates the advantages of using a variety of methods to study the processes involved in scientific and technological thinking. A provocative essay by Richard Feynman (Feynman & Leighton, 1985) described the dangers of "cargo cult" science, making an analogy to the South Sea islanders who—in hopes of bringing the departed World War II cargo planes back again—put a man in a hut with bamboo antennae on his head. Psychologists who assert that the only scientific way to study scientific thinking is by using experimental or computational techniques adapted from "harder" sciences are acting like cargo cult scientists. Sociologists who claim that only thick descriptions can be used to understand the processes involved in discovery and invention are making a similar error.

Experimental (*in vitro*) and computational (*in silico*) techniques certainly have an important contribution to make, as Dunbar and Fugelsang show in chapter 3. This volume has also demonstrated the value of fine-grained studies of scientific and technological problem solving—both historical (*sub specie historiae*) and contemporary (*in vivo*). Such case studies need to be complemented by experimental studies that attempt controlled comparisons of variables identified in those fine-grained studies and by computational simulations that model both the kind of thinking and the social interactions that might underlie the observed processes. The point is to achieve a kind of triangulation via several different methods that eventually support a cognitive theory of scientific innovation. Simulations can also be used to explore "what ifs"—whether different possible paths could have been taken to discoveries and inventions, as Gooding (1990) illustrated in his work on Faraday. Simulation methods also allow one to evaluate the variation in this set of possible paths, which may in turn suggest other parameters for further study by *in vitro* and *in vivo* methods (Gooding & Addis, 2003).

Different methods often develop in different contexts having distinctive assumptions, instrumentation, and discourses. Science studies scholars have produced many examples of circumstances in which different assumptions, practices, and discourses call for the development of new means of communicating between distinct cultures (Galison, 1997; Star & Griesemer, 1989). For example, the contrasting beliefs of Faraday and Ampère about the role of a priori assumptions produced very different approaches to the design and meaning of their experiments as well as different languages of description (Gooding, 1990; Steinle, 1997, 2003). In some cases, the divergence can become so great that communication between research traditions is compromised. The eventual synthesis of two traditions in high-energy physics—the image (bubble chamber) and logic (statistical) traditions—involved the development of composite languages, pidgins and creoles, to facilitate communication between research groups (Galison, 1997). This shows that multiple methods alone are not sufficient for triangulation. Results need to be shared and compared across different discourse communities. For cognitive studies, as Gorman (chap. 13) noted, this will require a creole, or reduced common language, and a network that encourages sharing.

In this volume we have tried to take the first step toward establishing such a creole for cognitive studies of science and technology by proposing a comparative framework, partly using the idea of problem spaces as outlined in chapter 4. This problem space framework is intended as a heuristic that researchers can use to try to compare the significance of their findings. If successful, this heuristic will encourage critical discussion of assumptions as well as results. Through such criticism it will, we hope, be transcended or incorporated into a superior framework.

The point of a problem space analysis is based on the idea that problem-solving heuristics require some "pruning" in order to reduce the size of the problem search tree. In the present context, did the success of the Wright brothers and the Rocket Boys (as described in chap. 11) depend on setting up a set of decomposable problem spaces? Is this decomposition likely to be especially useful in engineering tasks, where it is important to get the parts as independent of each other as possible? In *Sciences of the Artificial,* Simon (1981) described two clockmakers, Hora and Tempus, who became so successful that their work was interrupted by frequent phone calls. Hora assembled piecemeal clocks that had 1,000 parts. Whenever he set down a clock to answer the phone, it fell back into its pieces, and he had to reassemble it from scratch. Tempus, however, built subassemblies of 10 elements at a time. When he put down a clock to answer the phone he lost only the work on the current subassembly; he did not have to return to the very beginning. Tempus, like the Rocket Boys, had taken his problem and decomposed it into subcomponents. Simon argued that most complex systems are capable of at least partial hierarchical decomposition. However, as Allenby shows in chapter 14, in complex, closely coupled systems, modifying a part may cause unpredictable perturbations across the whole system (Perrow, 1984). When it comes to designing and managing complex technological systems, decomposition has heuristic value, but the whole is often greater than the sum of its parts.

Although problem space analysis is clearly useful in engineering tasks, is it helpful in the case of scientific thinking? We wish to open the question of whether the utility of heuristics such as problem decomposition is one of the areas in which invention and discovery actually differ from each other. Kurz and Tweney (1998) and Tweney (2001) have argued this point explicitly: that understanding the use of cognitive and epistemic artifacts within a model of scientific cognition is incompatible with a problem space analysis. In chapter 9, Gooding challenges the notion of decomposability presupposed by problem space analysis, and Nersessian also shows, in chapter 2, that there is something cognitively "special" about at least some aspects of scientific thinking.

Another goal of this volume is to present cutting-edge work on scientific and technological thinking in a way that might be useful to practicing scientists, inventors, and entrepreneurs. Gooding (chap. 9) suggests some visualization strategies that are widely used to construct and communicate interpretations and models in a range of scientific fields. Thagard (chap. 8) provides explicit but general heuristics for becoming a successful scientist, derived in part from the workshop on which this volume is based. Regarding invention, Hughes (chap. 12) and Allenby (chap. 14) emphasize the importance of considering the whole technological system, and Bradshaw (chap. 11) emphasizes the importance of decomposing problems.

Although this volume is a promising start, it is clear that we need more studies that combine case analysis with experiment and computational simulation, using a framework that facilitates comparison among different studies. The point is to keep the framework creative—a creole that is adaptive, not a rigid blueprint. Specifically, we recommend the following:

1. More fine-grained studies of the evolution of technological systems and scientific theories that focus on the role of different cognitive agents, including cognitive and epistemic artifacts, as well as multiple individuals in a distributed cognition context. We need to ask of such agents:
- How do they view the overall system?
- How do they solve problems within it?
- How do they relate to other cognitive agents?

This kind of study would combine the kind of close look at individual cognition done by Trickett et al. (chap. 5), Bradshaw (chap. 11), Tweney et al. (chap. 7), Ippolito (chap. 10), and others in this volume with the kind of network analyses recommended by Gorman (chap. 9). Here the focus would be on the relationship between spaces used in individual and small-group problem solving with the dynamics of the overall system. Ideally, multi-agent computer simulations could be used to explore what happens when abstracted versions of individual cognizers interact in various ways. Variables could potentially be isolated for experimental investigation, as in the simulations done by Gooding and Addis (Gooding & Addis, 1999a, 1999b, 2003). It remains to be seen whether resulting models of cognition will be "scalable" to the kinds of systems discussed by Hughes (chap. 12), Allenby (chap. 14), and others. We need more case studies highlighting the role of cognitive processes in what Collins and Evans (2002) called *reflexive historical systems,* that is, those systems that evolve partly in response to human awareness and decision making. The evolution of the galaxies identified by Trickett et al.'s (chap. 5) astronomers is one example of a system with a very long history that has no reflexive component. Global climate change, on the other hand, has now become a system that is reflexive.

2. More studies of tacit knowledge in science and invention. Cognitive scientists tend to study reflective scientists and inventors who keep good notebooks. Sociologists such as Collins, in contrast, look carefully at aspects of knowledge that are hard—perhaps even impossible—to articulate (Collins, 1982; Collins & Kusch, 1998; but see Galison, 1997; Galison, 1999, for a rebuttal). There is an extensive literature on implicit learning and knowledge (Dienes & Perner, 1999; Reber, 1993; Sternberg & Horvath, 1999) that needs to be related to invention and discovery (Gorman, 2002; Patel, Arocha, & Kaufman, 1999). First steps have been

taken by Tweney and Gooding, whose studies of Faraday attempt to recover what is not articulated, even by this most articulate of scientists. One of the challenges facing scholars in science and technology studies is that they must not only master expertise in a social science but also learn a domain of scientific or technological expertise (Collins & Evans, 2002). An alternative is to train some students undergoing graduate training in science and engineering to be participant–observers of their own cognitive processes. Shrager's diary (chap. 6) suggests the beginnings of a method that might be used to accomplish this goal. Shrager and Gorman are working with a graduate student entering nanotechnology to see if she can provide data on the process of becoming a nanoscientist. Our hope is that this process of reflection will also help her efforts to become a better scientist (Loh et al., 2001).

3. Investigation of the cognitive processes involved in trying to make discoveries and create technologies that are explicitly directed at benefiting society. Allenby's earth systems engineering management (chap. 14) is an example of a technoscientific system explicitly designed to improve the global environment. Architects of the new National Nanotechnology Initiative advocate focusing one societal dimensions of nanotechnology, including ethical issues. One important theme, related to mental models, is moral imagination, which consists of the ability to "go out toward people to inhabit their worlds, not just by rational calculations, but also in imagination, feeling, and expression" (Johnson, 1993, p. 200). There have been a number of case studies of the role of moral imagination in the development of new technologies (Gorman, Mehalik, & Werhane, 2000), but these cases have not been fine-grained enough to qualify as true *in vivo* studies. Recent research on the importance of "hot" cognition, especially positive emotions that encourage ethical behavior, also need to be part of future studies of scientific and technological thinking (Haidt, in press).

4. A multi-institutional effort, similar to a National Science Foundation center, that would plan and execute a strategic research program. Such an effort would involve multiple methods that would be closely coordinated. For example, a researcher studying several nanotechnology laboratories could team with an experimental psychologist at another institution and a computer scientist at a third to carry out a systematic comparison that would focus on different parts of the problem space. Dunbar's research on molecular biology laboratories and Nersessian's research on biotechnology, both reported in this volume (chaps. 3 and 2, respectively), illustrate the potential of this approach, but what we recommend is that cognitive scientists doing this kind of work collaborate strategically with others at other institutions who have different perspectives and methodological expertise.

To achieve true scientific and technological progress, we need to understand the process by which our species is transforming our world. We hope this book makes an important step in that direction by leading to breakthroughs in our understanding of discovery and invention.

REFERENCES

Collins, H. M. (1982). Tacit knowledge and scientific networks. In B. Barnes & D. Edge (Eds.), *Science in context: Readings in the sociology of science* (pp. 44–64). Cambridge, MA: MIT Press.

Collins, H. M., & Evans, R. (2002). The third wave of science studies. *Social Studies of Science, 32,* 235–296.

Collins, H. M., & Kusch, M. (1998). *The shape of actions. What humans and machines can do.* Cambridge, MA: MIT Press.

Dienes, Z., & Perner, J. (1999). A theory of implicit and explicit knowledge. *Behavioral and Brain Sciences, 22*(5), 735–808.

Feynman, R., & Leighton, R. (1985). *Surely you're joking, Mr. Feynman.* New York: Norton.

Galison, P. (1997). *Image & logic: A material culture of microphysics.* Chicago: University of Chicago Press.

Galison, P. (1999). Author's response. *Metascience, 8,* 393–404.

Gooding, D. (1990). *Experiment and the making of meaning: Human agency in scientific observation and experiment.* Dordrecht, The Netherlands: Kluwer Academic.

Gooding, D. C., & Addis, T. (1999a). Learning as collective belief-revision: Simulating reasoning about disparate phenomena. In *Proceedings of the AISB Symposium on Scientific Creativity* (pp. 19–28). Sussex, England: Society for the Study of Artificial Intelligence and the Simulation of Behaviour.

Gooding, D. C., & Addis, T. (1999b). A simulation of model-based reasoning about disparate phenomena. In P. Thagard (Ed.), *Model-based reasoning in scientific discovery* (pp. 103–123). New York & London: Kluwer Academic/Plenum.

Gooding, D. C., & Addis, T. (2003). *An iterative model of experimental science.* Manuscript submitted for publication.

Gorman, M. E. (2002). Types of knowledge and their roles in technology transfer. *Journal of Technology Transfer, 27,* 219–231.

Gorman, M. E., Mehalik, M. M., & Werhane, P. H. (2000). *Ethical and environmental challenges to engineering.* Englewood Cliffs, NJ: Prentice Hall.

Haidt, J. (in press). The emotional dog and its rational tail: A social intuitionist approach to moral judgment. *Psychological Review.*

Johnson, M. (1993). *Moral imagination.* Chicago: University of Chicago Press.

Kurz, E. M., & Tweney, R. D. (1998). The practice of mathematics and science: From calculus to the clothesline problem. In M. Oaksford & N. Chater (Eds.), *Rational models of cognition* (pp. 415–438). Oxford, England: Oxford University Press.

Loh, B., Reiser, B. J., Radinsky, J., Edelson, D. C., Gomez, L. M., & Marshall, S. (2001). Developing reflective inquiry practices: A case study of software, the teacher, and students. In T. Okada (Ed.), *Designing for science: Implications for everyday, classroom, and professional settings* (pp. 279–323). Mahwah, NJ: Lawrence Erlbaum Associates.

Patel, V. L., Arocha, J. F., & Kaufman, D. R. (1999). Expertise and tacit knowledge in medicine. In J. A. Horvath (Ed.), *Tacit knowledge in professional practice: Researcher and practitioner perspectives* (pp. 75–99). Mahwah, NJ: Lawrence Erlbaum Associates.

Perrow, C. (1984). *Normal accidents.* New York: Basic Books.

Reber, A. S. (1993). *Implicit learning and tacit knowledge: an essay on the cognitive unconscious.* New York: Oxford University Press.

Simon, H. A. (1981). *The sciences of the artificial.* Cambridge, MA: MIT Press.

Star, S. L., & Griesemer, J. R. (1989). Institutional ecology, "translations" and boundary objects: Amateurs and professionals in Berkeley's museum of vertebrate zoology, 1907–39. *Social Studies of Science, 19,* 387–420.

Steinle, F. (1997). Entering new fields: Exploratory uses of experimentation, *Philosophy of Science (PSA Proceedings)* (pp. S65–S74).

Steinle, F. (2003). The practice of studying practice: Analyzing research records of Ampère and Faraday. In F. L. Holmes, J. Renn, & H.-J. Rheinberger (Eds.), *Reworking the bench: Research notebooks in the History of Science* (pp. 93–117). Dordrecht, The Netherlands: Kluwer.

Sternberg, R. J., & Horvath, J. A. (Eds.). (1999). *Tacit knowledge in professional practice: Researcher and practitioner perspectives.* Mahwah, NJ: Lawrence Erlbaum Associates.

Tweney, R. D. (2001). Scientific thinking: A cognitive–historical approach. In K. Crowley, C. D. Schunn, & T. Okada (Eds.), *Designing for science: Implications for everyday, classroom, and professional settings* (pp. 141–173). Mahwah, NJ: Lawrence Erlbaum Associates.

Author Index

Note: Page number followed by *n* indicates footnote.

A

Addis, T., 3, *14,* 176, *214,* 346, 348, *350*
Agnetta, B., 32, *52*
Alberdi, E., 99, 104, 116, *117*
Allenby, B. R., 297, *300,* 305, 308, *340, 341*
Allott, M., 230, 231, 232, 234, 235, 236, 237, 243, 250, *252*
Alroy, J., 307, *340*
Altemus, S., 189–190, *217*
Alvarez, W., 290, *300*
Amsel, E., 86, *92*
Andersen, C., 88, *93*
Andersen, H., 37, *52,* 146, *155*
Anderson, J. R., 60, *77,* 175, *212,* 288, *301–302*
Andrews, C., 305, *342*
Angelev, J., 85, *93*
Anzai, Y., *13,* 288n, *300*
Arnheim, R., 233, *252*
Arocha, J. F., 347, *351*
Arthur, W. B., 339n, *340*
Austin, G. A., 5, *13,* 81, 87, *92,* 267, *272*
Ayliffe, L. K., 307, *342*
Azmitia, M., 89, *92*

B

Babbage, C., 281, *285*
Bacon, F., 57, *75*
Baird, D., 291–292, *300*

Baker, L. M., 58, 60, 73, *75*
Barabasi, A., 315n, *340*
Barr, S., 313n, *340*
Barry, R. G., 306, *341*
Barsalou, L. W., 25, 30, *52,* 56
Bates, E. A., 33, *53*
Beaulieu, A., 173n, 174n, *212*
Bechtel, W., 160, *170*
Bell, A. O., 223, 224, 232, 241, 242, 244, 245, 246, 247, 248, 250, 252–253, *253*
Bell, Q., 244, *253*
Bereiter, C., 220, 221, 233, 242, *253,* 256
Berger, P. L., 337, *340*
Berkes, F., 308, *340*
Berkhout, F., 305, *342*
Bertalanffy, L. von, 278, *285*
Bijker, W. E., 2, *13,* 293, *300,* 322n, *340*
Birnbaum, L., 268, *272*
Bishop, E. L., 240, *253*
Blanchette, I., 57, 58, 76, *77*
Bloch, M., 280, *285*
Bohrmann, G., 312n, *342*
Bostock, J., 147, *155*
Bottone, S. R., 149, *155*
Boutron, C. F., 306, *341*
Bowker, G. C., 292, *300*
Bradshaw, G. L., 2, 6, *14, 15,* 20, *54, 55,* 59, 61, *77,* 90, 94, 263, 264, 265, 266, *272*
Brem, S., 152, *155*

Brewer, W. F., 86, 95, 98, 117
Briggs, J., 230, 235, 243, 245, 249–250, 253
Brock, S., 86, 92
Bruner, J. S., 5, 13, 81, 87, 92, 228, 253, 267, 272
Bryson, M., 221, 233, 253
Bucciarelli, L. L., 37, 52
Bullock, M., 86, 92
Burge, T., 319, 340
Burtis, P. J., 233, 253

C

Call, J., 31, 32, 55
Callebout, W., 57, 76
Callon, M., 316, 340
Candelone, J., 306, 341
Caporael, L., 12, 13
Carey, L. J., 242, 253
Carey, S., 82, 84, 92, 95
Carlson, W. B., 2, 8, 13, 14, 18, 37, 53, 175, 212n, 213, 232, 254, 269, 270, 273, 282n, 285, 293, 300
Carmichael, L., 238, 253
Carpenter, B., 305, 341
Carraher, D. W., 26, 52
Carraher, T. D., 26, 52
Cartwright, N., 177, 178n, 213
Carver, S. M., 82n, 93
Case, R., 85, 92
Castellan, J. J., 89, 92
Castells, M., 324, 324n, 325n, 340
Catrambone, R., 30, 53
Cavicchi, E., 146, 151–152, 155
Cetina, K. K., 24, 52
Chapman, J. P., 73, 76
Chapman, L. J., 73, 76
Chase, W., 220, 253
Chen, X., 37, 52, 140, 155
Chi, M. T. H., 220, 221, 242, 253, 287n, 288, 301
Chinn, C. A., 98, 117
Choi, I., 25, 33, 54
Christensen, C. M., 1, 13
Churchland, P. M., 2, 13
Clancey, W. J., 25, 53
Clark, A., 174, 213, 321, 340
Clark, W. C., 305, 306, 342
Clement, J., 13, 288, 301
Clemow, L., 91, 92
Cohen, J. F., 305, 340
Cohen, M., 291–292, 300

Collins, H., 290, 301
Collins, H. M., 13, 287n, 288, 289, 291, 296, 297, 301, 348, 349, 350
Colson, I., 61, 76
Cooper, L. A., 174, 213
Costanza, M. E., 91, 92
Craig, D. L., 30, 53
Craver, C. F., 160, 171
Csikszentmihalyi, M., 219, 225, 229, 230, 253, 254
Cupchik, G. C., 235, 236, 238, 253

D

Daitch, V., 14
Dama, M., 99, 117
Darden, L., 61, 62, 67, 76, 152n, 155, 160, 171
Darley, J. M., 169, 171
Davies, G. E., 21, 53
Davies, J., 153, 157
Davis, S. N., 175, 215
Davy, H., 198, 207, 213
de Chadarevian, S., 173n, 190, 213
Dear, M. J., 339n, 340
deGroot, A., 220, 253
Delneri, D., 61, 76
DeMonbreun, B. G., 82, 93, 98, 116, 118
Dennett, D. C., 317, 341
Descartes, R., 319, 340
Dewey, J., 74, 75, 76
Diamond, J., 322, 323, 324, 324n, 340
Dienes, Z., 347, 350
Doherty, M. E., 6, 15, 57, 59, 60, 78, 81, 86, 87, 88, 90, 93–94, 94, 98, 118, 151, 156, 157
Donald, M., 25, 32, 53
Drakes, S., 2, 13
Dubey, M. K., 313n, 340
Dunbar, K. N., 5, 6, 13, 14, 18, 37, 53, 57, 58, 59, 60, 61, 73, 75, 75, 76, 77, 81, 83, 88, 92, 93, 98, 99, 116, 117, 121, 125, 126, 127, 135, 151, 152, 155, 160, 162, 170, 179, 213
Duncker, K., 233, 253
Durlauf, S. N., 339n, 340

E

Eagle, D., 238, 253
Edelson, D. C., 349, 350

Elliott, S. K., 313n, *340*
Elman, J. L., 33, *53*
Emteryd, O., 306, *342*
Enserink, M., 91, *92*
Epstein, R., 235, *253*
Epstein, S., 289, 292, 294, *301*
Ericsson, K. A., 68, *77*, 100, *117*, 220, 221,
    229, 242, *253*, *254*
Evans, P., 1, *13*
Evans, R., *13*, 287n, 288, 289, 291, 296,
    297, *301*, 348, 349, *350*
Eysenck, M. W., 230, 234, 238, 239, *254*

F

Faivre, I. A., 229, *254*
Fallshore, M., 228, *256*
Faraday, M., 138–139, 140, 142, 143,
    146–147, 151, *155*, 196–197,
    199, 200, 207, 208, 209, *213*
Farr, M. J., 220, 221, *253*
Fay, A., 73, *77*, 88, *93*
Feist, G. J., 58, *77*, 161–162, 169, *170*
Feltovich, P. J., 287n, 288, *301*
Ferguson, E. S., *213*
Ferguson, R. W., 152, *155*
Feshbach, M., 310, *341*
Feynman, R. P., *13*, 231–232, *254*, 345, *350*
Fiore, S. M., 228, *256*
Fisher, H. J., 154, *155*
Fitz, H. C., 298, *302*
Flannery, T. E., 307, *342*
Flower, L. S., 220, 221, 242, *253*, *254*
Flowers, J. H., 227, 228, 229, *254*
Fodor, J., 195n, *213*
Folke, C., 308, *340*
Forbus, K. D., 152, *155*
Fu, W. T., 99, *118*, 177, *217*
Fugelsang, J. A., 58, 59, 61, 73, 75, *77*
Fujimura, J., 11, *13*, 177, *213*
Fuller, S., 89, *93*

G

Gabrieli, J., 61, *77*
Galilei, G., 57, *77*
Galison, P., 2, 5, 11, *13*, *14*, 24, *53*, 162,
    170, 173n, 174n, 176, 177, *213*,
    215, 291, *301*, 346, 347, *350*
Gallagher, R., 305, *341*
Garbin, C. P., 227, 228, 229, *254*
Garcia-Mila, M., 88, *93*

Gardner, M., 176, *214*
Geertz, C., 28, *53*
Gentner, D., 152, *155*, 177, *214*
Gerlach-Arbeiter, S., 189–190, *217*
Getzels, J., 219, 225, *254*
Giaquinto, M., 174n, *214*
Gibby, R. E., 141, 143n, *157*
Gibson, E. J., 229, *254*
Gibson, J. J., 27, *53*
Giddens, A., 319, 324n, *341*
Giere, R. N., 2, *14*, 18, *53*, 160, *170*, 173n,
    174n, 177, 178, 192, *214*
Gilhooly, K. J., *14*, 288, *301*
Glaser, R., 88, 94, 220, 221, 242, *253*,
    287n, 288, *301*
Gleich, J., 4, *14*
Glenberg, A. M., 25, 30, *53*
Glover, G., 61, *77*
Gobet, F., 220, *256*
Goel, V., 61, *77*
Goethe, J. W. von, 339, *341*
Gold, B., 61, *77*
Goldenson, R. M., 229, *254*
Goldstein, E. B., 229, *254*
Golinski, J., 138, *156*
Gomez, L. M., 349, *350*
Goodfield, J., 245, *254*
Gooding, D. C., 2, 3, 4, 8, 9, *14*, *15*, 18,
    37, 38, *53*, 138, 140, 141, 151,
    153, 154, 154n, *156*, *157*, 160,
    170, 173n, 175, 176, 177, 178,
    179, 180n, 181n, 182, 192,
    197n, 198, 198n, 199, 200,
    214, 227, 232, 233, 237, *254*,
    288, 293, *301*, 346, 348, *350*
Goodman, D. C., 176, *214*
Goodman, N., 178, *214*
Goodnow, J. J., 5, *13*, 81, 87, 92, 267, *272*
Goodwin, C., 37, *53*, 177, *214*
Gorman, M. E., 2, 3, 5, 6, 8, *13*, *14*, 18,
    37, 47, *53*, 58, *77*, 91, *93*, 98,
    117, 152, *156*, 161–162, 169,
    170, 175, 212n, *213*, *214*, 232,
    254, 268, 269, 270, *273*, 282n,
    285, 287, 292, 293, 297, 300,
    301, 322n, *341*, 347, 349, *350*
Gould, S. J., 183, 184, 185, 186, *215*
Graedel, T. E., 305, 308, *341*
Graham, L. R., 292, *301*
Grammenoudi, S., 61, *76*
Green, A. J. K., *14*, 288, *301*

Greeno, J. G., 25, 26, 27, 29, 34, 47, 53
Gregory, R. L., 174, 181n, 215
Greinert, J., 312n, 342
Griesemer, J. R., 11, 15, 177, 216, 346, 351
Groezinger, E., 186, 191, 217
Grove, W., 147, 156
Gruber, H. E., 2, 8, 14, 15, 154n, 156, 175,
    215, 240, 250–251, 254
Grubler, A., 306, 319, 325, 339n, 341
Gunderson, L. H., 308, 309, 341, 342

H

Ha, Y., 59, 60, 77, 87, 93
Habermas, J., 319, 336, 341
Hacking, I., 162, 170, 178, 215
Haidt, J., 14, 349, 350
Hall, R., 37, 53
Hanson, H. P., 313n, 340
Hanson, N. R., 174n, 181n, 215
Hare, B., 32, 52, 55
Harnad, S., 318, 341
Harrison, A., 151, 157
Harrison, L. E., 324, 325, 341
Haugeland, J., 18, 53
Hayes, J. R., 220, 221, 225, 242, 254
Heering, P., 139n, 156
Hekkert, P., 174, 179, 217
Henderson, K., 173n, 174n, 177, 215
Hertz, J., 116, 118
Hickam, H. H., Jr., 259, 260, 261, 263, 265,
    267, 268, 270–271, 273
Hoddeson, L., 14
Hoffman, J. R., 21, 53
Hoffner, C. E., 151, 157
Hofstadter, D. R., 317, 341
Hogan, H. P., 238, 253
Holding, C. S., 308, 341
Holling, C. S., 309, 342
Holmes, F. L., 2, 14, 137, 151, 154, 156
Holton, G., 230, 234, 235, 254
Holyoak, K. J., 160, 170
Hong, S., 306, 341
Hooke, R., 227–228, 233, 254
Hopwood, N., 173n, 190, 213
Horvath, J. A., 347, 351
Hotchner, A. E., 236, 254
Houle, S., 61, 77
Hughes, T. P., 10, 14, 292, 294, 295, 301,
    308, 322n, 340, 341
Huntington, S. P., 324, 325, 341

Hussey, M., 238, 244, 248, 254
Hutchins, E., 25, 26, 28–29, 36, 42, 44,
    54, 138, 156

I

Ippolito, M. F., 153, 156, 177, 215, 226,
    229, 230, 231, 233, 237, 239,
    240, 242, 243, 250, 255, 288, 301

J

Jablonski, D., 307, 341
Jacobs, J., 339n, 341
Jacoby, S., 37, 55
Jager, J., 306, 341
James, F. A. J. L., 140, 156
James, W., 74, 77, 220, 229, 238, 255
Jasanoff, S., 298, 301
John-Steiner, V., 228, 233, 235, 255
Johnson, M., 25, 30, 33, 53, 54, 297, 301,
    349, 350
Johnson, T. R., 88, 93
Johnson-Laird, P. N., 177, 215, 234, 255
Jones, C., 173n, 174n, 215
Jones, R., 152, 156
Joram, E., 221, 233, 253

K

Kail, R. V., 239, 255
Kaiser, J., 307, 341
Kane, R. H., 329, 331, 336–337, 341
Kapur, S., 61, 77
Karmiloff-Smith, A., 33, 53, 227, 230,
    233, 255
Katalsky, A., 4
Kates, K. W., 305, 306, 342
Kauffman, S. A., 338, 341
Kaufman, D. R., 347, 351
Keane, M. T., 230, 234, 238, 239, 254
Keil, F. C., 82, 93
Keith, D. W., 305n, 313n, 314n, 341
Kemp, M., 174n, 215
Kitano, N., 326, 341
Klahr, D., 6, 12, 14, 37n, 54, 57, 58, 59,
    60, 73, 77, 81, 82n, 83, 83n, 88,
    89, 90, 91, 92, 93, 94, 151, 156,
    175, 215
Klauer, K., 73, 77
Klayman, J., 59, 60, 77, 87, 93
Kloor, K., 309, 341

Knorr, K. D., 98, *118*
Kolodner, J. L., 268, *273*
Korpi, M., 99, 104, 116, *117*
Koslowski, B., 86, 88, *93*
Kosslyn, S. M., 30, *54*, 174, 176n, *215*
Krems, J., 88, *93*
Krogh, A., 116, *118*
Kuhn, D., 85, 88, *93*
Kuhn, T. S., 62, *77*, 98, *118*, 222, 224, 240, 255, 288, *301*
Kulkarni, D., 2, *14*, 20, *54*, 59, 61, *77*, 98, *118*, 151
Kurz, E. M., 18, *54*, 151, *156*, 177, *215*, 347, *350*
Kurz-Milcke, E., 153, *157*
Kusch, M., 348, *350*
Kyander-Teutsch, T., 189–190, *217*

L

La Porte, T., 280, *285*
Lackner, K. S., 313n, *340*
Lakatos, I., 98, *118*
Lakoff, G., 25, 30, *54*
Landes, D. S., 324n, 325n, *341*
Lane, D. A., 339n, *340*
Langley, P. W., 2, 6, *14*, 20, *54*, 55, 59, 61, *77*, 90, *94*, 119, 128, *135*, 152, *156*
Langston, W. E., 25, *53*
Larkin, J. H., *15*, 288, *301*
Latour, B., 18, 20, 23, 24, 37, 43n, *54*, 119, 121, 123, 124, 128, *135*, 177, *215*, 289, *301*
Lausch, E., 312n, *342*
Lave, J., 25, 26, 27, *54*, 58, *77*
Law, J., 2, *14*, 334, *341*
Leaska, M. A., 238, *255*
LeGrand, H. E., 176, 192, 193, 194, *215*
Leighton, R. B., *14*, 231–232, *254*, *350*
Levidow, B. B., 152, *155*
Levine, J. M., 24, *55*, 89, *94*
Liebert, R. M., 85, *94*
Light, S. S., 308, 309, *341*, *342*
Linnaeus, C., 97, *118*
Lockmann, T., 337, *340*
Loh, B., 349, *350*
Lorenz, C., 86, *93*
Louis, E. J., 61, *76*
Lubchenco, J., 305, *343*
Luckmann, R., 91, *92*
Ludlow, P., 320, *342*
Luhrmann, T. M., 119, 130–131, *135*

Lynch, M., 37, *54*, 173n, 174n, *216*

M

MacFarlane, G., 67, *77*
Machamer, P., 160, *171*
Magnani, L., 174n, 197n, *215*, *216*
Mahoney, M. J., 82, *93*, 98, 116, *118*
Markman, A. B., 152, *155*
Marshall, S., 349, *350*
Martin, N., 320, *342*
Martin, T., 141, 142, *156*, 197n, *216*
Marx, K., 281, 282, 283, *285*
Mason, R., 192, *216*
Mathews, J. T., 305, 306, *342*
Matthews, D., 194, *217*
Mayr, E., 2, *14*
McCloskey, M., 84, *93*
McDermott, J., *15*, 288, *301*
McDonald, J., 87, *94*
McFarlane, A. C., 33, *55*
McNeill, J. R., 305, 306, 310, *342*
McNeillie, A., 223, 224, 241, 242, 244, 245, 246, 248, 250, *253*
McVoy, C., 298, *302*
Mears, R. P., 141, 143n, *157*
Medawar, P. B., 164, 165, 166, 167, 168, 169, *171*
Mehalik, M. M., 8, *14*, 282n, *285*, 292, *301*, 349, *350*
Melillo, J. M., 305, *343*
Merrill, G., 262, 268, 270, 271, *273*
Meyer, W. B., 305, 306, *342*
Miller, A. I., 2, *14*, 173n, 174n, 176, 180, 210, *216*, 232, *255*
Millgate, M., 231, *255*
Mitroff, I. I., 98, *118*
Montgomery, H., 235, *255*
Mooney, H. A., 305, *343*
Morgan, M., 177, *216*
Morris, W., 231, *255*
Morrison, M. S., 177, *216*
Musch, J., 73, *77*
Mynatt, C. R., 6, *14*, 57, 59, 60, 78, 81, 87, 88, 90, 93–94, 98, *118*, 151, *156*

N

Nagel, T., 119, 133, *135*
Naumer, B., 73, *77*
Nersessian, N. J., 18, 21, 30, 37, 38, 46, *53*, *54*, 59, 61, *77*, 98, *118*, 140, 152,

152n, 153, *156–157, 157,* 160,
171, 174n, 175, 177, 179, *216,* 231
Newell, A., 22, *55,* 83, *94,* 174, 175, 178,
181, 203, 210, 211, *216,* 222,
223, 255
Newstetter, W., 153, *157*
Nicolson, N., 224, 226, 232, 239, 241, 242,
243, 244, 246, 248, 249, 250, *255*
Nisbett, R., 25, 33, *54*
Noble, D. F., 325n, *342*
Noble, J. R., 240–241, *255*
Nonaka, I., 1, *15*
Norenzayan, A., 25, 33, *54*
Norman, D. A., 25, 28, 42, *54–55, 56,* 138,
153, *158*

O

Oberg, J., 305n, *342*
Oberlander, J., 176, *216*
Oblon, M., 8, *14,* 282n, *285*
Ochs, E., 37, *55*
Ochse, R., 250, *256*
Okada, T., 89, *94*
Okagaki, L., 86, *93*
Oliver, S. G., 61, *76*
Olley, J. M., 307, *342*
Olson, D. R., 86, *94*

P

Palmer, D., 330, *342*
Palmer, R. G., 116, *118*
Palumbi, S. R., 307–308, *342*
Pamplin, B. R., 290, *301*
Parisi, D., 33, *53*
Parkes, S., 145n, *157*
Paszek, D., 86, *94*
Patel, V. L., 347, *351*
Patterson, C. C., 306, *341*
Peng, K., 25, 33, *54*
Perkins, D. N., 223, 224, 225, 226, 229,
256, 257, 263, 269, 273
Perner, J., 86, *94,* 347, *350*
Perrow, C., 279, *285,* 347, *351*
Persson, M. W., 306, *342*
Pickering, A., 20, *55,* 177, 178, *216*
Pinch, T., 160, *170,* 296, *301,* 322n, *340*
Plunkett, K., 33, *53*
Poldrack, R., 61, *77*
Pool, R., 308, *342*
Popper, K., 160, *171*

Postel, S., 310, *342*
Potter, M., 228, *253*
Prabhakaran, V., 61, *77*
Prideaux, G. J., 307, *342*
Principe, L. M., 2, *15*
Puddephatt, R. J., 143n, *157*
Pylyshyn, Z. W., 176n, 195n, *213, 216*

R

Radinsky, J., 349, *350*
Raff, A., 192, *216*
Raghavan, K., 88, *94*
Ramón y Cajal, S., 162, 163, 164, 166,
167, 168, 169, *171*
Reber, A. S., 347, *351*
Redman, C. L., 307, *342*
Rees, E., 242, *253*
Reiner, M., 88, *94*
Reiser, B. J., 349, *350*
Renberg, I., 306, *342*
Renberg, I., 306, *342*
Resnick, L. B., 24, *55,* 89, *94*
Rheinberger, H.-J., 24, 42n, *55,* 153, *157*
Richards, A., 162, *171*
Richards, J. F., 305, 306, *342*
Richardson, R. C., 160, *170*
Richman, H. B., 220, *256*
Roberts, I., 61, *76*
Roberts, K. H., 280, *285*
Roberts, R. G., 307, *342*
Robinson, M., 90, *93*
Rochlin, G., 280, *285*
Rolland, R., 232, *256*
Rorty, R., 319, *342*
Rosenthal, M., 241, 249, *256*
Rowlands, M., 319, 320, 321, 323,
327–328, 342
Rudwick, M. J. S., 2, *15,* 138, *157,* 173n,
193, *216*
Ruffman, T., 86, *94*
Russell, B., 324n, *342*

S

Sagoff, M., 317, *342*
Sahdra, B., 132, 133, *135*
Samuels, M. C., 87, *94*
Sands, M., 231–232, *254*
Scardamalia, M., 220, 221, 233, 242, *253,*
256
Schaffer, S., 160, *170*
Schauble, L., 88, *94*

Schliemann, A. D., 26, *52*
Schooler, J. W., 228, *256*
Schraagen, J., 60, *77*
Schrope, M., 309, *342*
Schunn, C. D., 60, *77*, 81, 83n, 88, *94*, 99, 118, 151, *157*, 177, 193, *217*, 288, *301–302*
Scott, J. C., 292, *302*, 310, *342*
Seger, C., 61, *77*
Senge, P. M., 308, *342*
Shadish, W. R., 89, *93*
Shaklee, H., 86, *94*
Shapiro, J., 311, *342*
Shekerjian, D., 227, *256*
Shelley, C., 174n, *216*
Shepard, R. N., 174, *213*, 232, *256*
Shomar, T., 177, *213*
Shore, B., 25, 31, *55*
Shrager, J., 2, 12, *15*, 90–91, 91, *94*, 119, 120, 128, 129, *135*, 159
Shuerfeld, D., 186, 191, *217*
Shweder, R., 73, *78*
Siegel, D., 21, *55*
Siegler, R. S., 85, *94*
Simina, M. D., 268, *273*
Simon, D. P., *15*, 288, *301*
Simon, H. A., 2, *15*, 20, 22, 37n, *54*, *55*, 58, 59, 61, 68, *77*, 83, 83n, 89, 92, 93, *94*, 98, 100, *117*, *118*, 138, 139, 151, *156*, *157*, 219, 220, 222, 223, 233, 252, *253*, *255*, *256*, 268, *273*, 288, *301*, 347, *351*
Simon, H. L., 6, *15*, 20, *55*, 90, *94*
Sindermann, C. J., 169, *171*
Skinner, B. F., 89, *95*
Sklar, F. H., 298, *302*
Sleeman, D. H., 99, 104, 116, *117*
Smith, C., 21, *55*
Smith, J., 220, 221, 242, *254*
Smith, S. M., 160, *171*
Socolow, R. C., 305, 308, 311, *342*
Sodian, B., 82, *95*
Spitz, A. J., 91, *92*
Spitzmüller, C., 141, 143n, *157*
Stansky, P., 238, 239, *256*
Star, S. L., 11, *15*, 177, *216*, 292, 300, 346, *351*
Staszewski, J. J., 220, *256*
Steinle, F., 141, *157*, 346, *351*
Stenning, K., 176, *217*
Sternberg, R. J., 347, *351*

Stevens, A., 177, *214*
Stevens, R., 37, *53*
Stoddard, A. M., 91, *92*
Stratton, G. M., 229, 238, *256*
Stump, D., *214*
Suarez, M., 177, *213*, *216*
Suchman, L. A., 25, 27, *55*, 58, *78*
Suess, E., 312n, *342*
Sun, Y., 141, 143n, *157*

T

Takeuchi, H., 1, *15*
Teasley, S. D., 24, *55*, 89, *94*, 95
Teutsch, H. E., 186, 189–190, 190, 191, *216*, *217*
Thagard, P., 18, *55*, 59, 61, 62, *78*, 132, 133, *135*, 160, 162, 167, *170*, *171*, *216*
Thomas, V., 305, *342*
Thompson, E. P., 315n, *342*
Thompson, V., 73, *77*
Tomasello, M., 25, 31, 32, 33, *52*, *55*
Torralba, T., 37, *53*
Trafton, J. G., 99, *118*, 151, *157*, 177, 193, *217*
Trautmann, J., 224, 226, 232, 239, 241, 242, 243, 244, 246, 248, 249, 250, *255*
Trickett, S. B., 99, *118*, 151, *157*, 177, 193, *217*
Tschirgi, J. E., 84–85, *95*
Turner, B. L., 305, 306, *342*
Tversky, B., 174, *217*
Tweney, P., 59, *78*
Tweney, R. D., 2, 3, 6, 7, 8, 9, *15*, 18, 37, 38, 42n, *54*, *55*, 57, 59, 60, *78*, 81, 87, 88, 90, 93–94, 98, *118*, 140, 141, 142, 143n, 151, 152, 153, *156*, *157*, 175, 177, 196, *215*, *217*, 226, 227, 229, 230, 231, 232, 233, 237, 240, 243, *255*, *256*, 288, 294, *301*, *302*, 347, *350*, *351*

U

Umbach, D., 86, *93*

V

Vaid, J., 160, *171*

van der Kolk, B. A., 33, *55*
Van Leeuwen, C. I., 174, 179, *217*
Vera, A., 22, *55*
Verstijnen, X., 174, 179, *217*
Vine, F. J., 193, 194, *217*
Vitousek, P. M., 305, *343*
Vosniadou, S., 86, *95*
Vygotsky, L. S., 32, *55*

W

Walberg, H. J., 220, *257*
Wallace, D. B., 2, *15*, 238, *257*
Walter, A. A., 238, *253*
Ward, T. B., 160, *171*
Wason, P. C., 3, 4, 5, *15*, 59, 60, 78, 81, 87, 95, 98, *118*
Watson, J. D., 166, 167, 168, 169, *171*
Weber, M., 325n, *343*
Weber, R. J., 223, *257*, 263, 268, 269, *273*
Weisaeth, L., 33, *55*
Weisberg, R. W., 220, *257*
Werhane, P. H., 349, *350*
Westcott, M. R., 236, *257*
Westfall, R. S., 2, *15*, 175, *217*
Westly, F., 338n, *343*
White, L., Jr., 280, *285*
White, M. J., 91, *92*
Whittington, H. B., 183, 185, *217*
Wilkinson, C., 147, 148, 149, 150, *158*
Williams, L. P., 138, 140, 143n, *158*
Wilson, J. T., 193, *217*
Wise, M. N., 2, *15*, 21, *55*, 209, *217*
Wiser, M., 82, *95*
Wolff, P., 152, *155*

Wolpert, L., 162, *171*
Woods, D. D., 29, *55*
Woolf, V., 9–10, *15*, 220, 221, 223, 224, 226, 228, 232, 237–238, 239, 240, 241, 242, 243, 244, 245, 246, 247, 248, 249, 251, 252, *257–258*
Woolgar, S., 2, *15*, 18, 37, 43n, *54*, 119, 121, 123, 124, 128, *135*, 173n, 174n, 177, *215*, 289, 293, *301, 302*
Wu, Y., 298, *302*
Wurster, T. S., 1, *14*

Y

Yachanin, S. A., 6, *15*, 98, *118*
Yates, R. F., 259, *273*
Yeh, W., 30, *56*
Yoshida, H., 307, *342*

Z

Zaitchik, D., 82, *95*
Zanna, M. P., 169, *171*
Zee, R. V., 298, *302*
Zhang, J., 25, 28, *56*, 138, 153, *158*
Zhao, M., 61, *77*
Ziegler, A., 86, *92*
Zimmer, C., 312n, *343*
Zimmerman, C., 82n, *95*
Ziock, H. J., 313n, *340*
Zohar, A., 88, *93*
Zsigmondy, 138
Zytkow, J. M., 2, *14*, 20, *54*, 59, 61, *77*

Subject Index

Note: Page number followed by *n* indicates footnote.

A

Abductions, manipulative, 197n
Abstraction, 153, 176
  appropriate level of, 4
Accumulation, 198, 203
Action, as focal point of cognition, 25
Actor networks, 316–317, 319
Advanced Research Project Agency
  (ARPA), 294, 295
Affordances, 27, 30, 31
Agency, sense of personal, 33, *see also*
  Intentionality
Agricultural Revolution, 280–281
Agriculture, 298
AIDS activists and expertise, 289–290, 292,
  294, 295
Airplanes, 265, 266
Analogical transfers, promoting, 212
Analogies, 65, 152
Anomalies, 62, 97–98, 115–117, 192–194,
  see also Unexpected findings
  attending to, 106–107
    examples of, 107–109
  how scientists deal with
    approaches to investigating, 98–99
    study of, 99–116
  immediate response to, 109
    hypothesis elaboration, 104–108,
      112–114
    identifying features, 104, 109–110

    place in context, 105, 114–115
    proposing hypothesis, 105, 110–112
  noticing, 103, 105–106
  subsequent references to, 103–104, 107
Anthropogenic world, core dilemma of,
  333
Apted, Michael, 134
Aral Sea, 310
ARPANET, 294, 295
Artifacts
  cognitive, 42–44
  epistemic, 153
  technological, 38–39, 41, 48
Astronomy, *see* Anomalies
Attunement to constraints and
  affordances, 27
Automations, 281

B

"Bench top," 46
BigTrak, 90–91
"Biography of the object," 153
Biologist, molecular
  being and becoming a, 119–120,
    133–134
  commonsense perception and, 128–133
  diary of an insane cell mechanic, 120
  on "shallow" reason, 120–128
Biomedical engineering (BME) laboratory,
  34–36, 38, 39

as evolving distributed cognitive system, 40–50
Bioreactor, 44–46, 49
Biotechnology, 295–296, 327
"Blame the method," 121, 125, 126
Blood vessels, *see* Biomedical engineering (BME) laboratory
Bottom-up processes, 236
Boundary objects and boundary systems, 292–295, 298
British Industrial Revolution, 281, 282
Burgess shales, reanimation of the, 183–188

C

Calculator view, 91
Camera lucida, 183, 185, 197
Camera obscura, 195
Carbon cycle system, 311–313
Cardiovascular system, *see* Biomedical engineering (BME) laboratory
Cartesianism, 18, 22, 23, 319, 320, 323, *see also under* Science and technology studies
Case studies, 2, 98
Categorization, 116, 231, *see also* Classification schemes
Causal models, building of, 67–68
Causal (scientific) reasoning, *see also* Scientific thinking/reasoning
about expected and unexpected events, 70–75, *see also* Unexpected findings
in vitro, 58, 68–73
Causal scientific thinking and plausibility, 68–70
Cell combinations, studying various cells and, 84–88
Cellular biology, *see* Biologist
Choice, *see* Free will
Classification schemes, 97–98, 146, *see also* Categorization
Climate systems, 311–314
Clockmakers, 347
Clues, 225
Cogito, ergo sum, 323
Cognition, 24
consciousness and, 318
environmental perspectives on, 24–34
and the integration problem, 19, 33–34, 50
Newell's bands of, 210–211

as power, 322–323
technology and, 325
Cognitive accounts of science and technology, 17–18, 21
Cognitive artifacts, 28
Cognitive capabilities, 46
Cognitive-historical analysis, 36, 37
Cognitive partnerships, 47–50
Cognitive processes, 1–4, 26, 81
Cognitive science research, *see also* Research
aim, 28–29
Cognitive systems, 19, 26, 42, 318–319, 328n, *see also* Distributed cognitive systems
defined, 318
representations and processes, 21
Colloids, 138n, *see also* Faraday
Combinatory play, 234
Common-sense perception (CSP), 128–133
Communication, 294
levels of intergroup, 294
Component failure *vs.* poor component performance, 266
Concentration, 163
Concept formation, 37
Concept-learning task, 87
Conceptions, 231
Confirmation bias, 3, 98, 106
Confusions, 143, 146, 150, 152
Connections, making new, 160, 168
Consciousness, 318
Constraints, 27, 30, 309, 336
Construals, 197n
becoming concepts, 153–154
"Container" schema, 30
Continental drift, 192–194
Contributing expertise, 290–291
Converging Technologies (NBIC) for Human Performance, 295–297
Creative people, highly
habits of, 160–162, 167–170
Creativity
dimensions of, 203–305
perception and, 235
Cultural affordances, *see* Affordances
Cultural evolution, 31–32
Cultural models, 31
Culture, *see also* Sociocultural factors
and cognition, 31–33

D

Data analysis, *see* Anomalies
De Laval nozzle, 266–270
Decomposition, functional, 263, 266
Deflagrations, 140, 147–150
Design for environment, 308
Design improvement opportunities, 269
Design objectives and constraints, 309
Determinism, 329, 336, 337, *see also* Free will
Devices, 42–45
Diachronic cognitive systems, 36
Diaries, 8, *see also* Biologist; Faraday
Dimensional enhancement, 180–181,
    187–188, 194, 195, 204, *see also*
    Paleobiology
Dimensional reduction, 180–182, 187–188,
    194, 195, 197–199, *see also* Re-
    duction
Discovery, *see* Scientific discovery
Distributed cognition, 27–29
Distributed cognitive systems, 25
    evolving
        mixed method approach to investi-
            gating, 36–39
        research laboratories as, 34–50
Distributed mental modeling, 46–47
Distributed model-based reasoning, 43–45
Domain-general experiment space search,
    85, 87, 88–89
Domain-specific and domain-general evi-
    dence evaluation, 85–89
Domain-specific experiment space search,
    84–85, 87, 88–89
Domain-specific hypothesis space search,
    84, 88–89
    and evidence evaluation, 86–87
Domain-specific *vs.* domain-general knowl-
    edge, 82–83, 88–89
Dualism, Cartesian, *see* Cartesianism
Dyads, as primary unit of analysis, 12–13

E

Earth systems, 304
Earth Systems Engineering Management
    (ESEM), 11, 297–299, 304–314
    integrative cognitivism and, 314–339
Economic history, global, 305–306
Ecosystems, 298
Electric power, 282–284
Electrical discharge, luminous, 207, 208

striation of, 207
Electromagnetic fields, 231–232
Electromagnetic rotations and rotation
    motor, 151, 200–206
Electromagnetism, *see also* Faraday
    abstracting structure and theorizing
        process, 196–198
    experience and technique, 198–202
Electrostatic fields, 207
Embedded accounts of cognition, 19, 21–22
Embodied accounts of cognition, 19,
    21–22, 29–30
Embodied mental representation, 29–30
Emotional advice for scientists, 161–164,
    166–167
Emulations, 233
    construction of, 233–238, 247–251
    defined, 249
Engineering, *see* Earth Systems Engineer-
    ing Management
Environmentalsim, philosophical, 319, 320
Epistemic artifacts, 153
Equi-biaxial strain, 44
Equipment, 42–44
Ethics expertise, 296–297
Ethnographic research and analysis, 36–38
Everglades, 298–299, 308–310
Evidence evaluation, 84–89
Evolution, 31, 325
Ex vivo research, 5, 59–61
Excitement, 161–162
Experiment space search, 83–85, 87–89
Experimentation, 164, 166, 196
Expert-novice studies, 220–221, 288
Expertise, 288–289
    defined, 297
    and problem solving, 220–222, 225
    shared, 294
    types of, 289–291, 294, 296–297
    Virginia Woolf's development of,
        238–243
"Explanatory impulsion," 165
Externalism, 319–320
"Eye-hand-brain interaction," 151, 153

F

Faraday, Michael, 138–140, 151–154, *see
    also* Electromagnetism
    and the problem of gold, 140–142
    replicating deflagrations, 147–150
    replicating precipitates, 142–147

research, 8–9
sequence of his ideas, 145–146
Faraday-Tyndall effect, 138
Fleming, Alexander, 66–67
Florida Everglades, *see* Everglades
Flow loop, 39, 45
Focused-gambling strategy, 267–268
"Force" schema, 30
Free will, 315–319, 317n, 329–331,
    335–337, *see also* Agency;
    Intentionality
Freedom, 329–331
perverse, 330
Functional magnetic resonance imaging
    (fMRI), 60–61, *see also* Magnetic
    resonance imaging

G

Generativity, 195
Geophysics, 192–195
GOFAI ("Good Old Fashioned AI"),
    18–20, 23
Gold, 139–140
Faraday and the problem of, 140–142,
    see also Faraday
Gold wires, apparatus for exploding, 147,
    148
Golem, 296–297
Golem science, 297n
Group reasoning, 65–66
Groups, as primary unit of analysis, 12–13

H

Habits of highly creative people, 160–162,
    167–170
Hepatology, 186, 189–192, 195
Heuristics, 151, 250–251
problem representation, 226–238
"shallow," 121–127
Hickam, Homer, *see* Rocket Boys
Historical sciences, 297, 297n
Homing spaces, 224, 225
HOTAT ("hold one thing at a time"), 85
"Hunches," 153
Hydrologic projects, 310, *see also* Water
Hypothesis elaboration, 105–108, 112
elaborated *vs.* unelaborated hypotheses,
    104–105
source of elaboration, 105, 112–114
theoretical *vs.* visual, 105

Hypothesis space search, 83, 84, 86–89

I

Image-based thinking/reasoning, 173–174
Image schemas, 30
Imagery, 152, 153
Images, 211–212
and inference, 203–206
situating, 206–209
    in inference, 209–211
vs. mental models, 234
Impasse-circumvention opportunities,
    269
Impressionism, 238–239
In historico studies, 6, 59–62
In magnetico research, 5, 59–61
In silico research, 5–6, 59–62
In vitro studies, 5, 40, 43–44, 58–61
In vivo studies, 5, 40, 43–44, 58–61
Inceptions, 246
defined, 231
distillation of, 231–236
Inceptual processing, 226–238
defined, 226–227
of Virginia Woolf, 243–252
Industrial revolutions, 281–284
Inference, *see* Images; Visual inference
Information Revolution, 284
Information technology, 295–296, 307,
    327
Instantiate, 99
Instructionless learning, 90–91
Instruments, 42–44
Intact activity system, 27
Integration, 19, 33–34, 50, 67
Integrative cognitivism, 319–323, 322n,
    327, 328, 331, 338, 339, *see also*
    under Earth Systems Engineering
    Management
Intentionality, 315–319, 329–339, *see also*
    Agency
"congealed," 337n
interplay between system constraints
    and, 336
Interactional expertise, 289
Interactionist approach, 32
Internalism/internalist approach, 319, 320
Interpretation of data, 166, *see also* Unex-
    pected findings
Interpretive drift, 130–131
Invention, 10–11

J

Jacob's Room (Woolf), 224, 239–245, 248

K

Kepler's laws, 20
"Kew Gardens" (Woolf), 240–242
Klondike space, 224–227
Knowledge, 1
    local *vs.* global, 105
    types of, *see* Domain-specific *vs.* domain-general knowledge
Krebs cycle, 20

L

Labor, 281–284
Laboratory artifacts, sorting of
    by laboratory members, 42–43
Laboratory as "problem space," 41–43
"Laboratory Life" (Latour and Woolgar), 121
Lavoisier, Antoine, 137, 138
Leanchoilia, 184, 190, 191
Learning, *see* Problem solving
Life cycle assessment, 308
Light, *see* Faraday
Lived relations, 38
Liver function and structure, 186, 189–192
Load factor, 283
Local analogies, 65
Logic, 163

M

Machines, 281–282
Magnetic resonance imaging (MRI), 60–61, 291–292
Magnetism, *see* Faraday
Manipulations, 45
Manipulative abductions, 197n
Maps and mapping, 192–194
"Mark on the Wall, The" (Woolf), 240–242
Matching, 187–188, *see also* Paleobiology
Materials flow analysis, 308
Maxwell, James, 20–21
Mechanical conditioning, 45
Mechanical tester, 47–48
Medawar, Peter B., 164–169
Mental modeling, distributed, 46–47
Mental models, 293
    characterizations of, 234

Mental representations, 21, *see also* Imagery; Images
    as modal *vs.* amodal, 30
Metallic colloids, 138n
Mimesis, 32
Mind, 319, 328
    methods for discovering workings of the scientific, 58–61
Mind-body dualism, 319, 320, *see also* Cartesianism
Model-based claims and classification, 146
Models, 2–3, 178
    selectivity, 3–4
Modernism, 311
Molecular biologist, *see* Biologist, molecular
Moral, 337
Moral responsibility, 329, 331, 333, 334
Mrs. Dalloway (Woolf), 228, 249

N

Nanotechnology, 290–291, 295–296, 327
Narioia, 183
Nature-nurture debate *vs.* interactionist approach, 32
Negotiation, 154, 288
Network of enterprise, 240
Network states, three, 292–295, *see also* States 1, 2, and 3
Networks, actor, 316–317, 319
Neuroscience research, 32
Newell's bands of cognition, 210–211
Night and Day (Woolf), 239, 246–247
Notebooks, 7–8, *see also* Biologist
Novels and novelists, *see* Woolf, Virginia

O

Observation, 164
Observational practices, invention of, 154
Ontological dissonance, 327
Ontologies, 326–327, 333n
Opportunistic reasoning, 268
Optical mode of investigation, 140

P

Paleobiology, 183–188, 195
Paradigms, *see* Theories
Participant-observer studies, 119, 134
Pattern-structure-process (PSP) schema, 180–182, 195, 197, 206, 210–212

Penicillin, discovery of, 66–67
Perception, 335, see also "Seeing"
Perceptual inference, 209, see also Visual inference
Perceptual rehearsal, 227–231, 235–236, 243–246
    defined, 230, 243
Perceptual symbol systems, 30
Persistence, 160, 161, 163, 168
Phenomena, theories, and models, 176–178
Physical symbol system, 22
Physics, 288, see also Anomalies
Plasticity, cortical, 32
Plausibility, 68–69
Political domain, 291
Post-Impressionist art, 238, 239
Power sources, 282–284, 311–314, see also specific power sources
Primatology research, 31
Probability, 68–69
Problem finding, 219, see also Heuristics
nature and role of, 222–226
Problem re-representation, 227
Problem representation, 219–220, 226, 251–252
Problem representation heuristics, 226–238
Problem solving, 27, 90, 160–161
    expertise and writing, 220–222
    of experts vs. novices, 220–221
    nature and role of, 222–226
Problem space(s), 222–223, 225, 236, 346–347
    defined, 222
    laboratory as, 41–43
    search through a, 151
Procedural-declarative distinction, 132
Proximal reason, 127, 128
Public domain, 291

R

Rachet effect, 32
Ramón y Cajal, Santiago, 162–164, 167–169
Reasoning experiments, 3, 4
Reasoning processes, see also Causal (scientific) reasoning; Scientific thinking/reasoning
    types of, 66
Record keeping, 7–8
Reduction, 180–181, 187–188, see also Dimensional reduction; Paleobiology

Reductionism, see also under Science and technology studies
    cognitive, 21–22
    sociocultural, 22–23
Reflexive historical science, 297, 348
Relational trajectories, 36
Replication, see Faraday
Representational system, 317–318
Representations, 44, 327–328, see also Mental representations
    generation, 45, 211
    propagation, 45
Research, see also specific types of research
    in science, taxonomy of, 58–61
    types of studies, 5–7
Research methodology, 11–12
Researchers, see Cognitive partnerships
Risks, taking, 166
Rocket Boys, 259–263, 271–272, 347, see also Rockets
    coordinating multispace searches, 269–270
    efficiencies of invention, 263–268
    external expansion of the search space, 268–269
    model rockets developed by, 273–275
Rocket Boys (Hickam), 260n
Rockets, see also Rocket Boys
    design variables in model, 263–265
    mental models of, 270–271

S

Science, 92
    types of, 297n
Science and technology studies (STS), 17, 34, 37, 289
    Cartesian roots of cognitive and social reductionism in, 19–23
    rapprochement, 23–24, 50
    waves/movements in, 287–288
Scientific discovery, 5–10, 20
    activities involved in, 174
    beyond the myths of, 73–75
Scientific mind, see also Mind
    methods for discovering workings of, 58–61
Scientific thinking/reasoning, 57–58, see also Causal (scientific) reasoning
    fundamental feature of, 151
    integrative investigations of, 88
    processes in, 82–84

types of foci in psychological studies of, 82–84

Scientists, *see also under* Anomalies; Unexpected findings
  habits of successful, 160–162, 167–170
  rules for success, 166–167
  types of unsuccessful, 164
  what they do, 63–64

Second Industrial Revolution, 282–284

"Seeing," 154

"7up," 134

Shales, *see* Burgess shales

Sidneyia, 185

Similarities, recognition of, 210

Simulation, 30, 346, *see also* Emulations

Simulative model-based reasoning, 46–47

Situated cognition, 26–29, 35, 153

Sociability, 161, 162, 168

Social habits of successful scientists, 161, 162

Sociocultural accounts of science and technology, 17, 21, 50

Sociocultural factors, 26, *see also* Culture

Sociocultural processes, 26

South Florida Water Management District (SFWMD), 298

Spatial cognition, 173–175, *see also* Visual inference

"Spherical horses" approach, 25

Standardized packages, 177

States 1, 2, and 3 (systems), 11, 294, 326–327

Steam technology, 307

Stream of consciousness, 227, 238, 242, *see also* Woolf, Virginia

Stream of thought, 238

Striation of luminous electrical discharge, 207

Study of experience and expertise (SEE), 288–289, *see also* Expertise

Sub specie historiae research, 6

Suspended goals, 268

Symbol systems
  perceptual, 30
  physical, 22

Symbolic maps, 193

Synchronic cognitive systems, 36

Synergism, 165

Systematicity, 195

Systems
  defining, 278
  as primary unit of analysis, 12–13
  types of, 278–279

Systems-ordered world, 277–284
  deep ploughs, harnessed teams, and manors, 280–281
  labor, machines, and automation, 281–282

T

Taxonomic change in scientific concepts, 146

Teams, as primary unit of analysis, 12–13

Technological artifacts, 38–39, 41, 48

Technology, *see also* Science and technology studies
  evolution of, 325
  roles of, 325

Technology clusters, 306, 307

Textiles technology, 307

Theoretical elaboration of hypotheses, *see* Hypothesis elaboration, source of elaboration

Theories, 97–98
  use of, to avoid search, 266

To the Lighthouse (Woolf), 223–224, 243, 247

Top-down processes, 236

Trading zones, 291–292, 294, 298, 299

Transmission systems, 283

Tunneling process/system, 249

U

Unexpected findings, 70–75, *see also* Anomalies
  changes in reasoning with patterns of, 65–67
  expecting, 160, 161, 168
  frequency, 64–65
  how scientists assess, 67–68
  science and the, 61–62
  in vivo investigation of scientists and the, 62–68

"Unwritten Novel, An" (Woolf), 241, 242

V

Vascular constructs, 41, 42, *see also* Biomedical engineering (BME) laboratory

Vascular structures from modular sections, 186, 189–192

View application, 90–91

Visual elaboration, *see* Hypothesis elaboration, source of elaboration
Visual inference, 180–194, 209
  images and, 203–206
  and mental process, 194–198
  and Newell's bands of cognition, 210–211
  situating, 178–180
Visual inference tables, 203–206
Visual manipulations, 187–188, *see also* Paleobiology
Visualization, 176, *see also* Spatial cognition
  ways of exposing cognitive factors relevant to, 176
VOTAT ("vary one thing at a time"), 85, 267, 272

W

Wason, Peter
  2-4-6 task, 3, 4, 87

Water, 298, 307, *see also* Hydrologic projects; Steam technology
Watson, James D., 166–169
Waves, The (Woolf), 244, 247, 249
Will, 329n, 331, 332, *see also* Free will
  defined, 331
Wilson, Quentin, *see* Rocket Boys
Woolf, Virginia
  development of expertise, 238–243
  inceptual processing, 243–252
  writings
    Jacob's Room, 224, 239–245, 248
    "Kew Gardens," 240–242
    To the Lighthouse, 223–224, 243, 247
    "The Mark on the Wall," 240–242
    Mrs. Dalloway, 228, 249
    Night and Day, 239, 246–247
    "An Unwritten Novel," 241, 242
    The Waves, 244, 247, 249
Wright brothers, 265, 266
Writing problem solving, expertise and, 220